ISOZYMES AND ENZYME REGULATION IN CANCER

Japanese Cancer Association
GANN Monographs on Cancer Research

GANN Monographs on Cancer Research are intended for the publication of the proceedings of international conferences and symposia dealing with cancer or closely related research fields, and series of papers on specific subjects of importance to cancer research. Publication of each individual monograph is to be decided by the Editorial Board of the Japanese Cancer Association with the final approval of the Board of Directors of the said Association. It is hoped that series of GANN Monographs on Cancer Research will serve as an important source of international information in cancer research.

Tomizo Yoshida, M.D.

Editorial Board

Shichiro Ishikawa	Masayasu Kitagawa	Yoshiyuki Koyama
Toru Miyaji	Yasuaki Nishizuka	Shigeyoshi Odashima
Tetsuo Ono	Kunio Oota	Yoshio Sakurai
Haruo Sato	Hiroto Shimojo	Yoichiro Umegaki
Tadashi Yamamoto	Yuichi Yamamura	Tomizo Yoshida

Editor-in-Chief

Tomizo Yoshida

Editorial Staff

In Charge of Periodicals
Tetsuo Ono

In Charge of Monographs
Shigeyoshi Odashima

Technical Staff

Dorothy U. Mizoguchi   Hisaë Izumo

JAPANESE CANCER ASSOCIATION

GANN Monograph on Cancer Research 13

ISOZYMES AND ENZYME REGULATION IN CANCER

Edited By SIDNEY WEINHOUSE
TETSUO ONO

UNIVERSITY OF TOKYO PRESS

© UNIVERSITY OF TOKYO PRESS, 1972
UTP No. 3047-67707-5149
Printed in Japan.

All rights reserved. No part of this publication may be reproduced or transmitted in any form or by any means, electronic or mechanical, including photocopy, recording, or any information storage and retrieval system, without permission in writing from the publisher.

Published by
UNIVERSITY OF TOKYO PRESS
7-3-1 Hongo, Bunkyo-ku, Japan

November 10, 1972

PREFACE

Eleven years ago there grew out of discussions between President John F. Kennedy and Prime Minister Hayato Ikeda the U.S.-Japan Cooperative Program in Science. The purpose of this program, jointly administered by the National Science Foundation in the United States and by the Society for the Promotion of Science in Japan, is to foster international understanding through cultural, educational, and scientific exchange between individual scientists of both countries. Under the aegis of this program, a large number of symposia, seminars, and conferences have been held, several of which have been in the field of cancer. About six years ago, in November 1966, the first of such joint U.S.-Japan conferences on cancer was held in Kyoto under the title, " Biological and Biochemical Evaluation of Malignancy in Experimental Hepatomas." The proceedings of that conference were published by the editors of *GANN* and became GANN Monograph 1 of a continuing series now being issued as " GANN Monograph on Cancer Research." This initial monograph was edited jointly by Osamu Hayaishi and Harold P. Morris with Tomizo Yoshida as Editor-in-Chief.

This first cancer conference, held five years ago under the U.S.-Japan Cooperative Program, emphasized the liver and its neoplastic counterpart, the hepatoma, as a biological model for the study of carcinogenesis and for a broader understanding of the molecular basis of malignancy. The reasons for a special focus on the liver-hepatoma system are obvious. The versatility and diversity of the functions of the liver, as well as its size and accessibility, have made it a favored target for biochemical exploration, and there is little doubt that more is known about its enzymes, their synthesis and regulation, and the processes in which they take part than those of other tissues. Moreover, the liver cell is peculiarly susceptible to neoplastic transformation by chemicals, and the pioneering research by Tomizo Yoshida and Harold P. Morris have made available to cancer researchers a " spectrum " of many tumors ranging widely in their growth rates, degree of differentiation, and other biological properties generally associated with the term " malignancy." At this conference, much groundwork was laid for subsequent approaches to the molecular bases of neoplasia. The studies made it clear that certain alterations in enzyme patterns and metabolism underlie the biological diversity of hepatomas and pointed to aberration in gene expression as an inherent property of the malignant state.

The present monograph, published six years later, and titled " Isozymes and Enzyme Regulation in Cancer," is a subsequent stage in the exploration of questions and issues which developed from the initial symposium. It represents the proceedings of another cancer conference held in March 1971 in San Diego, California. Attended by some of the researchers who participated in the first conference, much attention was devoted to aberrations of gene expression. A recurrent pattern was the apparent retention or partial loss of liver-type enzyme species in slow-growing, well-differentiated hepatomas, but near or complete loss of those enzymes and their replacement in some instances by nonhepatic enzyme species in fast-growing, poorly differentiated hepatomas. A striking feature of the poorly differentiated tumors was the occurrence of certain isozymes found in fetal liver, but low or absent in adult liver. Thus, neoplasia appears to involve activation of genes normally repressed during embryonic development and the repression of genes normally expressed in the functional organ. Further understanding of these phenomena must await a deeper exploration of those factors which regulate gene expression in eukaryotic cells.

Other noteworthy observations reported at this conference were a striking correlation between growth rates of hepatomas and their ability to incorporate thymidine into DNA, and some significant effects of an implanted tumor on the metabolism and properties of host tissues.

The participants unanimously agreed that this conference was a uniquely beneficial experience in international cooperation toward common objectives. The conferences expressed their special thanks to Drs. Morris and Yoshida for the hepatomas which their researches have made available to the scientific community.

Thanks are also expressed to the U. S. and Japan sponsoring agencies for their generous support, to *GANN* and its Board for publishing the proceedings, and to Margaret Foti, Managing Editor of *Cancer Research*, and her staff for valuable editorial assistance with the manuscripts.

January, 1972

Sidney Weinhouse
Tetsuo Ono

CONTENTS

Preface v
Isozymes in Relation to Differentiation in Transplantable Rat Hepatomas
  ....Sidney WEINHOUSE, Jennie B. SHATTON, Wayne E. CRISS,
                              Francis A. FARINA, and Harold P. MORRIS   1
Regulation of Enzyme Synthesis Relating to Differentiation of Malignant
  Cells............Tetsuo ONO, Kiyoshige WAKABAYASHI, Kyoko UENOYAMA,
                              and Hideki KOYAMA   19
Disdifferentiation and Decarcinogenesis—the Isozyme Patterns in Malignant Tumors and Membrane Changes of Cultured Tumor Cells
  ....Takashi SUGIMURA, Taijiro MATSUSHIMA, Takashi KAWACHI,
  Kikuko KOGURE, Noritake TANAKA, Setsuko MIYAKE, Motoo HOZUMI,
                              Shigeaki SATO, and Hiroshi SATO   31
Molecular Correlation Concept: Ordered Pattern of Gene Expression in
  Neoplasia ....George WEBER   47
Dedifferentiation of Enzymes in the Liver of Tumor-bearing Animals
  ....Masami SUDA, Takehiko TANAKA, Susumu YANAGI,
          Shin-ichi HAYASHI, Kiichi IMAMURA, and Koji TANIUCHI   79
Isozymes in Selected Hepatomas and Some Biological Characteristics of a
  Spectrum of Transplantable Hepatomas ....Harold P. MORRIS   95
Thymidine Kinases of Neoplastic Tissues, Regenerating and Embryonic
  Liver, Marrow Cells, Potato, and *Tetrahymena* ....Setsuro FUJII,
  Takaki HASHIMOTO, Takahiko SHIOSAKA, Teruo Arima, Michiko MASAKA,
                              and Hiromichi OKUDA   107
Survey of Current Studies on Oncogeny as Blocked Ontogeny: Isozyme
  Changes in Livers of Rats Fed 3'-Methyl-4-dimethylaminoazobenzene
  with Collateral Studies on DNA Stability
  ....V. R. POTTER, P. Roy WALKER, and Jay I. GOODMAN   121
Abnormal Gene Expression on the Mode of Amino-Nitrogen Excretion in
  Rat Hepatomas from Phylogenic Aspects.........Nobuhiko KATUNUMA,
          Yasuhiro KURODA, Yoshiko MATSUDA, and Keiko KOBAYASHI   135
Relationship between Degree of Differentiation and Growth Rate of
  Minimal Deviation Hepatoms and Kidney Cortex Tumors Studied with
  Glutaminase Isozymes.........Nobuhiko KATUNUMA, Yasuhiro KURODA,
          Tasuku YOSHIDA, Yukihiro SANADA, and Harold P. MORRIS   143
Isozymes of Fructose 1,6-Diphosphatase, Glycogen Synthetase, and
  Glutamine: Fructose 6-Phosphate Amidotransferase
  ....Shigeru TSUIKI, Kiyomi SATO, Taeko MIYAGI, and Hisako KIKUCHI   153
Isozymes and Heteroglycan Metabolism of Hepatomas
  ....R. K. MURRAY, D. J. BAILEY, R. L. HUDGIN, and H. SCHACHTER   167
Isozymes of Branched Chain Amino Acid Transaminase in Normal Rat
  Tissues and Hepatomas ....Akira ICHIHARA and Koichi OGAWA   181

Regulation of the Levels of Multiple Forms of Serine Dehydratase and Tyrosine Aminotransferase in Rat Tissues
  Henry C. Pitot, Yoshifumi Iwasaki, Hideo Inoue, Charles Kasper, and Harvey Mohrenweiser ... 191

Mammalian Ribonucleotide Reductase and Cell Proliferation ... Howard L. Elford ... 205

Multimolecular Forms of Pyruvate Kinase and Phosphofructokinase in Normal and Cancer Tissues
  ... Takehiko Tanaka, Kiichi Imamura, T. Ann, and Koji Taniuchi ... 219

Biochemical Studies of the Preneoplastic State
  Hideya Endo, Masao Eguchi, Susumu Yanagi, Takehiko Torisu, Yukio Ikehara, and Tomoya Kamiya ... 235

Enzymology, Ultrastructure, and Energetics of Mitochondria from Three Morris Hepatomas of Widely Different Growth Rate
  ... Peter L. Pedersen ... 251

Glucose 6-Phosphate Dehydrogenase Isozymes in Cultured Morris Hepatoma Cells ... Mochihiko Ohashi and Tetsuo Ono ... 267

Diagnostic Value of Aldolase and Hexokinase Isozymes for Human Brain and Uterine Tumors ... Shigeaki Sato, Yoshihiro Kikuchi, Kintomo Takakura, Te Chen Chien, and Takashi Sugimura ... 279

Blood of Tumor-bearing Animals as a Cause of Metabolic Deviations
  ... Eiji Ishikawa and Masami Suda ... 289

Ferritin Isoproteins in Normal and Malignant Rat Tissues
  ... M. C. Linder, J. R. Moor, H. N. Munro, and H. P. Morris ... 299

Subject Index ... 315

ISOZYMES IN RELATION TO DIFFERENTIATION IN TRANSPLANTABLE RAT HEPATOMAS[*1]

Sidney WEINHOUSE, Jennie B. SHATTON, Wayne E. CRISS,[*2]
Francis A. FARINA,[*3] and Harold P. MORRIS

*Fels Research Institute and Department of Biochemistry, Temple University School of Medicine,[*4] and the Department of Biochemistry, School of Medicine, Howard University[*5]*

To determine the degree of retention and deletion of isozymes involved in normal liver function in liver neoplasms, a series of Morris hepatomas has been studied. Illustrative data on four isozyme systems reveal the following: With glucose-ATP phosphotransferases, aldolases, pyruvate kinases, and adenylate kinases, the highly differentiated, slow-growing hepatomas display the same isozyme pattern as normal adult liver; with decreased differentiation and increased growth rate, there is a variable loss of the "liver-specific" isozymes, gluokinase, aldolase B, pyruvate kinase II, and adenylate kinase III. With a series of poorly differentiated, fast-growing tumors, these liver-specific isozymes are virtually completely lost. In poorly differentiated hepatomas the liver-specific phosphotransferase, aldolase, and pyruvate kinase are each replaced by an isozyme which is very low in the adult liver. These findings reveal that retention of differentiated tissue function is not incompatible with the neoplastic transformation, but suggest that replacement of highly regulated isozymes by others not subject to host regulation may account for the lack of growth control of poorly differentiated tumors. These as well as other isozyme studies point to an instability of gene expression as a characteristic feature of the neoplastic cell.

There is an astonishing resemblance between poorly differentiated hepatomas and fetal liver in the isozyme patterns thus far studied. This observation may be added to a growing body of evidence for a "switching on" of fetal protein synthesis following or accompanying the "switching off" of gene products of the differentiated cell. These findings further suggest that impairment of

---

[*1] This work described has been supported by Grants CA-10916, CA-10439, and CA-10729 from the National Cancer Institute, and by Grant, P202 from the American Cancer Society. We also acknowledge the skillful technical assistance of Mrs. Billie P. Wagner and Mr. Albert Williams, and the help of Dr. David Meranze in the histological studies.

[*2] Present address: Department of Obstetrics and Gynecology, University of Florida, Gainesville, Florida, U.S.A.

[*3] Present address: Division of Clinical Chemistry, Temple University, School of Medicine, Philadelphia, Pa., U.S.A.

[*4] Philadelphia, Pennsylvania, U.S.A.   [*5] Washington, D.C. 20001, U.S.A.

gene control, rather than alteration of gene structure, may be a crucial factor in the neoplastic transformation.

At the present time, exciting new findings in virology and immunology overshadow many of the more traditional disciplines which share in the battle against cancer. While these now occupy the center of the stage, rightly so because of the possible immediacy of their application to the prevention or cure of human cancer, nevertheless there are certain other characters in this drama whose role may assume great importance in the future. One of these is the subject we have chosen to discuss at this meeting, that of isoenzymes and their regulation. This relatively new area of biochemistry holds great promise in furthering our understanding of such basic biological processes as gene activation and expression, disorders of which loom as possible etiological vectors in the cancer problem.

In this introductory presentation, it is appropriate to discuss briefly the theoretical background of our common field of interest. Although heterogeneity of enzyme structure had been occasionally suggested in the early years of biochemistry, it was not until simple and powerful methods of protein separation and identification were developed that the science of what we might call isoenzymology was created. The molecular heterogeneity of certain esterases and lactate dehydrogenase was demonstrated by Markert and Moller (31) hardly more than a decade ago. In the intervening years, the field has blossomed and borne much good fruit. An examination of the literature, which is now of overwhelming magnitude, reveals that the study of isozymes has become an unparalleled boon to the geneticist by furthering our understanding of genetic diseases and is assuming increasing practical utility in general medical diagnosis (28, 46, 52, 55).

TABLE I. Experimental Procedures for Identification of Isozymes

1. Kinetic	4. Isoelectric focusing
2. Electrophoresis	5. Immunological
3. Chromatographic	

Table I lists the experimental techniques that have been most often employed for the identification of multiple forms of enzymes. Perhaps no method is more generally useful than that of zone electrophoresis on a solid matrix such as starch or polyacrylamide gel, combined with specific enzymatic staining methods. Kinetic differences can be employed to good advantage when circumstances permit. We, as well as others, have taken advantage of this method to determine isozymes of glucose-ATP phosphotransferases (44, 45), aldolases (2), and lactate dehydrogenases (38) and are now applying it to pyruvate kinase isozymes.*

Methods of column chromatography, particularly the newly developed procedure of isoelectric focusing (22), are not easily adaptable to routine estimations but are extremely useful in specific applications. The same is true of immunological techniques, whose sensitivity and specificity have been thoroughly documented. Needless to say, combinations of these procedures complement and reinforce the individual methods.

---

* Unpublished work of F. Farina, S. Weinhouse, and H. P. Morris.

TABLE II. Molecular Basis for Multiple Forms of Enzymes

1. Multiple genes	4. Protein modification
2. Aggregation	5. Conformational isomerism
3. Partial proteolysis	

The voluminous isozyme literature has been adequately reviewed in recent books and monographs (*28, 46, 52, 55*), but we might consider briefly the nature and origins of isozymes. Basically, multimolecular forms may arise in two ways; either by synthesis at different gene loci, giving rise to structurally different polypeptides, or by modification of preexisting proteins (Table II). In the former instance, individual polypeptides may be active, as with the glucose-ATP phosphotransferases (*21*), or inactive subunits may combine in different proportions to form a series of isozymes. For example, the two subunits of lactate dehydrogenase combine to form 5 active tetramers (*21*) and the three subunits of aldolase A, B, and C likewise combine to form active tetramers consisting of combinations of A with B and C (*39*).

There are so many isozymic forms arising by modification of proteins that we can barely scratch the surface of this subject. If the isozyme concept can validly be extended to all of the allosteric enzymes whose kinetic properties are profoundly affected by combination with various substances, there would hardly be an enzyme that had no isozymic forms. Of the many well-recognized examples of this type, there are the glutamine synthetases of *Escherichia coli* studied by Stadtman (*47*), which differ in their attachment to AMP; the glycogen synthetases and phosphorylases which exist in phosphorylated and dephosphorylated forms; the proteases, which are activated by partial proteolysis; and the glutamate dehydrogenases, which have different substrate specificities depending on the state of aggregation.

With this background, we should like to explore with you some issues which arise from isozyme studies carried out in our laboratory during the past eight or nine years. These studies were all carried out with the pleasant and profitable collaboration of Dr. Harold P. Morris, who will provide a more detailed description of these tumors. The work we have chosen to discuss covers the study of isozyme patterns of four enzymes in liver under a variety of dietary and hormonal influences and in a " spectrum " of Morris tumors varying in growth rate and in degree of differentiation. The degree of differentiation is based on cellular morphology and tissue architecture from histological examinations carried out in our laboratory by Dr. David Meranze. For convenience, these have been divided into three groups, characterized as highly, well-, and poorly differentiated. In gross appearance, the highly and well-differentiated hepatomas resemble liver in color and consistency, the cells are large and contain abundant pale-staining eosinophilic cytoplasm, and they contain variable quantities of glycogen. The nuclei are round, with large nucleoli, and there are occasional double-nucleated cells. The cells are arranged in sheets, often in a lobular pattern. There are canaliculi, sometimes with bile pigment and sinusoids are conspicuous, with lining cells resembling Kuppfer cells. There may be bands of connective tissue, with macrophages containing hemosiderin, ceroid, lipid, and bile pigment. There are

TABLE III. General Properties of Morris Hepatomas (28, 44, 57)

Property	Degree of differentiation		
	High	Well	Poor
Growth rate	Very low	Low	Rapid
Chromosome number	Normal	Nearly normal	Abnormal
Chromosome karyotype	Normal	Nearly normal	Abnormal
Respiration	High	Moderate	Moderately low
Glycolysis	Low	Low	High
Enzyme pattern	Liver-like	Some deletions	Many deletions

many transitional stages between different hepatocellular carcinomas, based on architecture, cytology, and staining characteristics, often within a particular hepatoma. The choice between well- and highly differentiated is a matter of judgment based on the above-mentioned criteria.

The poorly differentiated hepatomas are grossly firm and gray or white in color. The cells are smaller, with great variations in size and pattern, and are laid down in crowded sheets. There is a complete loss of hepatocellular and lining cell pattern. Mitoses are frequent, with only rare canalicular formation and pigment deposition. A more detailed description of the gross and microscopic characteristics may be found in earlier reviews (33–35).

A summary of the properties of these hepatic tumors is shown in Table III. Closely correlated with the degree of differentiation is the growth rate. It is very low in the few highly differentiated tumors, extending from three or four months to a year for a transplant generation (33, 34). Growth rates are considerably faster, at two to six months, for well-differentiated tumors, and are extremely rapid, at one month or less, for the poorly differentiated hepatomas. Some of the highly differentiated tumors have the normal liver chromosome number and karyotype. Some of the well-differentiated tumors have the diploid number of chromosomes but differ slightly in karyotype. The poorly differentiated tumors deviate markedly in chromosome karyotype and number from those of rat liver. Respiration decreases moderately with loss of differentiation, but the striking feature of these tumors is the low or negligible glycolysis in the well-differentiated, in contrast with the usual high level of glycolysis in the poorly differentiated tumors (4, 54). These well-differentiated tumors, with their low glycolytic activity, make it clear that high lactic acid production is not an absolute requirement for tumor survival. These well-differentiated low-glycolyzing tumors can grow, albeit slowly, and they can also metastasize and ultimately kill their hosts. With decreased differentiation there is also a decrease or loss of certain enzymes which play a unique role in liver function, the so-called liver marker enzymes. We will elaborate on this aspect shortly.

Of the four enzymes whose isozyme patterns we plan to discuss, three are key enzymes of carbohydrate metabolism, the glucose-ATP phosphotransferases, the aldolases, and the pyruvate kinases. The fourth enzyme is adenylase kinase. Though not specifically involved in a metabolic pathway, it has the important function of maintaining the equilibrium among the three adenine nucleotides,

AMP, ADP, and ATP, substances which are either substrates, products, or effectors of a host of metabolic processes.

Glucose-ATP Phosphotransferases

This enzyme, as seen in Fig. 1, exists in multiple forms, with each normal tissue having its individual pattern. Liver has four forms, three of which, marked I, II, and III, are collectively called hexokinase. The fourth form is called glucokinase. In this figure is shown the relative intensities of the bands on starch gel, though it is important to point out that the intensity is not a reliable index of the activity. In fetal and neonatal liver, the preponderant forms are the three hexokinases, and the presence of glucokinase is undetectable by spectrophotometric assay, although a very faint band denotes its presence in extremely low activity.

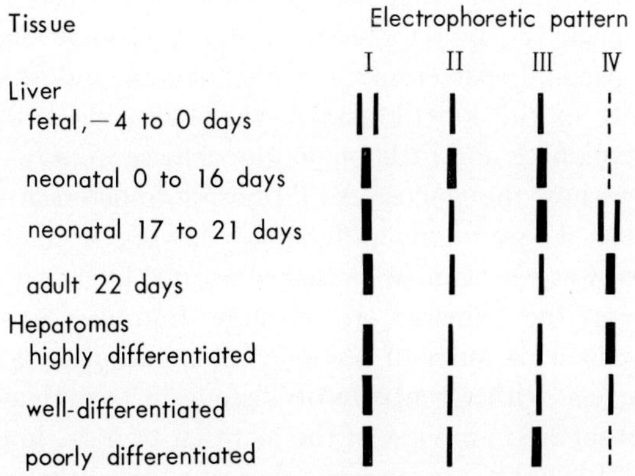

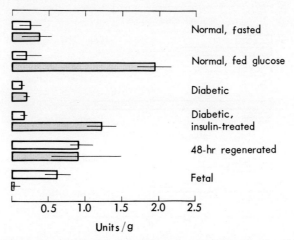

FIG. 1. Starch gel electrophoretic patterns of fetal and neonatal liver, and hepatoma glucose-ATP phosphotransferases

FIG. 2. Values for hexokinase (□) and glucokinase (▨) determined by kinetic assay

Hexokinase includes all three low K_m isozymes I, II, and III.

Not until 17 days after birth does the glucokinase make itself evident, and by 20 or 21 days after birth it is at the normal adult level. In the lower portion of this figure are patterns for the various classes of hepatomas. All three hepatoma types have the same three hexokinases. However, the glucokinase varies markedly with the state of differentiation.

Kinetic assays supplement the electrophoretic results in providing quantitative data, as shown in Fig. 2. The hexokinase isozymes, I, II, and III, collectively shown in the clear bars, are present at a low total activity which does not change with diet or hormonal conditions. On the other hand, the fourth isozyme, called glucokinase, is highly responsive to dietary and hormonal conditions, being low in fasted normal animals and high in carbohydrate-fed animals *(5, 10, 40, 44, 45, 53)*. It is also insulin-dependent, being extremely low in diabetes, and is restored by insulin injection *(10, 40)*. This isozyme has an important physiological function in hepatic glucose utilization. It has a very high K_m for glucose, and it is this property that is responsible for the fact that the liver takes up glucose only when the blood glucose concentration is high. During liver regeneration after partial hepatectomy, both hexokinase and glucokinase are high. We point especially to the kinetic assays which also show that fetal liver has essentially only hexokinase and little or no glucokinase *(5, 45)*.

Figure 3 shows how the glucose-ATP phosphotransferase activity changes in liver neoplasms. In a few highly differentiated, very slow-growing hepatomas the isozyme pattern was very similar to that of normal liver, with high glucokinase and low-to-moderate hexokinase. In a large number of well-differentiated tumors studied over a wide range of transplant generations, the glucokinase dropped to very low levels, with essentially no change in hexokinase *(41, 45)*. This is a very striking observation in view of the fact that tumors, in general, have high hexokinase levels. In contrast, a large number of poorly differentiated hepatocarcinomas do exhibit high hexokinase activity, but with little or no glucokinase.

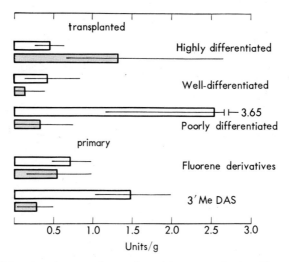

FIG. 3. Values in primary and transplanted hepatomas for hexokinase (□) and glucokinase (▨), determined by kinetic assay

Thus, loss of differentiation in hepatic tumors, which is accompanied by greatly increased growth rate, results in virtual loss of a glucose-ATP phosphotransferase which is physiologically functional and under regulation by diet and hormones, being replaced by high activities of three isozymes which are ordinarily low in normal liver. In the poorly differentiated Novikoff and 3924A hepatomas, all three hexokinase isozymes share in the marked rise, but predominant activity is present in Isozyme III. We call attention particularly to the striking resemblance of the isozyme pattern of the poorly differentiated tumors to that of the fetal liver.

Aldolase

Aldolase also exists in multiple forms which, like lactate dehydrogenase are tetramers of subunits with different primary structures distinguishable by immunological or kinetic criteria (*39*). In addition to aldolase A, which is the sole form of muscle aldolase, and aldolase B, which is the major form in liver, a new form termed aldolase C, as shown by Dr. Sugimura and others, has been found in brain, where it exists largely as an A-C hybrid with a large preponderance of A (*36, 39, 43*). The three forms are conveniently detected by starch gel electrophoresis, but the relative quantities of A and B subunits in liver and hepatomas can be assayed kinetically by the differences in their activity towards fructose 1, 6-diphosphate and fructose 1-phosphate.

Figure 4 contains composite data on liver and hepatomas from our own studies and data on fetal and perinatal liver by Rutter and Weber (*39*). In normal rat liver the almost exclusive form is aldolase B. In early fetal liver, up to about ten days before birth, the major form is not the liver-type aldolase B, but the nonhepatic aldolase A. At four days before birth there are approximately equal quantities of A and B, and the adult pattern is reached by one or two days before

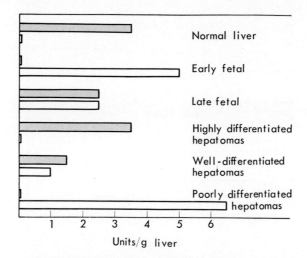

FIG. 4. Values for aldolase A (□) and aldolase B (▨) in liver and hepatomas, as determined by ratio of activities toward FDP and F1P

birth. Again, the highly differentiated hepatomas exhibit essentially only aldolase B activity.

In the well-differentiated, slow-growing tumors, the A form becomes evident, but the B form is still preponderant. However, in the poorly differentiated tumors, the B form has been essentially completely replaced by the A form. Similar observations have been made by George and Fanny Schapira in their extensive studies of aldolase isozymes in neoplasia (*42, 43, 48*). An interesting example of the persistence of the C form in brain tumors has been provided recently by Sugimura *et al.* (*48*), and these authors as well as Schapira *et al.* (*43*) found aldolase C to be present also in certain poorlydifferentiated hepatomas. Again we observe a striking similarity between the poorly differentiated tumors and the fetal liver.

Pyruvate Kinase

A crucial enzyme in the glycolytic pathway is the transphosphorylase which catalyzes the transfer of phosphate from phosphoenolpyruvate to ADP. This enzyme, like the glucose-ATP phosphotransferases, also exists in multiple forms. Two major forms are found in liver; type I, similar in kinetic properties to the muscle enzyme, and type II, which is the predominant form in normal liver (*9, 30, 49*). The latter is highly responsive to carbohydrates in the diet and has a number of distinctive kinetic properties, such as inhibition by ATP and activation by fructose diphosphate, which differentiate it sharply from the former. It is also reported by Weber *et al.* (*56*) to be low in diabetic rats, and its activity is restored by insulin treatment. Thus, it shares, in common with certain other liver

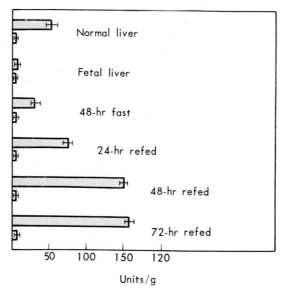

FIG. 5. Values for pyruvate kinase isozymes type I (□) and type II (▓) in rat liver, as determined by differential absorption on DEAE-cellulose

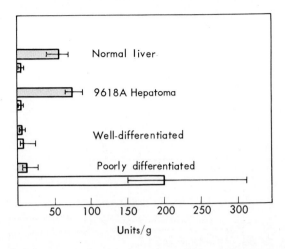

Fig. 6. Values for pyruvate kinase isozymes type I (□) and type II (■) in normal liver and in rat hepatoma, as determined by differential absorption on DEAE-cellulose

"marker" enzymes, specific functions that give the liver cells their unique metabolic capabilities. Figure 5 shows how the activities of forms I and II vary in liver. In normal liver form II is predominant, but as with other liver marker enzymes, is very low in fetal liver. The level of Isozyme II is lowered by about two-thirds by fasting and is restored readily by feeding carbohydrates. Throughout the dietary and hormonal manipulations, Isozyme I remains low and unchanged (14).

As shown in Fig. 6, a single, highly differentiated tumor, the 9618A, has the same isozyme pattern as liver, with a predominance of type II. However, a sharp distinction was observed between the well-differentiated and the poorly differentiated tumors. The former had very low levels of both isozymes, whereas the rapidly growing, poorly differentiated Novikoff and 3924A tumors had extremely high levels of an isozyme which has the properties of type I. Here we see again that with decreased differentiation there is a complete switch in isozyme pattern, with loss of a liver marker isozyme and its replacement by a nonhepatic type.

It now appears that the liver type is more complex and exists in multiple forms. Tanaka et al. (49) reported that normal liver has four pyruvate kinase isozymes detectable by starch gel electrophoresis. Taylor et al. (50) partially purified the major isozymes from rat liver, muscle, and the poorly differentiated 3924A hepatoma, and observed that the muscle and tumor isozymes differed in their electrophoretic migration on starch gel, in their stability, and in their susceptibility to SH reagents. More recently, Criss (7) in our laboratory separated the multiple forms of this enzyme by means of isoelectric focusing and obtained results depicted in Fig. 7. Liver and a highly differentiated hepatoma, the 9618A, exhibited similar patterns of four isozymes, one of which was identical with the single isozyme present in skeletal muscle. In a well-differentiated hepatoma, the 9633, the same four isozymes appeared, but they were accompanied by a fifth

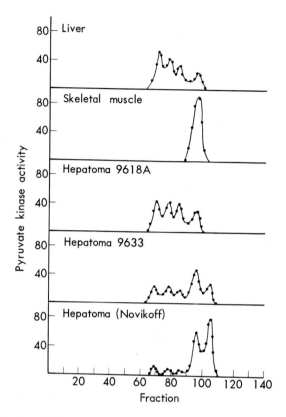

FIG. 7. Isoelectric focusing patterns of rat liver and hepatoma pyruvate kinase isozymes

form not detectable in normal liver. In the poorly differentiated Novikoff hepatoma, the first three were very low, with the preponderant activity in the muscle type and the new type. According to Criss, the new form is probably identical with the above-mentioned enzyme purified by Taylor et al. (50) from the 3924A hepatomas.

With pyruvate kinase, the resemblance of the poorly differentiated hepatomas to fetal liver is not altogether clear. Although the complex of liver-type isozymes evidently is markedly decreased, there is no evidence of the presence of another nonhepatic type in the fetal liver. It remains to be determined whether the nonhepatic type pyruvate kinase of fetal liver is the muscle type or the fifth form present in the poorly differentiated tumors.

Adenylate Kinase

An ATP-AMP phosphotransferase, responsive to diet and insulin, was found by us to be present in rat liver (3). This enzyme is high in fasting rat liver and is lowered markedly by glucose ingestion. It is also very high in fed diabetic rat liver and is lowered markedly by insulin treatment. Isoelectric focusing was employed by Criss et al. (8) in our laboratory to separate the liver enzyme into

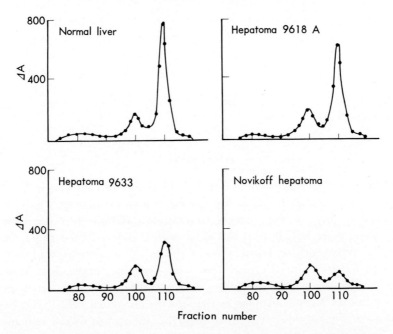

FIG. 8. Isoelectric focusing patterns of rat liver and hepatoma adenylate kinase isozymes

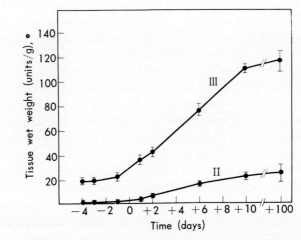

FIG. 9. Adenylate kinase isozymes II and III in fetal and neonatal rat liver

four distinct forms, the predominant one of which is the above-mentioned adaptive enzyme (Fig. 8). Two of these, Isozymes I and IV, are mere blips. Isozyme II is moderately high whereas Isozyme III, the adaptive isozyme, is predominant. Although all four forms are retained in the hepatomas, the activity of Isozyme III, which is unique to liver, drops sharply with decreased differentiation. Again, we see another example of the loss of a liver " marker " enzyme in poorly differentiated hepatic tumors.

Here, there is again observed a marked similarity to the fetal liver pattern (Fig. 9). According to Criss *et al.* (*16*), total adenylate kinase activity is low in fetal rat liver and, as shown in Fig. 9, the predominant form is isozyme III. The level of Isozyme III is relatively low, however, and begins to increase only after birth, to reach essentially the adult pattern only after ten days. Although Isozyme II also increases after birth somewhat, the level is substantially below that of Isozyme III.

DISCUSSION

Until recently, the prevailing cancer dogma regarded the cancer cell as a dedifferentiated cell, one in which the normal functional activities gave way to a new and competing function, namely, to replicate and multiply. A corollary view is that cancer cells, having thus lost their normal differentiated character have a new and common identity which they all share. While it may well turn out to be true that there is in all cancer cells some heritable, single characteristic that defines the neoplastic transformation, it is nevertheless evident from these and other studies on hepatic neoplasms that the parenchymal cell of the liver may yield tumors of great diversity. It is of further interest to us as biochemists that the diversity of morphology and tissue architecture has its counterpart in a molecular diversity.

The variable retention of liver marker isozymes in highly and well-differentiated hepatomas demonstrates that the neoplastic state is not incompatible with continued functional activity. Apparently, the genome of the neoplastic cell can be transcribed for enzymic synthesis unrelated to cell division even while it is replicated. Thus, the loss of liver marker isozymes, when it occurs in the lesser or poorly differentiated tumors, need not be attributable to the neoplastic transformation *per se*, but rather to an alteration in gene expression which may or may not accompany the neoplastic change.

Whatever may be the underlying mechanism of this phenomenon, it conceivably could play an important role in the growth and development of the tumor. Loss of highly regulated enzymes geared to liver function and their replacement by nonregulated enzymes geared to efficient utilization of fuel, may give such cells competitive advantages in growth rate which would overcome protective mechanisms of the host and enhance their replication. Even though certain regulatory mechanisms may be deleted from the cancer cell (and this deletion may be a key to its uncontrolled growth) there must, of course, be retained a high degree of internal metabolic regulation. Otherwise, cell replication or any orderly metabolic process would be impossible. Therefore, loss of enzyme regulation must be highly specific and, of course, only partial.

Resemblance to Fetal Liver

Perhaps the most significant aspect of the studies here described is the similarity of some of the isozyme patterns of the poorly differentiated hepatomas to those of fetal liver. These findings may be added to a large and ever-growing body of evidence that tumors may acquire fetal characteristics. We have already commented on this striking phenomenon (57), as have Schapira (42) and Sugimura et al. (48) in connection with their studies on the aldolases.

Abelev (1) has shown that α-fetoglobulin (fetal $α_1$-globulin) is synthesized and excreted by many human and experimental hepatomas. This protein is present in fetal liver and plasma of all animals and is the fetal counterpart of the serum albumin. It is never observed in normal adult liver or plasma. It can be readily detected by various immunological techniques, and its presence in serum of adult animals is indicative of the presence of liver neoplasms. Although it is also formed temporarily during liver regeneration following either surgery or treatment with liver toxins, it appears rarely, if at all, in nonmalignant liver disease such as hepatitis or cirrhosis, and is not present in cholangiocarcinoma. Histochemical evidence has further established the parenchymal cell as the site of formation of this protein.

Irlin et al. (26) found that transplantable mouse hepatomas also produce α-fetoglobulin, and Hull et al. (24) reported that stable cultures of primary chemical-induced monkey hepatomas produce not only α-fetoglobulin but also serum albumin and ferritin. Thus, these neoplasms carry out both fetal and adult functions.

α-Fetoprotein also appears in the yolk sac of the embryo and is present in about 50% of testicular and ovarian teratocarcinomas. It does not appear, however, in other embryonal cancers such as seminomas, chorionepitheliomas, neuroblastomas, nephroblastomas, and Wilm's tumors.

Studies by Gold et al. (19) demonstrated that all human cancers of the digestive tract contain common tumor-specific antigens. These antigens are absent from all other normal, diseased, or cancerous tissues, but are present in fetal gut, liver, and pancreas until the second trimester of gestation. These were named carcinoembryonic antigens, or CEA. Two types of subject had antibodies to these antigens in the plasma; women during all stages of pregnancy, and patients with nonmetastatic gastrointestinal tract cancer.

Another striking occurrence of a fetal protein is in virus-transformed cells. These cells produce components which can be detected immunologically; a T (or tumor) antigen in the nucleus, and surface antigens such as the tumor-specific transplantation antigen, TSTA, and the S (serological) antigen. Although the T antigen is coded for by the viral genome, it appears that the S antigen is coded for by the host cell (51). According to Duff and Rapp (11), SV-40 virus-induced hamster cells reacted specifically with sera from pregnant hamsters, thus indicating in these cells the presence of a derepressed embryonic antigen. It is not certain whether or not this is the S antigen. That the S antigen is a normal cell antigen unmasked by SV-40 transformation is also suggested by Hayry and Defendi (23). Another similar example is the observation by Burger (6) and by Inbar and Sachs (25) that SV-40 virus-transformed cells unmask sites on their cell surface which react with agglutinins or concanavalin A.

An interesting variant of the fetal isozyme pattern in cancer is the identification by Fishman et al. (17) of the so-called Regan isoenzyme of alkaline phosphatase as a placental antigen. Fiala and Fiala (50) also observed glutathionase in a series of chemically induced rat hepatomas. This enzyme was absent from normal rat liver but was present in fetal and neonatal rat liver.

Added to the examples already cited, there is a growing body of literature on bizarre aberrations of protein synthesis in cancer. Many of these are from the clinical literature. Lipsett (29) has cited over 100 cases of Cushing's syndrome associated with a variety of clinical nonpituitary neoplasms, principally bronchogenic carcinoma. This was apparently due to secretion of substances having the hormonal activity of ACTH. Severe hypoglycemia has been frequently reported as an accompaniment of various tumors of nonendocrine origin, and recently Miyabo et al. (32) found evidence of immunoreactive insulin in a gastric tumor associated with hypoglycemia. Goodall (20) and Eliel (12) have also reviewed the clinical literature which points to ectopic production and secretion of hormones from a variety of tumors of nonendocrine origin.

Evidently the gene readout mechanism can be distorted in certain neoplasms, so that genes normally completely repressed in the tissue of origin are unmasked for transcription and translation. These aberrations are all the more astonishing when we view them in the light of the extraordinary stability of the normal cell genome. The liver cell makes a wide variety of proteins but has never been known to make proteins that are characteristic of other cell types. If we accept the concept of a totipotent genome in every cell, then the repression in differentiated cells of all but the appropriate genes must verge on the absolute. There must be an unbelievably accurate control of gene transcription.

Is the derepression of otherwise repressed genes that is so often observed in cancer somehow trying to tell us something about the nature of the neoplastic transformation itself? How are the activities of these isozymes switched on and off, and how are these presumed alterations in gene action related to cell differentiation? Are we dealing with a disease of transcription? The steady state level of the adaptive enzymes in metazoan cells is established by complex homeostatic mechanisms involving not only substrate and hormone induction, operating at sites of synthesis, but also must involve degradative processes as well at sites of destruction. Until we know more about the molecular mechanism of this type of regulation, we can only speculate on the inadequate basis of what is known about gene regulation in microbial systems. Although we use the terminology borrowed from microbial studies, the concepts thus developed have not yet provided insight into the infinitely more complex problem of enzyme synthesis and destruction in liver and its neoplastic counterparts.

Another question to which we must address ourselves is how does the transformation of the normal parenchymal cell give rise to such diverse neoplasms? This question also cannot yet be answered and only leads to further questions. Is it the normal adult liver cell that is transformed? If so, this would explain the occurrence of well- and highly differentiated hepatomas, and one might expect on this basis that the poorly differentiated hepatomas arise by progression. However, our isozyme data reveal that loss of differentiation is a patterned process. Nonhepatic isozymes only appear when the hepatic isozyme is depleted. It seems unlikely, therefore, that these changes can be due to the random, independent series of deletions that are presumed to be involved in tumor progression (18), although this is not to say that some poorly differentiated hepatomas cannot arise by progression.

Does an undifferentiated stem cell undergo transformation followed by partial differentiation, as suggested by Pierce (37) for the teratocarcinoma? Can the neoplastic transformation occur at any stage of partial differentiation and then become " fixed " in the tumor? Finally, one might ask whether a " population " of parenchymal cells may become transformed, the ultimate neoplasm which develops depending on the immediate environment, the operation of immunologic mechanisms, and the relative proportion of well- and poorly differentiated cells. This hypothesis attains plausibility, particularly because primary liver neoplasms often consist of mixed cell types. Farber (13) has pointed out that chemical hepatic carcinogenesis is a prolonged, multi-step process which includes a reversible preneoplastic, hyperplastic nodule state.

If we have no ready answers to these questions, they should provide, nevertheless, sufficient issues for much lively and profitable discussion during the next few days.

REFERENCES

1. Abelev, G. I. α-Fetoprotein in oncogenesis and its association with malignant tumors. *Adv. Cancer Res.*, **14**, 295–354 (1971).
2. Adelman, R. C., Morris, H. P., and Weinhouse S. Fructokinase, triokinase, and aldolases in liver tumors of the rat. *Cancer Res.*, **27**, 2408–2413 (1967).
3. Adelman, R. C., Lo, C. H., and Weinhouse, S. Dietary and hormonal effects on adenosine triphosphate-adenosine monophosphate phosphotransferase activity in rat liver. *J. Biol. Chem.*, **243**, 2538–2544 (1968).
4. Aisenberg, A. C. and Morris, H. P. Energy pathways of hepatoma 5123. *Nature*, **191**, 1314–1316 (1961).
5. Ballard, F. J. and Oliver, I. T. Ketohexokinase isozymes of glucokinase and glycogen synthesis from hexoses in neonatal rat liver. *Biochem. J.*, **90**, 261–268 (1964).
6. Burger, M. M. A difference in the architecture of the surface membrane of normal and virally transformed cells. *Proc. Natl. Acad. Sci. U. S.*, **62**, 994–1001 (1969).
7. Criss, W. E. A new pyruvate kinase isozyme in hepatomas. *Biochem. Biophys. Res. Commun.*, **35**, 901–905 (1969).
8. Criss, W. E., Litwack, G., Morris, H. P., and Weinhouse, S. ATP-AMP phosphotransferase isozymes in rat liver and hepatomas. *Cancer Res.*, **30**, 370–375, (1970).
9. de Asua, L. J., Rozengurt, E., and Carminati, H. Some kinetic properties of liver pyruvate kinase. *J. Biol. Chem.*, **245**, 3901–3905 (1970).
10. DiPietro, D. L., Sharma, C., and Weinhouse, S. Studies on glucose phosphorylation in rat liver. *Biochemistry*, **1**, 455–462 (1962).
11. Duff, R. and Rapp, F. Reaction of serum from pregnant hamsters with surface of cells transformed by SV40. *J. Immunol.*, **105**, 521–523 (1970).
12. Eliel, L. P. Non-endocrine secreting neoplasm clinical manifestations. *Bull. Cancer*, **20**, 37–39 (1968).
13. Farber, E. Studies on the molecular mechanisms of carcinogenesis. *Miami Winter Symp.*, **2**, 314–334 (1970).
14. Farina, F. A., Adelman, R. C., Morris, H. P., Lo, C. H., and Weinhouse, S.

Metabolic regulation and enzyme alterations in the Morris hepatomas. *Cancer Res.*, **28**, 1897–1900 (1968).

15. Fiala, S. and Fiala, Q. E. Acquisition of an embryonal biochemical feature by rat hepatomas. *Experientia*, **15**, 889–890 (1970).
16. Filler, R. and Criss, W. E. Development of adenylate kinase isozymes in rat liver. *Biochem. J.*, **122**, 553–555 (1971).
17. Fishman, W. H., Inglis, N. R., and Green, S. Regan isoenzyme: a carcino-placental antigen. *Cancer Res.*, **11**, 1054–1057 (1971).
18. Foulds, L. "Neoplastic Development," Academic Press Inc., New York, Vol. 1, p. 45 (1969).
19. Gold, P. and Freedman, S. O. Specific carcinoembryonic antigens of the human digestive tract. *J. Exp. Med.*, **122**, 467–481 (1965).
20. Goodall, C. M. A review on para-endocrine cancer syndromes. *Int. J. Cancer*, **4**, 1–10 (1969).
21. Grossbard, L. and Schimke, R. T. Multiple hexokinases of rat tissues. Purification and comparison of soluble forms. *J. Biol. Chem.*, **241**, 3546–3560 (1966).
22. Haglund, H. Isoelectric focussing in natural pH gradients. *Sci. Tools (LKB Instruments)*, **16**, 2–7 (1967).
23. Hayry, P. and Defendi, V. Surface antigens of SV40-transformed tumor cells. *Virology*, **41**, 22–29 (1970).
24. Hull, E., Carbone, P. P., Gitlin, D., O'Gara, R. W., and Kelley, M. G. α-Fetoprotein in monkeys with hepatoma. *J. Natl. Cancer Inst.*, **42**, 1035–1044 (1969).
25. Inbar, M. and Sacks, L. Interaction of the carbohydrate-binding protein concanavalin-A with normal and transformed cells. *Proc. Natl. Acad. Sci. U.S.*, **63**, 1418–1425 (1969).
26. Irlin, I. S., Perov, S. D., and Abelev, G. I. Changes in the biological and biochemical properties of mouse hepatoma during long-term cultivation *in vitro*. *Int. J. Cancer*, **1**, 337–347 (1966).
27. Katzen, H. M. and Schimke, R. T. Multiple forms of hexokinase in the rat. *Proc. Natl. Acad. Sci. U.S.*, **54**, 1218–1225 (1965).
28. Latner, A. and Skillen, A. "Isoenzymes in Biology and Medicine," Academic Press Inc., London and New York (1968).
29. Lipsett, M. B. Humoral syndromes associated with non-endocrine tumors. *Ann. Int. Med.*, **61**, 733–756 (1964).
30. Llorente, P., Marco, P., and Sols, A. Regulation of liver pyruvate kinase and the phosphoenolpyruvate crossroad. *Eur. J. Biochem.*, **13**, 45–54 (1970).
31. Markert, C. L. and Moller, F. Multiple forms of enzymes: tissue, ontogenetic, and species specific patterns. *Proc. Natl. Acad. Sci. U.S.*, **45**, 753–763 (1959).
32. Miyabo, S., Fujimura, T., and Murakami, M. Gastric cancer containing insulin and associated with hypoglycemia. *Diabetes*, **17**, 286–289 (1968).
33. Morris, H. P. Studies on the development, biochemistry and biology of experimental hypatomas. *Adv. Cancer Res.*, **9**, 227–302 (1965).
34. Morris, H. P. and Wagner, B. P. Induction and transplantation of rat hepatomas with different growth rate (including minimal deviation hepatomas). *Methods Cancer Res.*, **4**, 125–152 (1968).
35. Nowell, P. C., Morris, H. P., and Potter, V. R. Chromosomes of "Minimal Deviation" hepatomas. *Cancer Res.*, **27**, 1561–1579 (1967).
36. Penhoet, E., Rajkumar, T., and Rutter, W. J. Multiple forms of fructose diphosphate aldolase in mammalian tissues. *Proc. Natl. Acad. Sci. U.S.*, **56**, 1275–1282 (1966).

37. Pierce, G. B. Teratocarcinoma; model for developmental concept of cancer. *Curr. Top. Dev. Biol.*, **2**, 223–246 (1967).
38. Rosado, A., Morris, H. P., and Weinhouse, S. Lactate dehydrogenase subunits in normal and neoplastic tissues of the rat. *Cancer Res.*, **29**, 1673–1680 (1969).
39. Rutter, W. J. and Weber, C. S. Specific proteins in cytodifferentiation. University of Texas M. D. Anderson Hospital 19th Annual Symposium on Developmental and Metabolic Control Mechanisms and Neoplasia, 195–218 (1965).
40. Salas, M., Vinuela, E., and Sols, A. Insulin dependent synthesis of liver glucokinase in the rat. *J. Biol. Chem.*, **238**, 3535–3538 (1963).
41. Sato, S., Matsushima, T., and Sugimura, T. Hexokinase isozyme patterns of experimental hepatomas of rats. *Cancer Res.*, **29**, 1437–1446 (1969).
42. Schapira, F. Isozymes et cancer. *Pathol. Biol.*, **18**, 309–315 (1970).
43. Schapira, F., Reuber, M. D., and Hatzfeld, A. Resurgence of two fetal-type aldolases (A and C) in some fast-growing hepatomas. *Biochem. Biophys. Res. Commun.*, **40**, 321–325 (1970).
44. Sharma, C., Manjeshwar, R., and Weinhouse, S. Effects of diet and insulin on glucose-ATP phosphotransferase of rat liver. *J. Biol. Chem.*, **238**, 3840–3845 (1963).
45. Shatton, J. B., Morris, H. P., and Weinhouse, S. Kinetic, electrophoretic, and chromatographic studies on glucose-ATP phosphotransferases in rat liver and hepatomas. *Cancer Res.*, **29**, 1161–1172 (1969).
46. Shugar, D. "Enzymes and Isoenzymes, Structure, Properties and Function," Academic Press Inc., London and New York (1970).
47. Stadtman, E. R. "The Enzymes," ed. by P. Boyer, Academic Press Inc., New York and London, p. 397 (1970).
48. Sugimura, T., Sato, S., and Kawake, S. The presence of aldolase C in rat hepatoma. *Biochem. Biophys. Res. Commun.*, **39**, 626–630 (1970).
49. Tanaka, T., Harano, H., Sue, F., and Morimura, H. Crystallization characterization and metabolic regulation of two types of pyruvate kinase isolated from rat tissues. *J. Biochem. (Tokyo)*, **62**, 71–91 (1967).
50. Taylor, C. B., Morris, H. P., and Weber, G. A comparison of the properties of pyruvate kinase from hepatoma 3924A, normal liver and muscle. *Life Sci.*, **8**, 635–644 (1969).
51. Tevethia, S. S., Diamandopoulos, G. T., Rapp, F., and Enders, J. F. Lack of relationship between virus-specific surface and transplantation antigens in hamster cells transformed by simian papovirus SV40. *J. Immunol.*, **101**, 1192–1198 (1968).
52. Vesell, E. S. Multiple molecular form of enzymes. *Ann. N. Y. Acad. Sci.*, **151**, 1–689 (1968).
53. Vinuela, E., Salas, M., and Sols, A. Glucokinase and hexokinase in rat liver. *J. Biol. Chem.*, **238**, 1175–1177 (1963).
54. Weber, G., Banerjee, G., and Morris, H. P. Comparative biochemistry of hepatomas. I. Carbohydrate enzymes in Morris hepatoma 5123. *Cancer Res.*, **21**, 933–937 (1961).
55. Wroblewski, F. Multiple molecular forms of enzymes. *Ann. N.Y. Acad. Sci.*, **94**, 655–1030 (1961).
56. Weber, G., Stamm, N. B., and Fischer, E. A. Insulin: Induction of pyruvate kinase. *Science*, **149**, 65–67 (1965).
57. Weinhouse, S. Respiration, glycolysis and enzyme alterations in liver neoplasms. *Miami Winter Symp.*, **2**, 462–480 (1970).

REGULATION OF ENZYME SYNTHESIS RELATING TO DIFFERENTIATION OF MALIGNANT CELLS[*1]

Tetsuo ONO, Kiyoshige WAKABAYASHI, Kyoko UENOYAMA, and Hideki KOYAMA

Cancer Institute[*2]

The relationship between growth rates and activities of liver marker enzymes was pursued using the wide spectrum of Morris hepatomas. Most of the liver marker enzymes, which are related to the differentiating functions of liver, are inversely correlated to the growth rates of hepatomas, and these enzyme activities decreased gradually in the hepatomas as the hepatomas progressed after many transplantations. However, there appeared another type of enzyme, such as carbamylphosphate synthetase and threonine dehydrase. These are also liver-specific enzymes and are related to differentiated functions of the liver, but showed remarkably higher activities in hepatomas than in normal liver; these high activities persisted even after the hepatomas had progressed during transplantations. Expression of the latter class of enzymes should be interpreted as derepression. As for the mechanism of regulation of enzyme activities in the first class by the growth rates of hepatomas, the role of extrachromosomal genomes was discussed and some circumstantial evidence presented.

Another system employed to explore the mechanism of regulation for the expression of differentiated functions was a cultured cell line which produced hyaluronic acid. This cell line was found among the hybrids of mouse mammary carcinoma cells and Chinese hamster fibroblast cells. Hyaluronic acid production was detected only in the rapidly growing state and ceased in the stationary phase. BUdr, an analog of thymidine, reversibly inhibited this function more than the cell growth. BUdr was found to decrease the hyaluronic acid synthetase in the cells and to induce alkaline phosphatase remarkably, but had no effect at all on acid phosphatase.

Studies on isozyme patterns of different enzymes in a wide spectrum of experimental tumors have revealed that changes of isozyme pattern in malignant cells could be explained as the results of dedifferentiation (*4, 15*) and/or disdifferentiation (*16*), rather than by mutation. On these bases, it has been suggested that cancer might be considered as a disease of differentiation rather than

---

[*1] This work was supported by grants for scientific research from the Ministry of Education and from the Princess Takamatsu Fund for Cancer Research.

[*2] Kami-Ikebukuro 1-37-1, Toshima-ku, Tokyo 170, Japan (小野哲生，若林清重，上野山恭子，小山秀機).

a mutation (11). Therefore, studies on the mechanism for the development of differentiation and the fixation of differentiated states in the cells are most relevant to elucidation of the mechanism of carcinogenesis as well as to discovery of the clue to controlling cancer.

Relationship between Growth Rates and Enzyme Activities of Minimal Deviation Hepatomas

The expressions of differentiated functions are generally explained by the derepression (6) or the switch-on of the genomes (17). Weber (18) has emphasized the close correlation between growth rates and enzyme activities in the Morris hepatoma spectrum. We also pursued the relationship between growth rates and activities of liver marker enzymes, using the wide spectrum of Morris hepatomas (12). The results are summarized in Table I, in which we have classified the enzymes we tested in the hepatomas into two groups, one group containing those enzymes regulated by growth rate of hepatomas, and the other containing these not regulated by growth rate. Enzymes in the upper half of the table are known as liver-specific enzymes, while the ones in the lower half have no relation to differentiation of liver. The inverse correlations between the enzyme activities of glucose-6-phosphatase, fructose diphosphatase, glutamate dehydrogenase, or ornithine transcarbamylase and growth rates of minimal deviation hepatomas were reported by the author (12). Also, we reported the decrease of these enzyme activities in the hepatomas as the hepatomas progressed after many transplantations (12).

TABLE I. Enzyme Activities of Morris Minimal Deviation Hepatomas

		Correlated with growth rate	Not correlated with growth rate
Relation with differentiation	+	G-6-Pase	Thr-dehydrase
		FDPase	Carbamylphosphate synthetase
		Glutamate dehydrogenase	
		OTC[a]	
	−	Glycolysis	G-6-P dehydrogenase
		Catalase	Tyr-transaminase
Mode of gene expression		(?)	(Derepression)

[a] OTC, ornithine transcarbamylase.

There appeared another type of enzyme, such as carbamylphosphate synthetase, which is shown in Fig. 1A, and threonine dehydrase, which is shown in Fig. 1B. They are also liver-specific enzymes and are related to differentiated functions of the liver, but showed remarkably higher activities in hepatomas than in normal liver. These high activities in hepatomas persisted even after the hepatomas had progressed during transplantations. In Fig. 1, the level of enzyme activity of normal rat liver is illustrated by a dashed line. As one can see, when there were enzyme activities in hepatomas, they were higher than in normal rat

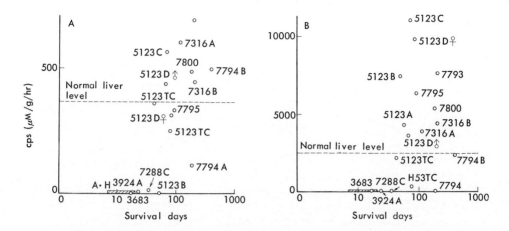

Fig. 1. Cases of no correlation between liver marker enzyme activities and growth rates of hepatomas
  A. Carbamylphosphate synthetase activities of hepatomas and survival days of host rats. B. Threonine dehydrase activities of hepatomas and survival days of host rats.

liver. Therefore, in these cases, it is apparent that they are derepressed in the expression.

Extrachromosomal Genes and Differentiation

Expression of the derepressed enzymes should be interpreted by the mechanism of enzyme regulation proposed by Jacob and Monod (6). However, how and why the enzymes of the first group are intimately regulated by the growth rate of hepatomas need answering. One possibility considered was that the genomes of the enzymes in this group are amplified and contained in the cell as extrachromosomal genomes. If the chromosome replication rate, that is, the growth rate, is too fast and the replication of these extrachromosomal genomes cannot keep pace with it, then they might be diluted out. As one of the extrachromosomal DNA, the presence in mitochondria is generally accepted (5), but its functions have not yet been fully revealed. Therefore, we tried to determine its function in relation to the expression of differentiated functions.

First, we concluded that mitochondrial genomes are really transcribed to messenger RNA in the cytoplasm, and these messenger RNA's are quite different from those produced in the nuclei. As was shown in Table II, the RNA's from mitochondria of rat liver hybridize well only with mitochondrial DNA, not with DNA from nuclei. In contrast to this, messenger RNA's from nuclei hybridize preferentially with nuclear DNA and not much with mitochondrial DNA.

Concerning the genomes and their expressions in mitochondria of hepatomas, we have compared the mitochondrial DNA and messenger RNA of Yoshida as-

TABLE II. Hybridization Efficiency of Mitochondrial and Nuclear Messenger RNA's with Mitochondrial and Nuclear DNA

Source of DNA	RNA	Hybridization efficiency (%)
Mitochondria	Mit.-RNA[a]	12.1[c]
	30SP-RNA[b]	1
Nuclei	Mit.-RNA	2.1
	30SP-RNA	15.6

[a] Mit.-RNA is chemically labeled mitochondrial RNA of rat liver.
[b] 30SP-RNA is prepared from the 30S-ribonucleoprotein particles extracted from nuclei and contains the rapidly labeled RNA produced in nuclei.
[c] The percentage of RNA hybridized out of input RNA.

TABLE III. Hybridization Efficiency (%) of Mitochondrial RNA's from Yoshida Ascites AH-130 and Rat Liver with Mitochondrial DNA of AH-130 and Rat Liver

Mit.-DNA \ Mit.-RNA	AH-130	Liver
AH-130	12.0	1.0
Liver	1.2	12.5

cites hepatoma AH-130 with that of normal rat liver by RNA-DNA hybridization. The results of that experiment are shown in Table III. Mitochondrial RNA's of ascites hepatoma hybridize only with mitochondrial DNA of hepatoma. On the other hand, liver mitochondrial RNA's hybridize well with mitochondrial DNA of liver, but not with that of hepatoma. Therefore, the differences in mitochondrial genomes and in their expression between ascites hepatoma and rat liver were quite clear. The question of interest and which need solving is whether these differences of mitochondrial genomes between normal liver and AH-130 are general phenomena between normal and malignant cells and have any meaning in the malignant transformation.

Possible Existence of Catalase Genome in Mitochondria

During the course of another experiment, we found that nascent catalase on rat liver polysomes had catalase activity. Our evidence to support this is as follows. We prepared polysomes from rat liver and purified it by Sephadex G-200 gel filtration. From the column, polysomes came out at the top, and later soluble catalase was eluted out and, in every case, we found some distinct catalase activity on the peak of the polysomes. The catalase activity on the polysomes could be completely abolished by the administration of Actinomycin-D. The time course of disappearance of catalase from the polysomes by Actinomycin-D is shown in Fig. 2. As shown in this figure, the half-life of catalase activity on the polysomes after Actinomycin-D administration is only 1 hr. This indicates that the turnover rate of messenger RNA for catalase is very rapid. In addition to Actinomycin-D, we found that ethidium bromide, which preferentially inhibits the transcrip-

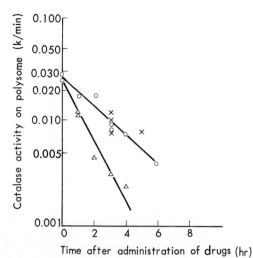

Fig. 2. Decrease of catalase activity on the polysomes of the rat after the administration of Actinomycin-D and ethidium bromide

○, ethidium bromide, 1.5 mg/100 g body weight; ×, ethidium bromide, 2.25 mg/100 g body weight; △, Actinomycin-D, 150 μg/100 g body weight.

tion in mitochondria but not of that in the nuclei (*19*), curtailed the supply of messenger RNA for catalase in the liver. Thus, catalase activity on the polysomes decreased by the rate indicated by the upper line in Fig. 2.

Third, we showed that the RNA extracted from mitochondria did contain the messenger RNA for catalase by assay system of protein synthesis *in vitro*. The messenger RNA for catalase was even more concentrated in mitochondria than in other RNA fractions.

The case of catalase is just one example, but it seems quite possible that further experiments would reveal the role of mitochondrial genomes in the expressions of differentiated functions, which are regulated by growth rates of the cells.

Differentiated Function of Cultured Cells

Another system employed in our laboratory to explore the mechanism of regulation for the expression of differentiated function was a cultured cell line which produced hyaluronic acid. Figure 3 illustrates the procedure we carried out to obtain the hybrids between mouse mammary carcinoma cell (FM3A) and Chinese hamster fibroblast cell (CHL), using UV-killed HVJ (Sendai virus) (*10*). The mouse cell chromosomes we used were mostly telocentric and 43 in modal number, and the Chinese hamster cell chromosomes were mostly biarmed and 22 in modal number. The chromosome constitution of the B-6 cell, one of the hybrids, was a mixed population of mouse and hamster chromosomes. There were 90 in all, and they seemed to be composed of nearly two sets of mouse chromosomes and one-half set of Chinese hamster origin. The culture medium of B-6 clone of hybrid was found to become more viscous day by day, and this viscous material was soon identified as hyaluronic acid (*7*). Since neither of

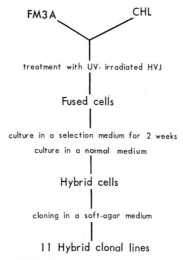

FIG. 3. Method of cell hybridization
 Normal medium: Eagle's MEM with 5% calf serum and 0.1% Bacto-peptone.
 Selection medium: Amethopterin, 1 μg/ml; hypoxanthine, 10 μg/ml; thymidine, 5 μg/ml; in the normal medium.

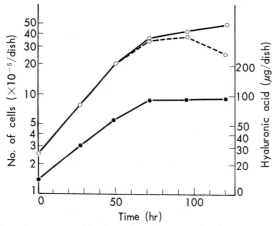

FIG. 4. Growth curves and hyaluronic acid synthesis of B-6 cells
 ○———○ growth curve of total cells in culture; ○- - - -○ growth curve of viable cells in culture; ●———● accumulation of hyaluronic acid in culture fluid.

the parental cells produced a significant amount of hyaluronic acid, and only the B-6 line among the 4 clones of hybrids tested produced a considerable amount of hyaluronic acid, it is evident that hyaluronic acid production was accidentally induced in B-6 cell by the hybridization procedure. However, we have not yet clarified the mechanism of this induction.

As shown in Fig. 4, hyaluronic acid production occurred only in the rapidly growing state and ceased in the stationary phase of cell growth (7).

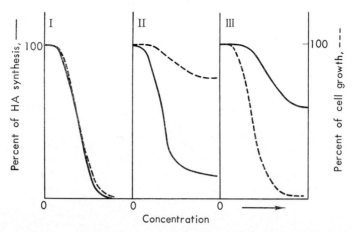

FIG. 5. Three types of effects on cell growth and hyaluronic acid synthesis of B-6 cells

The effects of hydroxyurea, an inhibitor of DNA synthesis, Actinomycin-D, an inhibitor of RNA synthesis, and cycloheximide, an inhibitor of protein synthesis, on B-6 cell growth and hyaluronic acid synthesis were compared. Cell growth was found to be inhibited more strongly than hyaluronic acid production by hydroxyurea. On the other hand, Actinomycin-D and cycloheximide equally inhibited cell growth and hyaluronic acid production. The effects of bromodeoxyuridine (BUdr), excess thymidine, and fluorodeoxyuridine (FUdr) on cell growth and hyaluronic acid synthesis were also tested. BUdr preferentially inhibited hyaluronic acid synthesis more than cell growth and, in contrast to BUdr, excess thymidine and FUdr inhibited cell growth more strongly than they did hyaluronic acid production (8).

Summarizing the results of these experiments, we classified the drugs tested into three groups, as shown in Fig. 5. The chemicals in the first group (I), Actinomycin-D, cycloheximide, Acridine Orange, and ethidium bromide, inhibited equally cell growth and hyaluronic acid synthesis. It is interesting to note that Acridine Orange and ethidium bromide, which are known to eliminate extrachromosomal genome, that is, episome in bacteria, fall into this group. The chemicals in the second group (II), such as BUdr, 5-bromodeoxycytidine, and 5-iododeoxyuridine, preferentially inhibited hyaluronic acid production more than cell growth. Recently, similar inhibitory effects of these drugs have been reported on the expression of differentiated functions in chick embryo cells, such as myogenesis in myoblasts (14), chondrogenesis in chondrocytes (1), melanin synthesis in retina cells (3), and hyaluronic acid synthesis in amnion cells (2). Furthermore, in mouse melanoma cells, both melanin synthesis and tumorigenicity were also found likely to be suppressed by BUdr (13). The chemicals of the third group (III), excess thymidine, hydroxyurea, and FUdr, which are all inhibitors of DNA synthesis, inhibited cell growth more strongly than hyaluronic acid synthesis. This seems to indicate that the continuous synthesis of metabolic DNA, or so-called messenger DNA, is not essential for the expression and maintenance of the differentiated function of hyaluronic acid synthesis.

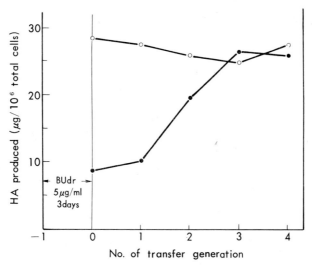

FIG. 6. Recovery of hyaluronic acid synthesis in B-6 cells after treatment with BUdr

○ hyaluronic acid (HA) synthesis in control culture; ● hyaluronic acid synthesis in culture treated with BUdr during the first three days.

The effect of BUdr in suppressing hyaluronic acid production was counteracted by the simultaneous addition of thymidine in amounts more than double that of the BUdr used (8), but production was not protected by any pyrimidines other than thymidine.

As in the cases of chondroitin sulfate production of chondrocytes (7) and myosin production by myoblasts (14), the effect of BUdr in suppressing hyaluronic acid synthesis is completely reversible. As shown in Fig. 6, by treatment with

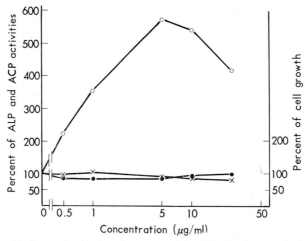

FIG. 7. Alkaline and acid phosphatase activities of B-6 cells treated with BUdr of various concentrations

○ percent of alkaline phosphatase (ALP) activity of B-6 cells treated with BUdr for three days; ● percent of cell growth after three days; × percent of acid phosphatase (ACP) activity after three days.

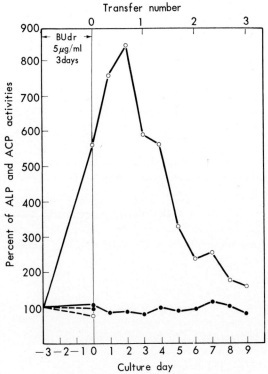

Fig. 8. Decrease of alkaline phosphatase activity induced by BUdr after removing BUdr from the medium
○ change of alkaline phospatase activity of B-6 cells; ● change of acid phosphatase activity of B-6 cells; —— treated with BUdr; --- control.

BUdr for three days, the production of hyaluronic acid of B-6 cells decreased to 30% of the control. Thereafter, the culture medium was replaced with one without BUdr and the recovery course of hyaluronic acid synthesis was investigated. By the first transfer, the recovery of hyaluronic acid production was very little, but by two more transfers it was gradually and completely restored to the untreated control level. In one transfer we cultured the cells for three days. Therefore, complete recovery of hyaluronic acid synthesis takes about ten generations, which is too long to allow us to assume the recovery process to be a simple replacement by thymidine of BUdr incorporated into DNA.

We also assayed the hyaluronic acid synthetase enzyme activity, that is, hyaluronic synthesis from UDP-N-acetylglucosamine and UDP-glucuronic acid, of cells treated for three days with BUdr of various concentrations. It was found that the decrease of synthetase activity parallelled the retardation rate of hyaluronic acid production.

As a control of hyaluronic acid synthetase, we assayed acid phosphatase as well as the alkaline phosphatase activities of these cells. As shown in Fig. 7, BUdr treatment exhibited almost no effect upon acid phosphatase activity but, unexpectedly, alkaline phosphatase was remarkably induced by BUdr (9). The optimal concentration of BUdr for the induction of alkaline phosphatase was found also to be around 5 g/ml, as in the case of inhibition of hyaluronic acid synthesis. Figure 8 shows the time course of the changes in alkaline phosphatase activity of B-6 cells treated for three days with BUdr, which was thereafter removed. The induction of alkaline phosphatase by BUdr did not stop immediately after removing it from the medium, but continued two more days and then gradually decreased to the original level. During this course, no significant changes in acid phosphatase by BUdr were observed. The induction of alkaline phosphatase in the cell by BUdr was suppressed only by thymidine and not by other pyrimidine nucleosides or pyrimidines as observed in the inhibition of hyaluronic acid synthesis.

So far, we have not yet clarified the mechanism of BUdr in modulating gene expressions, especially of differentiated functions. However, both hyaluronic acid synthesis and alkaline phosphatase are likely to reside on the cell membrane, and thus it is interesting to note that BUdr specifically modifies the functions of membrane.

In conclusion, we would like to mention that our hybrids between mouse carcinoma cell and Chinese hamster normal cell still retain their tumorigenicity to the mouse (9).

REFERENCES

1. Abbott, J. and Holtzer, H. The loss of phenotypic traits by differentiated cells. V. The effect of 5-bromodeoxy-uridine on cloned chondrocytes. *Proc. Natl. Acad. Sci. U.S.*, **59**, 1144–1151 (1968).
2. Bischoff, R. and Holtzer, H. Inhibition of hyaluronic acid synthesis by BUdr in culture of chick amnion cells. *Anat. Rec.*, **160**, 317 (1968).
3. Coleman, A. W., Kunkel, D., Werner, L., and Coleman, J. R. Cellular differentiation *in vitro*: perturbation by halogenated deoxyribonucleosides. *J. Cell Biol.*, **39**, 27a (1968).
4. Farina, F. A., Adelman, R. C., Lo, C. H., Morris, H. P., and Weinhouse, S. Metabolic regulation and enzyme alterations in the Morris hepatomas. *Cancer Res.*, **28**, 1897–1900 (1968).
5. Humm, D. G. and Humm, J. H. Hybridization of mitochondrial RNA with mitochondrial and nuclear DNA in agar. *Proc. Natl. Acad. Sci. U.S.*, **55**, 114–119 (1966).
6. Jacob, F. and Monod, J. Genetic repression, allosteric inhibition, and cellular differentiation. *In* " Cytodifferentiation and Macromolecular Synthesis," ed. by M. Locke, Academic Press Inc., New York, pp. 30–64 (1963).
7. Koyama, H., and Ono, T. Initiation of a differentiated function (hyaluronic acid synthesis) by hybrid formation in culture. *Biochim. Biophys. Acta*, **217**, 477–487 (1970).
8. Koyama, H. and Ono, T. Effect of 5-bromodeoxyuridine on hyaluronic acid synthesis of a hybrid between mouse and Chinese hamster cell lines in culture. *J. Cell Physiol.*, **78**, 265–272 (1971).
9. Koyama, H. and Ono, T. Induction of alkaline phosphatase by 5-bromodeoxyuridine in a hybrid line between mouse and Chinese hamster in culture. *Exp. Cell Res.*, **69**, 468–470 (1970).
10. Koyama, H., Yatabe, I., and Ono, T. Isolation and characterization of hybrids between mouse and Chinese hamster cell lines. *Exp. Cell Res.*, **62**, 455–463 (1970).

11. Markert, C. Neoplasia: a disease of cell differentiation. *Cancer Res.*, **28**, 1908–1914 (1968).
12. Ono, T. Enzyme patterns and malignancy of experimental hepatomas. *GANN Monograph*, **1**, 189–205 (1966).
13. Silagi, S. and Bruce, S. A. Suppression of malignancy and differentiation in melanotic melanoma cells. *Proc. Natl. Acad. Sci. U.S.*, **66**, 72–78 (1970).
14. Stockdale, F., Okazaki, K., Nameroff, M., and Holtzer, H. 5-Bromodeoxyuridine: effect on myogenesis *in vitro*. *Science*, **146**, 533–535 (1964).
15. Suda, M., Tanaka, T., Sue, F., Harano, Y., and Morimura, H. Dedifferentiation of sugar metabolism in the liver of tumor-bearing rat. *GANN Monograph*, **1**, 127–141 (1966).
16. Sugimura, T. Decarcinogenesis, a newer concept arising from our understanding of the cancer phenotype. *In* " Chemical Tumor Problems," ed. by W. Nakahara, Japan Society for the Promotion of Science, Tokyo, pp. 269–284 (1970).
17. Sugimura, T., Matsushima, T., Kawachi, T., Hirata, Y., and Kawabe, S. Molecular species of aldolases and hexokinases in experimental hepatomas. *GANN Monograph*, **1**, 143–149 (1966).
18. Weber, G. The molecular correlation concept: studies on the metabolic pattern of hepatomas. *GANN Monograph*, **1**, 151–178 (1966).
19. Zylber, E., Vesro, C., and Penman, S. Selective inhibition of the synthesis of mitochondria-associated RNA by ethidium bromide. *J. Mol. Biol.*, **44**, 195–204 (1969).

DISDIFFERENTIATION AND DECARCINOGENESIS[*1]
The Isozyme Patterns in Malignant Tumors and Membrane Changes of Cultured Tumor Cells

Takashi SUGIMURA,[*2] Taijiro MATSUSHIMA,[*2] Takashi KAWACHI,[*2]
Kikuko KOGURE,[*2] Noritake TANAKA,[*2] Setsuko MIYAKE,[*2]
Motoo HOZUMI,[*2] Shigeaki SATO,[*3]
and Hiroshi SATO[*4]

*Biochemistry Division, National Cancer Center Research Institute,[*2]
Department of Molecular Oncology, Institute of Medical
Science, University of Tokyo,[*3] and Sasaki Institute[*4]*

Studies on aldolase isozymes in hepatomas revealed that the liver-type aldolase, aldolase B, disappeared in fast-growing hepatomas of the rat, and the muscle-type aldolase, aldolase A, appeared instead. Slow-growing hepatomas contained aldolases A, B, and their hybrids. In some strains of fast-growing hepatomas, aldolase C, which is specific to normal brain and nerve in adult rats, appeared. These changes in aldolase isozymes were explained as a " switch-on " and " switch-off " of genes. The term " disdifferentiation " was proposed to express the abnormal patterns of gene expression in malignant tumors.

Normal rat liver contains hexokinase isozymes I, II, III, and IV. Fast-growing hepatomas have hexokinases I and II. A slow-growing substrain derived from a fast-growing line possessed hexokinase III in addition to hexokinases I and II. This represents tumor reversal at the molecular level. The term " decarcinogenesis " was proposed for tumor reversion.

Experiments on tumor reversal were carried out *in vitro*. A cell line which had lost its transplantability *in vitro* was established from a mouse mammary tumor cell line, FM3A. The properties of the cell membrane of this cell line, which had undergone decarcinogenesis, were altered as revealed by its increased agglutinability with phytohemagglutinins (PHA).

In human gastric carcinoma also, abnormal differentiation was revealed by the existence of sucrase in the gastric mucosa.

The importance of regarding cancer as an epigenetic change in somatic cells is emphasized.

[*1] This work was supported by grants from the Ministry of Education and the Ministry of Health and Welfare of Japan, and from the Seminar on Metabolic Regulation (Amino Acid and Protein).

[*2] Tsukiji 5-1-1, Chuo-ku, Tokyo 104, Japan (杉村　隆, 松島泰次郎, 河内　卓, 木暮喜久子, 田中宣威, 三宅節子, 穂積本男).

[*3] Shirokanedai 4-6-1, Minato-ku, Tokyo 108, Japan (佐藤茂秋).

[*4] Kanda-Surugadai 2-2, Chiyoda-ku, Tokyo 101, Japan (佐藤　博).

Many investigations on various biochemical parameters have been carried out in an attempt to find critical differences between normal and cancer cells. It has been generally accepted that cancer cells, which originate from various normal tissues, show biochemical patterns rather similar to those of embryonic tissues in spite of the different biochemical patterns in the normal tissues from which they originate (14). Consequently, it has been concluded that cancer cells are in the state of dedifferentiation. However, this concept does not seem sufficient to explain the specific biochemical features of cancer cells since the undifferentiated state of embryonic tissues is in a normal and well-regulated condition. Emphasis on the similarity of the cancer and embryonic phenotypes did not offer any biochemical clue to explain the principal biological characteristics of cancer cells, such as autonomous growth and metastasis. Too much energy has been spent on simple comparison of numerous kinds of enzymes and substances in normal and cancer cells without settling on definite objective points. Under these circumstances, studies on the molecular species of isozymes seem to provide a new approach to a rational understanding of the biochemical features of cancer cells.

In the following will be summarized the experimental results from our laboratory, which started from studies on aldolase and hexokinase isozymes in hepatoma cells and led us to the idea that cancer is caused by an epigenetic change in somatic cells. An epigenetic change seemed more likely to be reversible than a genetic change, and this encouraged us to search for indications of reversion of cancer cells. Our work also includes studies on abnormal differentiation of the gastric mucosa in man, which may be related to gastric carcinoma.

Aldolase Isozymes—Disdifferentiation Proposal

The pioneer work on the difference in the molecular species of aldolases in normal liver and hepatomas in man and rats was reported by Schapira *et al.* (47)

TABLE I. Properties of Crystalline Aldolases from Muscle and AH-130

Property measured	Muscle aldolase	AH-130 aldolase
K_m for FDP (M)	6.1×10^{-5}	5.6×10^{-5}
K_m for F1P (M)	1.2×10^{-2}	1.3×10^{-2}
FDP/F1P activity ratio	52	30
FDP activity remaining after carboxypeptidase treatment (%)	8	6
Activity toward F1P remaining after carboxypeptidase treatment (%)	97	94
Inhibitory effects of adenine nucleotides	ATP>ADP>AMP	ATP>ADP>AMP
Inhibition after treatment with anti-A antibody (%)	100	100
Inhibition after treatment with anti-B antibody (%)	0	0

in 1963. At a previous conference held in Kyoto, Japan, in 1965, under the US-Japan Science Cooperation Program, we also reported on differences in the molecular species of aldolases in fast-growing Yoshida ascites hepatomas, slow-growing Morris hepatomas, and normal liver, represented by very high (30 to 50), moderate (3 to 5), and low (1 to 2) ratios of the activities of FDP to F1P (54).

Since then, we have succeeded in demonstrating the existence of aldolase A in many strains of Yoshida hepatomas and aldolases A, B, and hybrids between aldolases A and B in Morris hepatomas 7316A and 7793 (32). Aldolase was purified in crystalline form from Yoshida ascites hepatoma, AH-130, which contained glucose 6-phosphatase and was supposed to be of hepatic origin (33). As shown in Table I, many enzymic properties of the aldolase of this hepatoma are quite similar to those of muscle aldolase from rats, and the amino acid compositions of the two enzymes are the same. Similar research has been carried out by the groups of Rutter (45), Schapira (38, 47), Weinhouse (1, 8), Horecker (13), and Endo (21). They all concluded that the liver-type aldolase, aldolase B, disappeared in fast-growing hepatomas and aldolase A appeared instead. We have expressed this change as a " switch off of the gene for aldolase B " and " switch on of the gene for aldolase A " (32).

Aldolases A and B and their hybrids are well separated by electrophoresis on a cellulose acetate membrane, and we found that fast-growing hepatomas have aldolase A almost exclusively while slow-growing hepatomas contain aldolases A and B, and their hybrids.

More than 50 strains of ascites hepatomas have become available through the industriousness of Yoshida and his associates (58, 59) in Japan. On analysis of the aldolases in these hepatomas, we detected bands of A_3C_1 and A_2C_2 hybrid molecules of aldolase on cellulose acetate membrane electrophoresis, in addition to aldolase A in strains AH-143A and AH-66F, as shown in Photo 1. The A_4 and A_3C_1 molecules of AH-143A were purified by the methods of Nicholas et al. (37) and Penhoet et al. (41) using DEAE-cellulose column chromatography. The activities of A_4 and A_3C_1 were inhibited by anti-aldolase A in the same way as the A_4 and A_3C_1 molecules of normal rat brain (55). The appearance of aldolase C in some strains of rat hepatoma was also reported by Schapira et al. (48). Aldolase C or hybrids of aldolases C and A were found only in the brain and nerve tissues of normal adult rat, although electrophoresis of embryonic liver showed a weak band of the A_3C_1 hybrid (48) in addition to that of aldolase A. The various patterns of aldolase molecules found in liver and hepatomas are represented schematically in Fig. 1.

As reported by Ichihara and Ogawa (39), branched-chain amino acid transaminase isozyme III, found only in the brain tissue of rats, appeared in a rat hepatoma. Tanaka et al. (57) also mentioned that phosphofructokinase isozyme IV, usually found only in normal liver, appeared in all kinds of tumors, including Walker carcinosarcoma. Katunuma et al. (24) reported that the glutaminase isozyme pattern in hepatomas is not the liver type and is also dissimilar to the isozyme pattern in embryonic liver. Our data, together with those of other investigators, indicate that gene expressions, which are represented by isozyme phenotypes, do not fully support the conclusion that the phenotypes of cancer cells

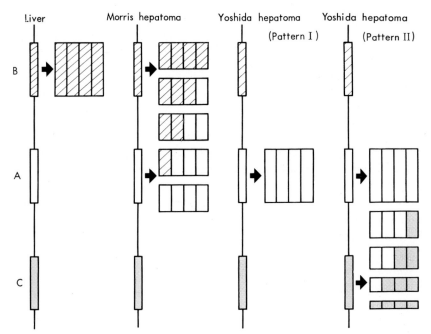

Fig. 1. Schematic representation of gene expressions of aldolase isozymes in liver and hepatomas

A, gene for aldolase A; B, gene for aldolase B; C, gene for aldolase C.

are similar to those in an undifferentiated state. It seems that cancer cells have properties which do not correspond to those in any differentiated state found in normal or embryonic tissues. The combination of expressed and unexpressed genes in cancer cells is probably not identical with that in any normal tissues. Based on the idea that the cancer phenotype might be defined as an abnormality of differentiation, we proposed the term disdifferentiation. Markert (31) suggested that cancer is a disease of differentiation, while Potter (43) considered that oncongeny represents blocked ontogeny. If Potter used the term "blocked" in the sense of abnormal discontinuation of development in a coordinated status of gene expression in the successive program of normal ontogenic development, these three terms actually mean the same thing.

Hexokinase Isozymes — Decarcinogenesis Proposal

There have been many reports on hexokinase isozymes in the liver and in hepatomas by González et al. (12), Katzen et al. (25), Weinhouse et al. (8, 50, 51), and others (16, 26). We mentioned the difference between the molecular species of hexokinase isozymes in Yoshida ascites hepatomas, Morris hepatomas, and normal liver at the conference in Kyoto in 1965 (54). Later, we studied the hexokinase isozymes in more detail again, separating them by cellulose acetate membrane electrophoresis, and we reported the hexokinase patterns of fast-growing Yoshida hepatomas and slow-growing Morris hepatomas. Normal rat liver has hexokinases I, II, III, and IV, in order of increasing electrophoretic mobility to the anode at pH 8.6. Their K_m values for glucose are 10^{-5}, 10^{-4}, 10^{-6}, and 10^{-2} M, respectively, and hexokinase IV is the so-called glucokinase.

In general, fast-growing hepatomas contained only hexokinases I and II, the latter always predominating. Slow-growing hepatomas showed a definite band of hexokinase III and a faint band of hexokinase IV, in addition to hexokinases I and II (46).

Yoshida sarcoma cells were also investigated. Yoshida sarcoma was thought for a long time to have originated from mesenchymal tissue. However, Yoshida suggested in his recent review on hepatomas that it may have originated from hepatic cells (58). Yoshida sarcoma can be transplanted by injection of a single cell into the peritoneal cavity of a rat; in other words, it can be cloned by the *in vivo* transplantation technique. By transplanting Yoshida sarcoma cells repeatedly into the peritoneal cavities of rats sensitized with leukemia or Yoshida ascites hepatoma cells and then retransplanting the cells into rat peritoneal cavities, Hiroshi Sato obtained various subclones of Yoshida sarcoma, which grow more slowly in rats and take longer to kill the animals. As shown in Photo 2, the original Yoshida sarcoma contained hexokinases I and II, but a substrain of Yoshida sarcoma, LY5, possessed hexokinase III also. In a more slowly growing substrain established from Yoshida hepatoma, AH-109, hexokinase III appeared concomitantly with conversion of the cells to a slow-growing type, while the original fast-growing strain showed only hexokinases I and II.

It is evident that the gene for hexokinase III is active in normal liver and in slow-growing Morris hepatomas but inactive in fast-growing hepatomas. It should be emphasized that the inactive gene for hexokinase III was re-activated in the slow-growing substrain derived from the fast-growing hepatoma. The reappearance of hexokinase III represents a reversion of the isozyme phenotype in the hepatoma in the direction of that of normal liver.

We found the isozyme patterns of lactic acid dehydrogenase (LDH) in slow-growing substrains of Yoshida sarcoma to be different from the pattern in the original Yoshida sarcoma. As shown in Photo 3, LY5 has a pattern similar to that of rat heart, while LY7 has a pattern similar to that of rat liver. The LDH patterns of LY5, LY7, and the original Yoshida sarcoma are quite constant and do not change with time after tumor transplantation.

Based on the understanding that cancer cells are produced through disdifferentiation, it seemed reasonable, therefore, to expect that reversion of the cancer phenotype is possible. For this kind of change in cancer phenotypes, we proposed the term decarcinogenesis, indicating tumor reversal (52). The relationship of the carcinogenic process of disdifferentiation to decarcinogenesis is illustrated schematically in Fig. 2. Decarcinogenesis does not necessarily mean complete reversion to the normal type but could, in some cases, result in types other than the normal or malignant cell types. Several years ago Katsuta *et al.* (23) proposed the term " third cell " for a cell derived from a hepatoma which had lost its transplantability after successive culture *in vitro*.

Neuroblastoma cells undergo differentiation to neuroblasts *in vivo* (7) and *in vitro* (2, 11, 49). Myeloid leukemia cells can differentiate to matured granulocytes or phagocytes *in vitro* under defined conditions (20). Erythroblastic leukemia

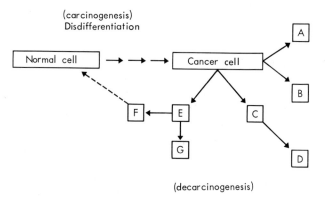

Fig. 2. Concept of decarcinogenesis

cells of mice can undergo maturation to red cells when placed in a millipore filter bag in the peritoneal cavity of mice (*10*). There are many other reports which substantiate the possible reversion of tumor cells. Plant tumors also undergo reversion (*3*). Braun (*4*) made a comprehensive survey and cited many instances of tumor reversal in man and experimental animals, such as mice, rats, frogs, and salamanders. Thus it no longer seems strange to expect tumor phenotypes to be reversible. The opposite concept of irreversibility of the tumor phenotype arose erroneously from experiments on tumors as masses of cancer cells. This concept of irreversibility was accepted by many investigators because the possible existence of a small fraction of reverted cells was not examined, and their growth was naturally masked by the growth of unreverted malignant cells constituting the majority of the tumor mass. To study these reverted cells we must deal with individual cancer cells like individual bacteria. In the following section, we will discuss *in vitro* experiments on decarcinogenesis.

Loss of Transplantability of Mouse Mammary Carcinoma Cells

A cell line of mouse mammary tumor with a high plating efficiency in soft agar was established by Nakano (*36*). We isolated a cell clone (1-8) which had spontaneously lost its transplantability from the original FM3A cell line (*18*). Intraperitoneal injection of 10^3 cells of the original FM3A cell line (1-6) into C3H/He mice always resulted in tumor formation, whereas injection of 10^6 cells of the 1-8 clone did not result in tumor formation. Koyama and Ishii (*27*) in the Institute succeeded in establishing a cell line (M6) which had lost the transplantability of the original FM3A cells by treating them with the carcinogen 4-nitroquinoline 1-oxide.

The properties of one of the transplantable clones (1-6) of FM3A and a subclone (1-8) which had undergone decarcinogenesis and lost transplantability are shown in Table II. The two strains have the same saturation density, mean generation time, and plating efficiency. The shape of the colonies and the number of chromosomes in the two clones are also the same. Fortunately we found that, after decarcinogenesis, cells of either 1-8 or M6 showed the property of

TABLE II. Properties *in vitro* of Original Clone (1-6) and Clone (1-8) Showing Decarcinogenesis

Property measured	Clone 1-6	Clone 1-8
Generation time (hr)	24	25
Saturation density ($\times 10^5$ cells/cm^2)	3	3
Cloning efficiency (%)	83	74
No. of chromosomes	67	67

TABLE III. Agglutination Titers of Clones with Various Transplantabilities

Clone	Transplantability	Titer with PHA from		
		Wheat germ	*Ricinus communis*	*Phaseolus vulgaris*
1-614	+	16	256	128
1-82	−	32	1,024	256
FM3A/B	+	32	256	128
M6	−	64	1,024	256

Agglutinin solutions (16 mg/ml) were diluted serially with 0.05 M phosphate buffer (pH 6.9) in 0.1 M NaCl. The cell suspension (0.1 ml, 2×10^5 cells) was mixed with 0.1 ml of agglutinin solution and the degree of cell agglutination was determined microscopically after 1 hr. Numerals indicate the maximum dilution causing agglutination. Clones 1-614 and 1-82 are subclones of 1-6 and 1-8, respectively, and grow on a glass surface, while clones FM3A/B and M6 grow in suspension.

aggregation in more dilute solutions of PHA from various kinds of plants than did FM3A cells, as shown in Table III and Photo 4. This means that after decarcinogenesis the architecture of the cell membrane differs from that of the original FM3A cells. It is well known that cells which have been transformed by viruses or chemical carcinogens aggregate more easily with PHA than do normal cells. This was explained as due to alteration of the membrane structure, probably by exposure of reactive sugar residues of glycoprotein in the membrane to PHA (5), since trypsinization enhanced the agglutinability of normal cells by PHA. It is interesting that after decarcinogenesis the cells showed more agglutinability than did the cancer cells. This is represented schematically in Fig. 3.

Thus by applying cells to a Falcon plastic "Microtest" plate with holes in it, we can very quickly detect cell lines with membrane alteration, which is associated with loss of transplantability probably due to change in surface antigens (17). Kinetic studies on decarcinogenesis of cells are now possible. Cells from individual colonies in soft agar were directly transferred in buffered saline to a hole in the plastic Microtest reaction plate. A solution of PHA of a defined concentration was added to holes containing cells from different colonies. After shaking the plate gently at room temperature, the frequency of occurrence of decarcinogenesis was easily determined. A change in agglutinability with PHA was also observed in the cell line (M6) which had undergone decarcinogenesis.

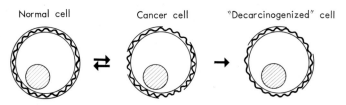

Fig. 3. Alteration of cell membrane structure accompanying carcinogenesis and decarcinogenesis

There have been many reports indicating that loss of transplantability occurs during cultivation of cancer cells *in vitro* (*6, 9, 15, 19, 23, 28, 30, 34, 40, 42, 44, 56*). We would like to point out that we have no reason to exclude the possibility that these changes do not also occur *in vivo*.

Abnormal Differentiation of Gastric Mucosa Associated with Human Gastric Tumors

We tried to find cases of abnormal differentiation in man. It is well known that gastric carcinoma in elderly people is frequently associated with intestinal metaplasia in the gastric mucosa (*22, 29, 35*). Intestinal metaplasia is defined as the histopathological appearance of intestinal epithelium in the gastric mucosa. This intestinal metaplasia attracted our attention, since it is a reflection of abnormal differentiation of normal gastric mucosa. Intestinal metaplasia was diagnosed histologically. Previously, there have been various histochemical studies using color reactions on enzymes, such as leucine aminopeptidase and alkaline phosphatase, both of which are normally found in the intestinal epithelium but not in the gastric epithelium. Unfortunately, these two enzymes are not specific to intestinal tissues but are also found in other tissues.

We recently devised a very quick, simple, and rational method to detect intestinal metaplasia (*53*). We used a method to detect sucrase, since sucrase is localized in the microvilli of the surface epithelium of the intestine but is not found in other tissues, including the gastric mucosa. Sucrase splits sucrose, yielding glucose and fructose. The presence of glucose is easily demonstrated with Tes-tape, which is used for the detection of glucose in the urine of diabetic patients. Tes-tape contains glucose oxidase, peroxidase, and dye. Glucose oxidase converts glucose to gluconate, yielding hydrogen peroxide which oxidizes the dye with peroxidase, so that the Tes-tape changes from yellow to green.

Excised stomach from human patients was quickly brought to the laboratory, washed well with cold saline, spread on a thick glass plate, and a solution of 5% sucrose sprayed over the whole mucosal surface. The glass plate was kept in an incubator at 37° for 5 min to allow hydrolysis of sucrose. Then the mucosa surface was quickly covered with pieces of Tes-tape. In the area where sucrase was present, the Tes-tape turned green. This area, showing a positive reaction for sucrase, coincided surprisingly well with the area where intestinal metaplasia was demonstrated histologically.

Photo 5 shows a stomach with carcinoma in an advanced stage from a 63-year-old woman. The area with cancer gave a negative reaction and was surrounded by a region of intestinal metaplasia, giving a positive reaction. With this method, carcinoma can be detected at an early stage as a spot giving a negative reaction in a zone giving a positive reaction.

Sucrase is located in the cell membrane, and it is well known that the membrane is altered during the carcinogenic process. Thus, the appearance of sucrase due to abnormal differentiation and its disappearance after malignant conversion may indicate that the abnormal differentiation and the consequent changes in the properties of the cell membrane are connected with malignant transformation.

In newborn rats, lactase was observed in the intestine, but it disappeared during weaning and sucrase appeared instead. We have some times found lactase also in the gastric mucosa of elderly patients with stomach cancer. Studies on molecular species of enzymes during intestinal metaplasia suggest the existence of more complicated abnormalities of differentiation than those detected in histopathological studies.

DISCUSSION

Studies on molecular species of isozymes have shed new light on cancer phenotypes. We are beginning to find qualitative differences between normal cells and cancer cells. The switch on and off of genes revealed in studies on aldolase might also occur with many other genes, either in a coordinated or uncoordinated way, during the process of carcinogenesis. In the normal state, the gene expression of enzymes and enzyme inhibitors and repressors is probably well coordinated. Unbalanced patterns of gene expression may result in the loss of various cellular control mechanisms, including very important enzyme systems for chromosome replication and for formation of the cell membrane. We propose that cancer is produced by abnormal differentiation (disdifferentiation) of somatic cells.

From studies on the molecular species of hexokinase, we have learned that a gene which is turned off in a hepatoma may be turned on again. This is an example of "tumor reversion," observed on a defined biochemical parameter. With regard to tumor reversion, we observed spontaneous and induced loss of tumor transplantability of a mouse mammary tumor *in vitro*. This loss of transplantability was apparently associated with change in the properties of the membrane structure revealed by change in agglutinability with PHA.

There are two main explanations for the carcinogenic process, a genetic change or an epigenetic change. As pointed out clearly in the Report of the National Panel of Consultants on the Conquest of Cancer for the Committee on Labor and Public Welfare, United States Senate (April 27, 1970), if cancer is a genetic disease, the cure of cancer can be achieved only by complete and selective destruction of cancer cells or by genetic engineering such as enzymic synthesis of a selected molecule of DNA and integration of this into chromosomes of cancer cells. On the contrary, if cancer is a reflection of an epigenetic change in somatic cells, a better method of treatment would be to promote the reversion process.

Now it is well known that viral genes may be integrated into chromosomes. This integration itself and/or gene products of viral genes may interfere with normal expression in normal cells and thus cause malignant transformation.

Finally, we would like to emphasize the case of abnormal differentiation in human gastric mucosa represented by the appearance of sucrase, an enzyme specific to the intestine, which was observed in the gastric mucosa in areas of intestinal metaplasia and which is thought to be related to the production of stomach cancer.

Further research on disdifferentiation should be fruitful.

REFERENCES

1. Adelman, R. C., Morris, H. P., and Weinhouse, S. Fructokinase, triokinase, and aldolases in liver tumors of the rat. *Cancer Res.*, **27**, 2408–2413 (1967).
2. Blume, A., Gilbert, F., Wilson, S., Farber, J., Rosenberg, R., and Nirenberg, M. Regulation of acetylcholinesterase in neuroblastoma cells. *Proc. Natl. Acad. Sci. U.S.*, **67**, 786–792 (1970).
3. Braun, A. C. The plant tumor cell as an experimental tool for studies on the nature of autonomous growth. *Can. Cancer Conf.*, **4**, 89–98 (1961).
4. Braun, A. C. Reversal of tumor growth. *In* "The Cancer Problem, A Critical and Modern Synthesis," Columbia University Press, New York and London, pp. 134–159 (1969).
5. Burger, M. M. A difference in the architecture of surface membrane of normal and virally transformed cells. *Proc. Natl. Acad. Sci. U.S.*, **62**, 994–1001 (1969).
6. Cailleau, R. and Costa, F. Long-term *in vitro* cultivation of some mouse ascites tumors: Ehrlich ascites carcinoma. *J. Natl. Cancer Inst.*, **26**, 271–281 (1961).
7. Everson, T. C. and Cole, W. H. Spontaneous regression of neuroblastoma. *In* "Spontaneous Regression of Cancer," W. B. Sanders, Philadelphia and London, pp. 88–163 (1966).
8. Farina, F. A., Adelman, R. C., Chai, H. L., Morris, H. P., and Weinhouse, S. Metabolic regulation and enzyme alterations in the Morris hepatomas. *Cancer Res.*, **28**, 1897–1900 (1968).
9. Foley, G. E. and Drolet, B. P. Loss of neoplastic properties *in vitro*. I. Observations with S-180 cell lines. *Cancer Res.*, **24**, 1461–1467 (1964).
10. Furusawa, M., Ikawa, Y., and Sugano, H. Development of erythrocyte membrane specific antigen(s) in clonal cultured cells of Friend virus-induced tumor. *Proc. Japan Acad.*, **47**, 220–224 (1971).
11. Goldstein, M. N., Burdman, J. A., and Journey, L. J. Long-term tissue culture of neuroblastomas. II. Morphologic evidence for differentiation and maturation. *J. Natl. Cancer Inst.*, **32**, 165–199 (1964).
12. González, C., Ureta, T., Sanchez, R., and Niemeyer, H. Multiple molecular forms of ATP: hexose 6-phosphotransferase from rat liver. *Biochem. Biophys. Res. Commun.*, **16**, 347–352 (1964).
13. Gracy, R. W., Lacko, A. G., Brox, L. W., Adelman, R. C., and Horecker, B. L. Structural relations in aldolases purified from rat liver and muscle and Novikoff hepatoma. *Arch. Biochem. Biophys.*, **136**, 480–490 (1970).
14. Greenstein, J. P. "Biochemistry of Cancer," Academic Press Inc., New York (1954).
15. Guerin, L. F. and Kitchen, S. F. Adaptation of 6C3HED tumor cells to culture *in vitro*. *Cancer Res.*, **20**, 344–346 (1960).
16. Gumaa, K. A. and Greenslade, K. R. Molecular species of hexokinase in hepatomas and ascites-tumour cells. *Biochem. J.*, **107**, 22p (1968).

17. Hozumi, M., Miyake, S., Mizunoe, F., Sugimura, T., Irie, R., Koyama, K., Tomita, M., and Ukita, C. Surface properties of non-tumorigenic variants of mouse mammary carcinoma cells in culture. *Int. J. Cancer*, **9**, 393–401 (1972).
18. Hozumi, M. and Nakamura, K. Spontaneous reduction of transplantability in cultured mammary carcinoma cells. *GANN*, **61**, 409–414 (1970).
19. Hsu, T. C. and Klatt, O. Loss of malignancy by Novikoff hepatoma cells *in vitro*. *J. Natl. Cancer Inst.*, **22**, 313–339 (1959).
20. Ichikawa, Y. Further studies on the differentiation of a cell line of myeloid leukemia. *J. Cell Physiol.*, **76**, 175–184 (1970).
21. Ikehara, Y., Endo, H. and Okada, Y. The identity of the aldolases isolated from rat muscle and primary hepatoma. *Arch. Biochem. Biophys.*, **136**, 491–497 (1970).
22. Järvi, O. and Laurén, P. On the role of heterotopias of the intestinal epithelium in the pathogenesis of gastric cancer. *Acta Pathol. Microbiol. Scand.*, **29**, 26–44 (1951).
23. Katsuta, H., Takaoka, T., and Kaneko, K. Studies on rat ascites hepatoma cells in tissue culture. IX. On the decrease in malignancy of tissue culture strain cells (JTC-1 and-2) by transfer into fluid-suspension culture (in Japanese). *Proc. Jap. Cancer Assoc., 18th Meet.*, 177–178 (1959).
24. Katunuma, N., Tomino, I., and Sanada, Y. Differentiation of organ specific glutaminase isozyme during development. *Biochem. Biophys. Res. Commun.*, **32**, 426–432 (1968).
25. Katzen, H. M. and Shimke, R. T. Multiple forms of hexokinase in the rat: Tissue distribution, age dependency, and properties. *Proc. Natl. Acad. Sci. U.S.*, **54**, 1218–1225 (1965).
26. Knox, W. E., Jamder, S. C., and Davis, P. A. Hexokinase, differentiation and growth rates of transplanted tumors. *Cancer Res.*, **30**, 2240–2244 (1970).
27. Koyama, K. and Ishii, K. Induction of non-transplantable mutant clones from an ascites tumor. *GANN*, **60**, 367–372 (1969).
28. Kuroki, T. Studies on *in vitro* growth of Yoshida sarcoma cells. III. Virulence of the long-term cultured cells. *GANN*, **57**, 379–389 (1966).
29. Laurén, P. The two histological main types of gastric carcinoma: Diffuse and so-called intestinal type carcinoma. An attempt at a histo-clinical classification. *Acta Pathol. Microbiol. Scand.*, **64**, 31–49 (1965).
30. Macpherson, I. Reversion in hamster cells transformed by Rous sarcoma virus. *Science*, **148**, 1731–1733 (1965).
31. Markert, C. L. Neoplasia: A disease of cell differentiation. *Cancer Res.*, **28**, 1908–1914 (1968).
32. Matsushima, T., Kawabe, S., Shibuya, M., and Sugimura, T. Aldolase isozymes in rat tumor cells. *Biochem. Biophys. Res. Commun.*, **30**, 565–570 (1968).
33. Matsushima, T., Kawabe, S., and Sugimura, T. Properties of crystalline fructose diphosphate aldolase purified from rat tumor. *Abstr. 10th Int. Cancer Congr.*, 256 (1970).
34. Midorikawa, O. Functional and phenotypic changes in cancer cells (in Japanese). *Proc. Jap. Cancer Assoc., 27th Meet.*, 6 (1968).
35. Morson, B. C. Carcinoma arising from areas of intestinal metaplasia in the gastric mucosa. *Brit. J. Cancer*, **9**, 377–385 (1955).
36. Nakano, N. Establishment of cell lines *in vitro* from a mammary ascites tumor of mouse and biological properties of the established lines in a serum containing medium. *Tohoku J. Exp. Med.*, **88**, 69–84 (1966).
37. Nicholas, P. C. and Bachelard, H. S. The separation, partial purification and

some properties of isoenzymes of aldolase from guinea-pig cerebral cortex. *Biochem. J.*, **112**, 587–594 (1969).
38. Nordmann, Y. and Schapira, F. Muscle type isozymes of liver aldolase in hepatomas. *Eur. J. Cancer*, **30**, 247–250 (1967).
39. Ogawa, K., Yokojima, A., and Ichihara, A. Transaminase of branched chain amino acids. VII. Comparative studies on isozymes of ascites hepatoma and various normal tissues of rat. *J. Biochem. (Tokyo)*, **68**, 901–911 (1970).
40. Paraf, A., Moyne, M. A., Duplan, J. F., Scherrer, R., Stanislawski, M., Bettane, M., Lelievre, L., Rouze, P., and Dubert, J. M. Differentiation of mouse plasmocytomas *in vitro*: Two phenotypically stabilized variants of the same cell. *Proc. Natl. Acad. Sci. U.S.*, **67**, 983–990 (1970).
41. Penhoet, E., Kochman, M., Valentine, R., and Rutter, W. J. The subunit structure of mammalian fructose diphosphate aldolase. *Biochemistry*, **6**, 2940–2949 (1967).
42. Pollack, R. E., Green, H., and Todaro, G. J. Growth control in cultured cells: Selection of sublines with increased sensitivity to contact inhibition and decreased tumor-producing ability. *Proc. Natl. Acad. Sci. U.S.*, **60**, 126–133 (1968).
43. Potter, V. R. Recent trends in cancer biochemistry. *Can. Cancer Conf.*, **8**, 9–30 (1969).
44. Rabinowitz, Z. and Sachs, L. Reversion of properties in cells transformed by polyoma virus. *Nature*, **220**, 1203–1206 (1968).
45. Rutter, W. J., Blostein, R. E., Woodfin, B. M., and Weber, C. S. Enzyme variants and metabolic diversification. *Adv. Enzyme Regulation*, **1**, 39–56 (1963).
46. Sato, S., Matsushima, T., and Sugimura, T. Hexokinase isozyme patterns of experimental hepatomas of rat. *Cancer Res.*, **29**, 1437–1446 (1969).
47. Schapira, F., Dreyfus, J. C., and Schapira, G. Anomaly of aldolase in primary liver cancer. *Nature*, **200**, 995–997 (1963).
48. Schapira, F., Reuber, M. D., and Hatzfeld, A. Resurgence of two fetal-types of aldolases (A and C) in some fast-growing hepatomas. *Biochem. Biophys. Res. Commun.*, **40**, 321–327 (1970).
49. Seeds, N. W., Gilman, A. G., Amano, T., and Nirenberg, M. W. Regulation of axon formation by clonal lines of a neural tumor. *Proc. Natl. Acad. Sci. U.S.*, **66**, 160–167 (1970).
50. Sharma, R. M., Sharma, C., Donnelly, A. J., Morris, H. P., and Weinhouse, S. Glucose-ATP phosphotransferases during hepatocarcinogenesis. *Cancer Res.*, **25**, 193–199 (1965).
51. Shatton, J. B., Morris, H. P., and Weinhouse, S. Kinetic, electrophoretic, and chromatographic studies on glucose-ATP phosphotransferases in rat hepatomas. *Cancer Res.*, **29**, 1161–1172 (1969).
52. Sugimura, T. Decarcinogenesis, a newer concept arising from our understanding of the cancer phenotype. *In* " Chemical Tumor Problems," ed. by W. Nakahara, Japan Society for the Promotion of Science, Tokyo, pp. 269–284 (1970).
53. Sugimura, T., Kawachi, T., Kogure, K., Tanaka, N., Kazama, K., and Koyama, Y. A novel method for detecting intestinal metaplasia of stomach with Tes-tape. *GANN*, **62**, 237 (1971).
54. Sugimura, T., Matsushima, T., Kawachi, T., Hirata, Y., and Kawabe, S. Molecular species of aldolases and hexokinases in experimental hepatomas. *GANN Monograph*, **1**, 143–149 (1966).
55. Sugimura, T., Sato, S., and Kawabe, S. The presence of aldolase C in rat hepatoma. *Biochem. Biophys. Res. Commun.*, **39**, 626–630 (1970).

56. Takahashi, M. Alteration of virulence of MN ascites sarcoma after continuous culture *in vitro*. *GANN*, **54**, 295–310 (1963).
57. Tanaka, T., An, T., and Sakaue, Y. Studies on multimolecular forms of phosphofructokinase in rat tissues. *J. Biochem. (Tokyo)*, **69**, 609–612 (1971).
58. Yoshida, T. Comparative studies of ascites hepatomas. *Methods Cancer Res.* **6**, 97–157 (1971).
59. Yoshida, T. and Sato, H. Ascites tumor-Yoshida sarcomas and ascites hepatoma(s). *Natl. Cancer Inst. Monograph*, **16**, (1964).

EXPLANATION OF PHOTOS

Photo 1. Appearance of A-C hybrid molecules of aldolase in Yoshida ascites hepatomas.

Photo 2. Hexokinase isozyme patterns of rat liver, Yoshida sarcoma, and its substrain, LY 5. The patterns of Yoshida ascites hepatoma, AH-109A, and its substrain are also represented.

Photo 3. LDH isozyme patterns of normal rat liver and heart, Yoshida sarcoma and its substrains.

Photo 4. Aggregation with PHA of cells showing decarcinogenesis. (M6) (left), and original cells (FM3A/B) (right).

Photo 5. Demonstration of intestinal metaplasia using the sucrase test on a stomach with cancer. Macroscopic appearance (left) and result with Tes-tape (right).

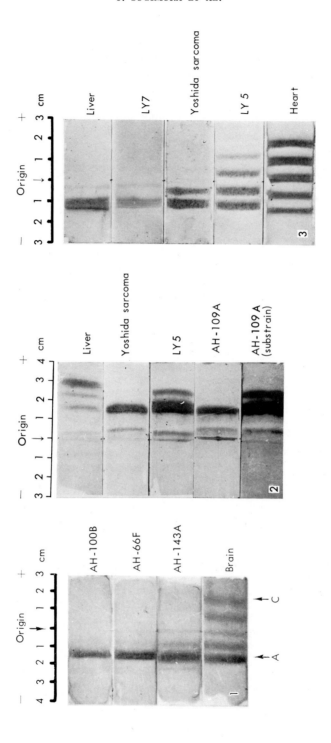

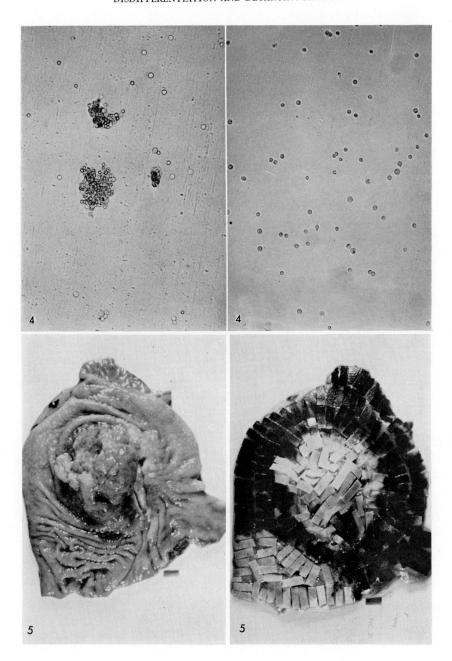

MOLECULAR CORRELATION CONCEPT: ORDERED PATTERN OF GENE EXPRESSION IN NEOPLASIA

George WEBER

*Department of Pharmacology, Indiana University School of Medicine**

This paper discusses recent advances achieved in my laboratories with application of the molecular correlation concept. The implications of the molecular correlation concept for our understanding of the pattern of gene expression in normal, differentiating, regenerating, and neoplastic liver are analyzed.

The results of this investigation provide evidence for the following conclusions.

(1) In the spectrum of hepatomas of different growth rates there is an ordered pattern of gene expression which can be understood in terms of imbalance of opposing key enzymes in synthetic and degradative pathways.

(2) The metabolic imbalance in neoplastic cells is linked with the tumor growth rate. Thus, there is a link between the metabolic and proliferative expression of the genes in the cancer cell.

(3). The metabolic imbalance confers a selective biological advantage on the neoplastic liver cells by the progressive predominance of the catabolic pathway over the anabolic one in carbohydrate metabolism and by the gradual predominance of the synthetic pathway over the decrease in the catabolic pathway in pyrimidine and nucleic acid metabolism. Furthermore, by the decrease in the activity of ornithine carbamoyltransferase, the competition of the urea cycle for aspartate and carbamyl phosphate utilization is gradually switched off, permitting an increase in the channeling of these precursors to nucleic acid biosynthesis.

(4) The existing evidence allows us to utilize the molecular correlation concept to analyze and search for new areas of metabolic imbalance, since the concept has predictive power in this field.

(5) The metabolic imbalance that is discovered in the hepatoma cells, in carbohydrate, pyrimidine, nucleic acid, and ornithine metabolism is specific to neoplasia, since no similar pattern of metabolic imbalance is present in regenerating or in differentiating liver.

(6) Correlations of various metabolic alterations and tumor growth

---

* Indianapolis, Indiana 46202, U.S.A.

The following abbreviations are used: ATC, aspartate transcarbamylase; DHO, dihydroorotase; FDPase, fructose 1,6-diphosphatase; GK, glucokinase; G-6-Pase, glucose 6-phosphatase; HK, hexokinase; OCT, ornithine carbamoyltransferase; PFK, phosphofructokinase; PEP-CK, phosphoenolpyruvate carboxykinase; PK, pyruvate kinase; Py carboxylase, pyruvate carboxylase; Tdr, thymidine.

rate are not restricted to liver tumors, as they have been shown also to apply to renal tumors, mammary cancer, and other neoplasms of different growth rates. Therefore, it appears that the approaches of the molecular correlation concept are valid not only for hepatomas but also for other types of neoplasms.

Objectives and Scope of This Presentation

The purpose of this paper is to discuss recent advances achieved in my laboratories with application of the molecular correlation concept and to analyze some of the implications of these discoveries.

My report at the first U.S.-Japan meeting on hepatomas concerned the identification of the biochemical pattern of the hepatomas in terms of the conceptual and experimental approaches of the molecular correlation concept. My paper also outlined in detail the evidence for the discovery that the activities of key enzymes and certain metabolic pathways correlate with hepatoma growth rate (*33*). At that time, there appeared to be little or no support for this concept among the other participants of the meeting. However, in subsequent years, more and more investigators have come to employ this approach and to interpret data from the point of view of the molecular correlation concept. As a result of research in my laboratory (*1, 2, 8, 13, 22–24, 26, 29–45, 51*) and in other centers (*3, 5, 7, 15, 17, 18, 21, 25*), extensive information is available which I consider good evidence in support of our original proposal for identifying the metabolic imbalance present in the hepatoma cells.

In the following presentation I will report the newest advances made in our laboratories, and the current status and the implications of the molecular correlation concept will be examined in terms of a number of relevant questions.

Is gene expression ordered or random in hepatomas of different growth rates and in differentiation? If gene expression is ordered, is it possible to identify a definitive enzymatic and metabolic pattern? If a biochemical pattern is identified, can it be understood in terms of imbalance of opposing key enzymes and synthetic and degradative pathways? If there is a metabolic imbalance in neoplastic cells, is it possible to identify the linking of gene expression in terms of the correlation of metabolic and proliferative pattern in tumor cells? Does the metabolic imbalance and its linking with tumor growth rate confer a selective biological advantage on the cancer cell? Does the existing evidence allow us to examine whether the molecular correlation concept is capable of providing predictive power in the investigation of the cancer cell?

Since the answers are in the affirmative, the question should be posed whether the pattern of metabolic imbalance uncovered in hepatomas is specific to neoplasia or whether it might be only a reflection of the metabolic pattern of rapidly growing or differentiating tissues. Finally, I would like to examine whether the metabolic pattern elucidated in the hepatoma spectrum is restricted to liver tumors or whether it applies to neoplasms other than hepatomas. The objective of these last two questions is to analyze the specificity and the general applicability of the metabolic and enzymatic imbalance identified in the hepatoma spectrum.

I will start with a very brief analysis of the chief points proposed by the molecular correlation concept and will examine some of the main conceptual and experimental approaches employed. In subsequent parts of this paper, the individual questions raised in the Introduction will be met.

Molecular Correlation Concept

In our research, we anticipated the presence of a meaningful pattern in the metabolism of the cancer cell. We stated earlier that "a careful comparison of the metabolism, histology, and biological behavior may bring a new understanding of the role of biochemical alterations in the pathological behavior of various liver tumors" (*40*). The same paper presented the first attempt to describe the biochemistry of a hepatoma in terms of a meaningful metabolic imbalance and enzyme lesions. The liver tumor used for detailed studies was the very rapidly growing, very malignant Novikoff hepatoma. We expected that a more complete picture of the biochemistry of liver tumors might be achieved if other types of liver tumors with different degrees of malignancy were available. " Since various liver tumors differ in their histological structure, cellular population, biological behavior, and growth rate, it is not unexpected that the biochemical lesions or alterations which underlie morphological and biological differences will be present in varying qualitative or quantitative extent. Such a concept agrees well with the common medical experience of finding many variations of the same disease, from subclinical through mild or severe manifestations to the rarely encountered, full-blown case in which all symptoms and signs are present to their maximum development" (*40*). Thus, we postulated the operation of a meaningful, ordered and correlated expression of morphological, biological, and biochemical behavior in neoplastic tissues which is linked with the expression of their malignancy and growth rate. In order to test the validity of this concept and the existence of such a linked pattern at the molecular level, a prerequisite was the availability of a spectrum of tumors of the same cell type, in which the molecular signs of cancer could be studied in a graded and quantitative manner (*31, 34, 45*). When the spectrum of liver tumors of different growth rates became available (*15*), the prerequisite was fulfilled for a biological system where the progression of the neoplastic process could be conveniently studied (*31, 34*).

We approached the analysis of gene expression in the spectrum of hepatomas with the molecular correlation concept. Since we outlined elsewhere in detail the various theoretical and practical aspects of this concept, I refer the reader to earlier papers (*31–35, 41, 44*) and to Table I, which summarizes the approach.

It is important to emphasize that the biochemical parameters are grouped into three classes according to their relationship with the growth rate and differentiation of the neoplasms.

The molecular correlation concept takes into account the fact that the critical change produced in a cancer cell must be inheritable and that the progression in the neoplastic process should be reflected in a progressive metabolic imbalance. Thus, there should be a linking between the gene expression controlling cell proliferation and the relevant biochemical pattern. We believe that any concept

TABLE I. Molecular Correlation Concept

A conceptual and experimental approach: To elucidate and interpret the molecular basis of altered gene expression and metabolic pattern in cancer cells

Relationship of metabolic parameters with growth rate	Significance of change for neoplasia
Class 1. Correlates with growth rate	Essential
Class 2. Altered in the same direction in all tumors	Ubiquitous
Class 3. No relationship	Irrelevant, coincidental

proposed for the analysis and understanding of the metabolism and behavior of the cancer cell should entail the discovery of a specific metabolic pattern that distinguishes cancer cells from normal and other pathological conditions, be readily testable, and provide a biochemical explanation for the selective biological and replicative advantages that the inheritable neoplastic change confers on the cancer cells. Such a concept should have power to predict biochemical alterations and metabolic imbalances in yet unexplored metabolic areas and should deal with the question of applicability of the molecular pattern discovered in one type of tumor (hepatoma spectrum) to other neoplasms.

Ranking of the Hepatomas in Regard to Growth Rate and Differentiation

Ae a brief overview, the significant properties of the Morris hepatoma spectrum are summarized in Table II.

Without overlooking other important aspects of neoplasia (invasiveness, metastasis formation), we have emphasized tumor growth rate in these studies because (1) it can be measured with some precision with an array of biological, cytological, and biochemical methods; (2) it enters into clinical diagnosis; and (3) it can be employed in the evaluation of treatment including chemotherapy. With this approach we have been conducting systematic studies of gene expression in analyzing the linkage of enzymatic and metabolic imbalance with cell replication rate. Evidence was presented in our previous publications that demonstrates

TABLE II. Significant Properties of Morris Hepatoma Spectrum

Properties	Parameters	Extent of change from normal		
		Slight	Intermediate	Extensive
Biological behavior	Growth rate	Low	Medium	Rapid
Morphology	Differentiation	Near normal	Medium	Poor
Genetic Apparatus	Chromosome number	Normal	Increased	High
	Chromosome karyotype	Normal	Nearly normal	Abnormal
Energy Generation	Respiration	Normal	Moderate	Moderately low
	Glycolysis	Low	Normal or increased	High
Replication and functions	Imbalance of opposing pathways of synthesis and degradation	Moderate	Pronounced	Extensive

TABLE III. Measuring Methods for Ranking Tumors of the Hepatoma Spectrum by Growth Rate

---

Growth rate: Measurable with precision with one or several of the following methods which all give similar ranking:

Biological—Tumor size and volume
            Tumor weight
            Average time between transplantations
            Time required to kill host

Cytological—Mitotic counts

Biochemical—Study of thymidine (Tdr) metabolism

In vivo (a) Injection of Tdr: Extraction and counting of DNA
In vitro (b) Incubation of tissue slices with Tdr: Determining the ratio of incorporation into DNA and degradation to CO_2
In vitro (c) Using Tdr: in autoradiography

Utility—Readily measurable by any biochemist

---

the usefulness of relating the biochemical pattern to the growth rate of the different tumor lines. Since our investigations were designed to study the correlation of gene expression with growth rate in hepatomas, the measurement of tumor proliferation rate is important. Various aspects of this approach were discussed previously in detail (*41, 44*), and these are summarized in Table III.

Advantages of Relating the Biochemical Pattern to Growth Rate Rather Than to the State of Differentiation

As it has become clear that there is a metabolic pattern that can be correlated with hepatoma growth rate, it has been suggested by some workers that these

TABLE IV. Advantages of Relating the Biochemical Pattern in Hepatomas to Growth Rate Rather Than to Differentiation

Variable	Growth rate	Degree of differentiation
Measurable with precision	Yes	No
Provides quantitative results	Yes	No
Makes statistical evaluation possible	Yes	No
Enters into clinical diagnosis	Yes	Yes
Enters into clinical prognosis	Yes	Yes
Indicator of tumor response to therapy	Yes	No
Responds to stimulation or inhibition	Yes	No
Biochemists can determine it routinely	Yes	No

Differentiation:
1. Difficult to quantitate or measure with precision
2. It is frequently a subjective evaluation
3. Does not seem to respond to therapeutic influence in patients; clinically no redifferentiation occurs
4. Biochemists would have to rely on pathologists for grading

Conclusion: Growth rate should be used as the basis of comparison

biochemical correlations should be made rather with the state of differentiation or dedifferentiation of the tumors. I do not see any conceptual or methodological advantage in this, and the disadvantages of referring to the state of differentiation and the advantages of utilizing growth rate measurements are assembled in Table IV.

Growth rate can be measured with some precision with the techniques available to biochemists. However, attempts could also be made to ascertain the state of differentiation and to eventually work out a precise quantitation of the histological grading. There is good agreement here, since both growth rate and histological grading provide essentially the same ranking order for all tumors. The scientific approach will always weigh in the direction of the more precise, quantitative approaches, which the measurement of growth rate can currently provide (41, 44).

Source of Failure to Detect Biochemical Correlations with Growth Rate

In our publications we outlined the various steps to be taken to provide appropriate controls, selection of key enzymes and metabolic pathways, standardization of biochemical methods, and the expression of enzymatical and biochemical results to be used in the application of the molecular correlation concept (31–35, 41, 44). In the course of the past decade a number of investigators did not detect biochemical correlations subsequently discovered to exist. It seems useful to summarize in Table V the most common errors which result in the failure to detect biochemical correlations with tumor growth rate.

TABLE V. Errors Leading to Failure to Detect Biochemical Correlations with Growth Rate

Conceptual errors	1.	Not selecting for study key regulatory enzymes which are the ones expected to show correlation
	2.	Selecting enzymes present in excess
	3.	Assuming that every biochemical parameter ought to correlate with growth rate
Technical errors	4.	Not adapting assay method to liver and hepatoma systems
	5.	Not assaying under optimal substrate, cofactor, and linear kinetic conditions
	6.	Not establishing proper extraction methods for the enzyme
	7.	Not recognizing the presence of isozymes that behave differently
Errors in ranking	8.	Not ranking tumors in their order of growth rate
	9.	Assaying tumors of similar growth rates, minimizing biochemical differences seen which can be detected in a wider spectrum
	10.	Not investigating a sufficient number of tumors

Molecular Correlation Concept and Examination of the Pattern of Gene Expression

We said previously that the molecular correlation concept takes into account the fact that the metabolic pattern, which is near the essence of neoplastic change, is inheritable.

Those who doubted the presence of any meaningful biochemical pattern in tumors ignored or rejected the validity of the metabolic correlations reported, and some investigators relied chiefly on observations of the apparent diversity in the activity and behavior of certain enzymes. There are two appropriate comments. The first will recall our observations that enzymes that are present in excess do not usually correlate with the growth rate of hepatomas, or, for that matter, with the alteration of the overall activities of metabolic pathways either in diabetes or under hormonal stimulation. We group such enzymes into Class 3, along with other parameters that do not correlate with growth rate. As we emphasized previously, the objective in most scientific endeavors in this area should be to uncover the metabolic pattern in face of apparently irrelevant and coincidental alterations. The changes that may occur in the overall transcription of the genome in neoplastic cells might also lead to the expression of previously dormant functions and properties. The presence of unusual functions and the emergence of new functions normally undetected in certain cell types represent important evidence which further confirms that in neoplasia we are dealing with alterations in gene expression. Table VI lists some of the nonendocrine tumors that have been observed to produce hormones or hormone-like substances (19).

TABLE VI. Tumors Associated with Specific Ectopic Hormone Syndromes

Syndrome	Hormones	Sites
Cushing's syndrome	ACTH	Lung (oat cell, bronchial adenoma), thymus, pancreas, thyroid (medullary)
Hyperparathyroidism	PTH	Kidney, lung (squamous), pancreas, ovary, many squamous sites
Inappropriate antidiuresis	Arginine-vasopressin (ADH)	Lung (oat cell)
Erythrocytosis	Erythropoietin	Cerebellum (hemangiomas), liver, uterus
Zollinger-Ellison syndrome	Gastrin	Pancreas (non-β-islet-cell) adenomas
Gynecomastia (adults) Precocious puberty	Gonadotrophins	Lung (large cell), liver
Hyperthyroidism	Thyrotrophin	Trophoblast, lung
Hypoglycemia	Unknown	Retroperitoneal mesoderm, liver

For example, lung cells are not known to be capable of secreting gonadotrophins, yet certain lung tumors produce this hormone. Such phenomena are most readily interpreted currently as activation or derepression of genic expression. The concept that neoplasia entails an alteration in gene expression provides a unifying mechanism for relating neoplastic transformation of the cells and its phenotype in terms of metabolic imbalance and tumor growth rate. The "diversity" which occurs is a sign that, in addition to the essential features of the altered neoplastic gene expression, some non-essential and coincidental changes might also take place. It seems to me, however, that it is a conceptual error to concentrate on the coincidental aspects of gene expression instead of on the essential imbalance in cancer cells that can be correlated with biological behavior.

Detection of Ordered Pattern of Gene Expression and Its Linking with Tumor Growth Rate

The molecular correlation concept postulated the operation of an ordered pattern of gene expression in neoplasia. It suggested that it should be possible to discover and identify the essential pattern of gene expression among a number of possibly coincidental alterations that may be associated with the core of altered gene expression that underlies neoplasia. In this paper, we will focus on the implications of the molecular correlation concept in the areas of carbohydrate, pyrimidine, nucleic acid, and ornithine metabolism.

Is gene expression random or ordered in differentiation and in neoplasia? If it is ordered, is it manifested in the gradual emergence of a metabolic pattern which characterizes various stages of differentiation and different degrees of neoplastic transformation? If there is a definite metabolic pattern, is it indicated in the selective expression of genes or groups of genes determining the production and operation of functionally related key enzymes? Is gene expression irreversibly fixed in the liver of adult rat or does it respond to regulatory modulation? In the hepatomas, is there a linking of gene expression as manifested in progressive metabolic imbalance with the gene expression as manifested in different cell proliferation rates?

Since the molecular correlation concept suggests that the biochemical pattern that correlates with hepatoma growth rate should be the most clearly detectable in the behavior of key enzymes, the following sections will examine the evidence for the existence of a pattern of gene expression in differentiation and in neoplasia by studying the key enzymes of carbohydrate, pyrimidine, nucleic acid, and ornithine metabolism.

Carbohydrate Metabolism: Pattern of Gene Expression in Differentiating Liver and in Hepatomas of Different Growth Rates

1. Pattern of gene expression in differentiation

In order to examine the possibility of the existence of a pattern in gene expression in carbohydrate metabolism, the activities of the key gluconeogenic and glycolytic enzymes were compared (46) in the liver of developing and adult rats. The results of these studies are summarized in Fig. 1, which shows that there is a characteristic developmental pattern present. The activities of the key gluconeogenic enzymes are low before term, and they rise rapidly after birth. The enzyme activities decrease to the level observed in the adult rat during the 3–4 week postpartum period. In contrast, the activities of the key enzymes of glycolysis rise during differentiation, reaching adult level during the same period (46). The behavior of the bifunctional enzymes does not conform to any pattern. This is in line with my view that the pattern of gene expression can be most readily identified by examining the key enzymes of the opposing metabolic pathways. The three groups of enzymes behave according to a pattern that can be predicted if the Functional Genic Unit concept is applied to the analysis of sequential gene expression in differentiation (48).

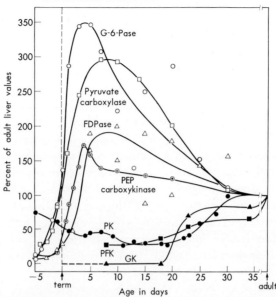

Fig. 1. Developmental pattern of key enzymes of gluconeogenesis and glycolysis in liver during postnatal differentiation

Thus, the evidence indicates that there is a definite pattern of gene expression in carbohydrate metabolism during differentiation.

2. *Pattern of gene expression in adult liver; gene expression can be modulated*

In order to test whether gene expression has reached an unalterable balance after differentiation is completed in the adult liver, we examined the effects of hormonal conditions on opposing key enzymes of carbohydrate metabolism. Two key enzymes selected from the opposing groups of gluconeogenic and glycolytic enzymes were analyzed as indicators of the sequential behavior in dif-

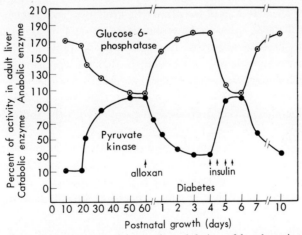

Fig. 2. Sequential gene expression and modulation of key hepatic gluconeogenic and glycolytic enzymes during differentiation and in hormonal imbalance

Repression and derepression of anabolic and catabolic enzymes in differentiation, diabetes, and insulin treatment.

ferentiation and in the endocrine control of gene expression (46). Figure 2 shows that during differentiation the glucose-6-phosphatase activity decreases and the pyruvate kinase activity increases, reaching the levels observed in normal adult liver. The experiment showed that when the animals are made alloxan-diabetic and the insulin level is markedly decreased, the activity of the gluconeogenic enzyme increases and that of the glycolytic one decreases to about the levels found in the newborn rat liver. As we showed previously, insulin functions as a suppressor of the synthesis of the key gluconeogenic enzymes and as an inducer of the key glycolytic ones (47, 49). In agreement with the earlier results, administration of insulin in the diabetic rats caused the activity of glucose-6-phosphatase to decrease (49) and that of pyruvate kinase to increase to normal liver levels (50). When the treatment of diabetic rats with insulin was discontinued, the activities of the opposing key enzymes returned to the levels observed in the diabetic rats. Thus, insulin coordinates intermediary metabolism, in part at least, through influencing in an antagonistic fashion gene expression for the opposing key enzymes of gluconeogenesis and glycolysis.

Then, the evidence indicates that the homeostatic balance of gene expression in the adult liver can be drastically altered by changing the hormonal status of the adult animal.

3. Pattern of gene expression in hepatomas in carbohydrate metabolism; progressive metabolic imbalance linked with hepatoma growth rate

Previous work showed that in the hepatoma spectrum in parallel with increasing tumor growth rate the key gluconeogenic enzymes decreased, whereas the key glycolytic enzymes increased (Fig. 3). These results, outlined in detail elsewhere, support the molecular correlation concept (33, 44). In recent studies, in order to investigate how closely the behavior patterns of these enzymes and hepatoma growth rate are linked, the ratios of the activities of opposing key enzymes were calculated and correlated with hepatoma growth rate. Figure 4 indicates that the ratios of the activities of the glycolytic/gluconeogenic enzymes (hexokinase/ glucose-6-phosphatase; phosphofructokinase/fructose-1, 6-diphosphatase) correlate closely with hepatoma growth rate over a wide range (41). From the behavior of the key enzymes, one can anticipate that the opposing pathways of glycolysis and gluconeogenesis should also correlate with the growth rate of the neoplasm. Experimental work demonstrated that the aerobic glycolysis did increase and gluconeogenesis decreased in parallel with the increase in hepatoma growth rate (26).

Then, the key enzymes and the opposing pathways in carbohydrate metabolism exhibit a characteristic pattern of behavior in differentiation, regulation, and neoplasia. The metabolic pattern in the hepatoma spectrum reveals the development of a progressive imbalance in gene expression, linked with the tumor growth rate.

These results show that in carbohydrate metabolism there is a pattern of gene expression as manifested in the behavior of key enzymes of anabolism and catabolism. The behavior of the key enzymes follows a pattern that is in line with the predictions deducible from the functional genic unit concept and the molecular

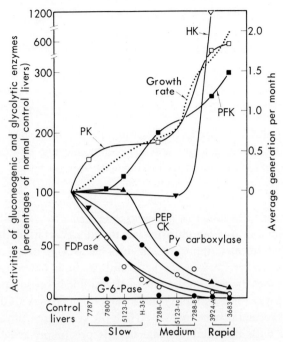

FIG. 3. Behavior of key gluconeogenic and glycolytic enzymes in hepatomas of different growth rates

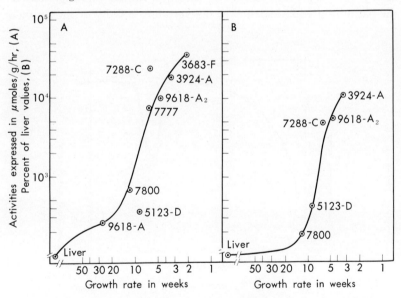

FIG. 4. Correlation of ratios of key glycolytic/gluconeogenic enzymes with hepatoma growth rate

Activities originally measured in μmoles/g/hr were converted to percentages of values observed in normal liver, and ratios were then calculated.
  A. Hexokinase/glucose 6-phosphatase.
  B. Phosphofructokinase/fructose 1,6-diphosphatase.

correlation concept. Now, it is relevant to examine the question whether the general pattern and the principles elucidated for the behavior and regulation of carbohydrate metabolism apply to pyrimidine and nucleic acid metabolism.

Pyrimidine and Nucleic Acid Metabolism: Pattern of Gene Expression in Differentiating and Regenerating Liver and in Hepatomas of Different Growth Rates

In view of the fact that cell replication and its disturbed control are in the center of the neoplastic change, it is especially relevant to examine the pattern of nucleic acid metabolism and its linking with differentiation and cell proliferation rate.

1. *Pattern of gene expression in differentiation*

Investigations in our laboratories led to the expectation that, similar to the pattern observed for the opposing enzymes in carbohydrate metabolism, there would be a pattern for the key synthetic and catabolic enzymes of DNA metabolism. Our studies focused on the fate of thymidine because, for this purecursor, a synthetic pathway (Tdr to DNA) and a catabolic pathway (Tdr to CO_2) operate in the liver and certain other organs. When the synthetic utilization of thymidine was contrasted with the catabolic fate of this precursor, a definite pattern was observed in the differentiating liver. As Fig. 5 shows, in the young rats the catabolic utilization of thymidine progressively increased, whereas the synthetic utilization progressively decreased. In the adult rat the very high activity of the catabolic pathway predominated over the very low activity of the synthetic path-

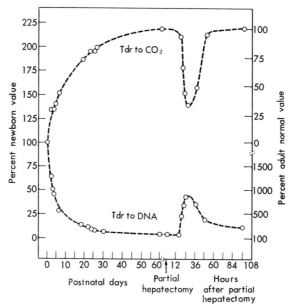

FIG. 5. Opposing behavior of synthetic and degradative utilization of thymidine in liver during differentiation and in regeneration

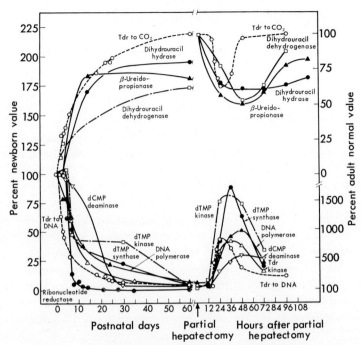

FIG. 6. Behavior of opposing pathways and key enzymes of nucleic acid metabolism in differentiation and regeneration

way. We inquired into the question whether gene expression in the adult liver is fixed or subject to modulation. In the case of DNA metabolism, the appropriate stimulus to examine this is partial hepatectomy, which induces a self-limited rapid proliferation of the liver. As Fig. 5 shows, there was a marked increase in the utilization for synthetic purposes and a concurrent drop in catabolic use of thymidine. These results are compatible with the view that the opposing pathways of thymidine utilization are regulated in an antagonistic fashion. The operation of a clear-cut pattern of gene expression was revealed in thymidine metabolism by studies in differentiation and regeneration (46).

When it is suggested that the opposing behavior of thymidine metabolic pathways represents alterations in the pattern of gene expression, what one implies is that the gene products, the specific enzyme proteins that are produced, depend on a sequential derepression or repression in the respective pathways. Thus, these results could be interpreted as implying a sequential derepression of the genes governing the production of the enzymes of the catabolic pathway and a gradual repression of those of the synthetic pathway of thymidine during differentiation in liver. In order to test the validity of this interpretation, we carried out investigations on the behavior of some of these enzymes under these conditions and assembled these results, with data recalculated from the literature for other enzymes. The results, given in Fig. 6, indicate that the behavior of the opposing overall metabolic pathways is closely reflected in the behavior of the activities of the enzymes involved. During differentiation there is a gradual increase in the activities of the catabolic pathway, and the catabolic enzymes, dihydrouracil dehydrogenase, β-ureidopropionase, and dihydrouracil hydrase, increase parallel with the overall pathway. The activity of thymidine phosphorylase (not shown in this figure) exhibits a pattern basically similar to that of the other catabolic enzymes (46). On the other hand, the synthetic utilization of thymidine into DNA rapidly decreases in differentiation, and the activities of the enzymes involved, ribonucleotide reductase, dCMP deaminase, dTMP kinase, dTMP synthase, Tdr kinase, and DNA polymerase, decrease parallel with the behavior of the overall synthetic pathway. Thus, during sequential gene expression, the activity of the opposing pathways and that of the key enzymes of thymidine utilization behave in a mirror picture fashion. When the process of differentiation is completed by 40 to 60 postnatal days the activities of the enzymes and of the overall pathways of the synthetic utilization are very low and those of the catabolic one are high. These results are compatible with the suggestion that a coordinated behavior pattern for gene expression operates manifesting opposing changes in the activity of antagonistic pathways and enzymes during differentiation. Moreover, studies in our laboratories have shown that Actinomycin treatment inhibits the rise of the degradative pathway of thymidine in the liver during the first five postpartum days (8). These observations further support the interpretation suggested for the behavior of this pathway in differentiation.

2. Pattern of gene expression in regeneration

In the adult liver the activity of the synthetic pathway of thymidine is very low, whereas that of the catabolic one is very high. Consequently, the ratio of the activity of Tdr to DNA/Tdr to CO_2 is very low. The resting state of DNA metabolism in the adult is, then, characterized by the overwhelming predominance of the activity of the catabolic pathway over the synthetic utilization of thymidine. One could assume from this metabolic relationship that gene expression in the fully differentiated adult animal liver is now settled in a fixed balance, but there is experimental evidence that the potential for altering the homeostatic equilibrium of gene expression is retained in the adult liver. This is shown by the fact that partial hepatectomy brings into operation a yet unknown physiological signal system that is capable of unleashing the potential of gene expression (Figs. 5 and 6). In the regenerating liver, gene expression is altered in such a way that it results in a sharp rise in the incorporation of thymidine to DNA and a decrease in the catabolic utilization of this precursor until about 24–36 hr after hepatectomy. Subsequently, the opposing pathways return to near normal range at or about 96 hr. These observations are interpreted by suggesting that the changes result in a repression of the derepressed pathway of thymidine catabolism, and in a derepression of the repressed pathway of the utilization of thymidine for synthesis of DNA (46). If this interpretation in terms of gene expression is a valid one, it should lead to discovery of similar behavior in the activities of the key enzymes involved in the two pathways. We have tested this prediction by compiling from the literature and from our laboratory results dealing with the behavior of these enzymes in regenerating liver. Figure 6 compares the overall metabolic pathway and the key enzymes and indicates a close parallelism between these processes (46).

Thus, these results support the concept that there is a pattern of gene expression that closely coordinates the opposing pathways and the key enzymes of thymidine metabolism (8, 46).

3. *Pattern of gene expression in pyrimidine and nucleic acid metabolism; progressive metabolic imbalance linked with hepatoma growth rate*

We have suggested that the metabolism of UMP in liver may be considered as synthetic and degradative pathways opposing each other (Fig. 7). Viewing pyrimidine metabolism in terms of antagonistic pathways in turn might provide an analytic and predictive approach (46). Current evidence indicates that in the adult liver the degradative pathway of UMP has a much higher activity than that of the anabolic pathway. In consequence, we would expect that in the cancer cells the overall synthetic pathway and the key enzymes would increase, and concurrently the activity of the catabolic pathway might remain unchanged or, more likely, would be decreased.

The two initial enzymes of the anabolic pathways of UMP, aspartate transcarbamylase and dihydroorotase, increase in parallel with the growth rate of hepatomas (27, 28). Recent investigations in our laboratories were directed to the analysis of the degradative pathways of UMP and thymidine (8, 13, 22, 24, 46). In the catabolic pathway that degrades uridine or thymidine to CO_2 and other products, dihydrouracil dehydrogenase (EC 1.3.1.2) was identified as the rate-limiting enzyme (4, 9). The behavior of the key enzyme, dihydrouracil dehydrogenase, was studied in hepatomas of different growth rates and compared with that of the enzyme in normal resting and in rapidly growing, regenerating liver. In the degradative pathway, the phosphorylase was also of interest because it is the enzyme that directly opposes thymidine kinase, which channels thymidine into

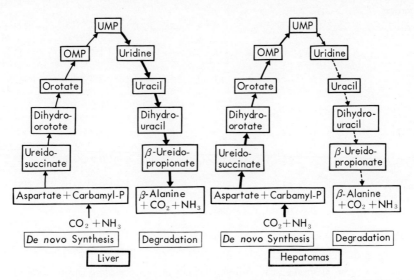

FIG. 7. Behavior of opposing pathways of synthesis and degradation of UMP in liver and in hepatomas

Correlation of enzymatic basis of imbalance in pyrimidine metabolism with hepatoma growth rate.

synthetic utilization. Since thymidine phosphorylase is not a rate-limiting enzyme and its activity is high, the molecular correlation concept would suggest that this enzyme would not decrease in parallel with hepatoma growth rate but would markedly decrease in all tumors.

TABLE VII. Correlation of Dihydrouracil Dehydrogenase Activity with Hepatoma Growth Rate

Tissues	No. of observations	Transplant generation	Growth rate	Dihydrouracil dehydrogenase activity	
				μmoles/g/hr	μmoles/cell/hr $\times 10^{-10}$
Normal liver (ACI/N)					
Control for 3924A	15			1.9 ± 0.2	95 ± 13
Control for 3683F	3			2.9 ± 0.1	129 ± 5
Normal liver (Buffalo)					
Control for 44	4			2.8 ± 0.4	118 ± 19
Control for 47-C	2			2.9 ± 0.1	137 ± 3
Control for 7787	2			2.4 ± 0.4	123 ± 10
Control for 9618A	5			2.0 ± 0.1	112 ± 6
Control for 9618B	4			1.8 ± 0.1	82 ± 3
Control for 7777	3			2.2 ± 0.2	116 ± 35
Control for 9618A2	3			2.6 ± 0.1	87 ± 2
Normal liver (Wistar)	4			2.2 ± 0.3	93 ± 13
Sham-operated (Wistar)	4			2.3 ± 0.3 (100)	108 ± 18 (100)
24-hr regenerating liver (Wistar)	4			1.7 ± 0.2 (74)	87 ± 9 (81)
Hepatomas					
44	3	6	9	2.5 ± 0.3 (89)[a]	111 ± 15 (94)
47-C	4	5	8	1.5 ± 0.1 (52)[a]	87 ± 7 (64)[a]
7787	2	15	7	2.3 ± 0.8 (94)	102 ± 38 (80)
9618A	6	6	5.8	1.1 ± 0.1 (55)[a]	55 ± 5 (49)[a]
9618B	4	5	4.5	1.2 ± 0.1 (69)[a]	59 ± 6 (73)[a]
7777	4	66	1.3	0.9 ± 0.2 (41)[a]	45 ± 16 (39)[a]
3924A	14	277	1.0	0.5 ± 0.05 (29)[a]	36 ± 5 (38)[a]
3683F	4	280	0.5	0.3 ± 0.04 (9)[a]	12 ± 1 (10)[a]
9618A2	4	19	0.4	0.3 ± 0.04 (12)[a]	13 ± 1 (15)[a]

The data are given as means $\pm$ SE with percentages of corresponding control liver values in parentheses. The activities per cell are to be multiplied by the exponent given to arrive at the actual values. Growth rate is expressed as the mean transplantation time given in months between inoculation and growth to a size of approximately 1.5 cm diameter. From Queener et al. (22).

[a] Values statistically significantly different from the respective controls.

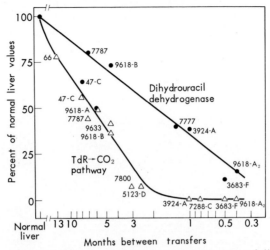

Fig. 8. Correlation with hepatoma growth rate of the activity of dihydrouracil dehydrogenase and of the activity of the overall degradative pathway of thymidine to CO_2

Spearman's rank correlation coefficients, $r' = 6\ d^2/[n(n^2-1)]$, were calculated for the relationships shown in the figure. For the eight tumor series, dihydrouracil dehydrogenase activity *versus*. Tdr to CO_2 pathway activity gave $r' = 0.79$. In both cases the correlation was significant $P = <0.05$.

The kinetic behavior of these two enzymes was investigated in our laboratories. The affinity of the dehydrogenase to substrate, to coenzyme, and the pH optima have been reported (22). As Table VII shows, in the spectrum of hepatomas, dihydrouracil dehydrogenase activity decreased parallel with the increase in tumor growth rate.

Since this enzyme is the rate-limiting one in the degradation of Tdr to CO_2, the behavior of the overall pathway and this enzyme were compared in the hepatomas. The data given in Fig. 8 show a correlation between the growth rate of the hepatomas and the decrease in dihydrouracil dehydrogenase activity and overall catabolic pathway of thymidine. These results are in line with the rate-limiting role proposed for dihydrouracil dehydrogenase, and the close linking of this enzyme activity with tumor proliferation rate provides further evidence in support of the molecular correlation concept (22).

More recent studies worked out the kinetic conditions of the phosphorylases involved in pyrimidine metabolism (24). Since special interest centers on thymidine phosphorylase (EC 2.4.2.4) because it directly opposes the activity of thymidine kinase, detailed examination was also carried out on this enzyme in the hepatomas. The results indicated that thymidine phosphorylase was significantly decreased in all hepatomas, irrespective of the growth rate (24). This behavior is in line with the predictions of the molecular correlation concept, which classifies this enzyme in Group 2 (Table I).

4. Correlation of behavior of the activity of opposing pathways of synthetic and catabolic utilization of thymidine with hepatoma growth rate

Table VIII lists the hepatomas in order of increasing growth rate and the

TABLE VIII. Correlation of Activity of Synthetic and Degradative Pathways of Thymidine with Hepatoma Growth Rate

Tissues	Transplant generation	Growth rate (months)	Thymidine into DNA	Thymidine to CO_2	Thymidine into DNA / Thymidine to CO_2
Liver[a]			100	100	100
Hepatomas					
9618-A	6	5.8	280	49.0	615
9618-B	6	4.5	292	36.0	810
7800	49	3.0	370	6.9	5,750
5123-D	81	2.5	706	6.5	13,500
3924-A	274	1.0	1,890	0.094	2,210,000
7288-C	101	0.8	2,360	0.050	5,620,000
7777	70	0.7	4,520	0.064	8,700,000
3683-F	272	0.5	3,900	0.045	11,500,000
9618-A_2	15	0.4	3,180	0.041	13,900,000

Data are expressed as percentages of normal liver values.[a] The growth rate is the mean transplantation time given in months between inoculation and growth to a size of approximately 1.5 cm diameter.[b] From Ferdinandus et al. (8).

[a] Normal liver values: thymidine into DNA = $11,330 \pm 250$; thymidine to CO_2 = $1,065,000 \pm 17,000$ in dpm/g/hr.

[b] Spearman's rank correlation coefficients, $r' = 1 - 6\sum d^2/n(n_2-1)$, were calculated from the difference, d, between rankings with $n=9$ for the orders of the 9 tumors in increasing growth rate and the activities of the pathways for ranking of growth rate against Td to DNA activity, $r' = 0.933$; for Tdr to CO_2 activity, $r' = 0.983$; and for Tdr into DNA/Tdr to CO_2 activity, $r' = 1.0$. All these values were significant ($P = <0.05$).

results on the incorporation of Tdr into DNA, the degradation of Tdr to CO_2, and the ratio of the activities of the two pathways (8). This spectrum spans a wide variation in tumor growth rate, including some of the very slow-growing and some of the fastest proliferating liver neoplasms. It is significant from the point of view of the linking of growth rate with biochemical events that even in the slowest growing hepatomas the incorporation of Tdr into DNA increased nearly 3-fold and, concurrently, the degradation of Tdr to CO_2 decreased to 49%. The ratio increased 6-fold over the value in the normal liver of the control rats. The results clearly show that in parallel with the increase in tumor growth rate there was a rise in the activity of the synthetic pathway and a decline in that of the catabolic pathway of Tdr. In the most rapidly growing hepatoma in this series, 9618-A_2, the incorporation into DNA increased 31-fold, whereas the degradative pathway declined to 0.041% of values observed in normal control rat liver. In consequence, the ratio increased nearly 140,000-fold (8).

Figure 9 shows the gradually widening gap between the synthetic and degradative utilization of Tdr, the resultant predominance of the synthetic over the catabolic pathway, and the relationship to hepatoma growth rate of the ratios of Tdr into DNA/Tdr to CO_2.

This pattern suggests a close linkage in the expression of the replicative potential of the genome with the extent of the progressive imbalance in DNA metab-

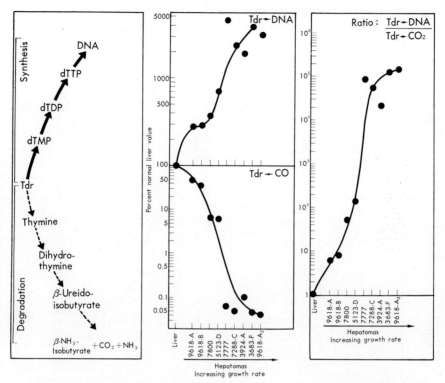

Fig. 9. Correlation of the activities of opposing pathways of thymidine utilization and the thymidine into DNA/thymidine to CO_2 ratio with hepatoma growth rate

olism, as reported earlier for carbohydrate metabolism (*33, 44*). The results further indicate the importance of the ratios of activities of opposing enzymes and metabolic pathways as sensitive indicators of the link between replicative, transcriptive, and translative expression of the altered genome in neoplasia.

5. *Correlation of the behavior of the activity of opposing pathways of thymidine and DNA metabolism with hepatoma growth rate*

The interpretation we suggested, that the alterations in the gradual imbalance in the activities of opposing metabolic pathways might involve altered gene expression (*8, 46*), is supported by considerable evidence describing a pattern of correlation with hepatoma growth rate for a number of key enzymes of nucleic acid metabolism (*46*). These results in the literature indicate that the following synthetic enzymes correlate positively with hepatoma growth rate: Tdr kinase, dTMP kinase, dTMP synthase, dCMP deaminase, aspartate transcarbamylase, dihydroorotase, ribonucleotide reductase, and DNA polymerase. On the other hand, there was a negative correlation with the increase in tumor growth rate for the degradation of thymine, thymidine, uracil, and the activity of dihydrouracil dehydrogenase. These correlations were discussed elsewhere, and Fig. 10 shows these relationships of key enzymes of nucleic acid metabolism with hepatoma growth rate (*46*).

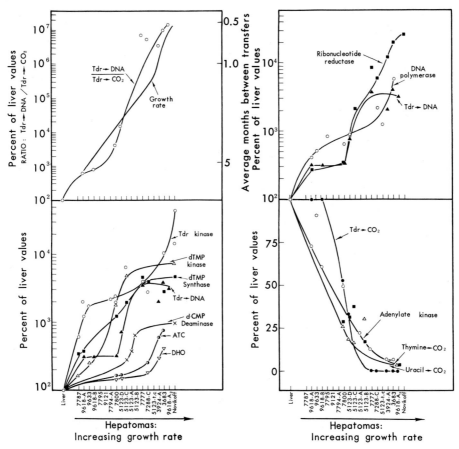

Fig. 10. Correlation of DNA metabolic pathway and enzyme activities with hepatoma growth rate

It is of interest to examine the extent of alterations that the various metabolic pathways and enzymes exhibit in relation to hepatoma growth rate. Figure 10 indicates that some metabolic parameters increase very considerably in the rapidly growing tumors, while other ones change relatively little. In order to illustrate this fact, Fig. 11 gives a comparison of the extent of correlation with hepatoma growth rate for the enzymes ribonucleotide reductase and aspartate transcarbamylase and for the thymidine ratio (46). Figure 11 shows that the rise in the ratio of Tdr to DNA/Tdr to CO_2 is the most striking for the extent of alteration of any biochemical parameter studied so far in the hepatoma spectrum. The thymidine ratio correlates with the hepatoma growth rate over a 100,000-fold range and should therefore have considerable interest as an indicator and a measure of tumor malignancy and growth rate. In examining Fig. 11, it may be noted that ribonucleotide reductase, which shows the greatest extent of increase, has very low activity, whereas aspartate transcarbamylase, which has a comparatively high activity, exhibits a much smaller increase (46). These observations, indicating a widely different behavior for different enzymes, suggest that a relationship might exist between the normal activity of the liver enzyme and the extent of alteration in

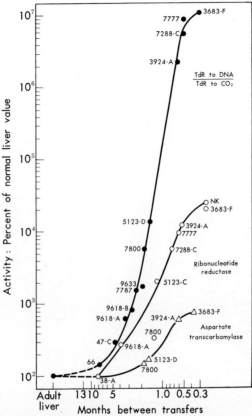

FIG. 11. Comparison of the extent of increase with hepatoma growth rate of the thymidine ratio, ribonucleotide reductase, and aspartate transcarbamylase activities

the activity of the enzyme in the hepatoma spectrum. Now I will examine this possibility.

6. *Comparison of activities and behavior of enzymes involved in the synthetic and catabolic pathways of pyrimidine and DNA metabolism in liver and hepatomas*

Since the molecular correlation concept suggests a linking of gene expression in terms of metabolic imbalance and tumor growth rate, it should be possible to utilize this concept for the prediction of presence and extent of metabolic imbalance and behavior of enzyme activity on the basis of the growth rate of the hepatomas and in view of the known activity of the enzyme in normal liver. Thus, in order to test the predicting power of the concept, it is necessary to know the activity and metabolic role of the enzymes involved. We have assembled from the literature and from our own work the activities of the enzymes of pyrimidine and DNA metabolism and recalculated them in order to express the activities in a uniform and comparable fashion (46). These data are summarized in Table IX, and the examination of the activity of the enzymes and their behavior in the rapidly growing tumors does suggest the possible presence of a relationship which may allow predictions and that can be experimentally tested.

TABLE IX. Comparison of Activities of Enzymes Involved in the Anabolic and Catabolic Pathways of Pyrimidine and DNA Metabolism

Enzymes	Livers		Hepatomas (rapidly growing)	
	Normally fed	Regen. 24 hr	Novikoff	3683-F
Anabolic enzymes	pmoles/mg protein/hr	% liver	% liver	
Ribonucleotide reductase	3		25,300	20,800
DNA polymerase	56	540		5,810
dTMP synthase	180	1,140	4,510	2,860
dTMP kinase	420	1,620	7,100	
Tdr kinase	2,400	680	14,000	10,200
Uridine kinase	6,200	210		
dCMP deaminase	12,000	350	900	750
OMP pyrophosphorylase	20,000	380		
OMP decarboxylase	71,000			
CP synthase	94,000	96		
Aspartate transcarbamylase	123,000	127	480	796
Dihydroorotase	246,000	108		418
Nucleoside diphosphate kinase	1,200,000			200
Catabolic enzymes				
Dihydrouracil dehydrogenase	26,000	78		9
β-Ureidopropionase	144,000	58		
Dihydrouracil hydrase	276,000	66		
Tdr phosphorylase	234,000	106		31

Modified from Weber et al. (46). In order to achieve a comparison, all data were recalculated in pmoles of substrate metabolized/mg protein/hr. In calculations, the values of 0.2 g protein/g wet weight of tissue for homogenates and 0.08 g protein/g wet weight for supernatant fluids were used.

The activities of the synthetic enzymes are extremely low, particularly those that play a role at critical places in the biosynthetic process, such as ribonucleotide reductase, which is at a strategic point at the reduction of nucleosides, DNA polymerase at the end stage of the reaction sequence, and the enzymes involved in thymidine utilization specifically for DNA synthesis, thymidine kinase and dTMP kinase. The enzyme dTMP synthase is also at a strategic point and has low activity. In contrast, the enzymes present at the early steps of pyrimidine biosynthesis, aspartate transcarbamylase and dihydroorotase, have nearly 100,000-fold higher activities than ribonucleotide reductase. From Table IX it appears that the synthetic enzymes which have the lowest activities in the resting liver are the ones that increase to the highest extent in rapidly growing neoplasms such as the 3683-F or the Novikoff hepatoma. Thus, ribonucleotide reductase, DNA polymerase, and dTMP kinase increase to approximately 20,800, 5,800, and 2,800%, respectively. In contrast, aspartate transcarbamylase and dihydroorotase increase to 800 and 400%, respectively.

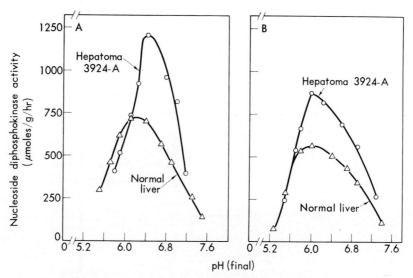

Fig. 12. Comparison of nucleoside diphosphokinase activity in normal liver and in rapidly growing hepatoma 3924-A
  A. Substrate: IDP. B. Substrate: GDP.

From these apparent relationships, one may predict that an enzyme with lower activity than ribonucleotide reductase should increase to an even higher extent than the reductase, or, conversely, an enzyme with higher activity than dihydroorotase should show a smaller extent of rise. An opportunity for testing this prediction can also be made by examining the behavior of an enzyme that is known to have very high activity, nucleoside diphosphokinase. The kinetic conditions of this enzyme were worked out very carefully in my laboratory and an assay system was established for liver and hepatomas where the only limiting factor was the amount of enzyme protein present. The behavior of nucleoside diphosphokinase in normal liver and in very rapidly growing hepatoma 3924-A is compared in Fig 12. This figure shows that in the hepatoma the activity is about twice as high as that observed in the liver of normal control rats. This observation is in agreement with the predictions made.

The pattern of altered gene expression that was discovered in the hepatoma spectrum is integrated in Fig. 13, which indicates a state of imbalance in the synthesis and degradation in pyrimidine and nucleic acid metabolism as observed in the rapidly growing liver tumors.

Ornithine Metabolism in Liver and in the Hepatoma Spectrum

Ornithine is an interesting amino acid, since it appears to be not a precursor of any major protein but may be considered a key metabolite for which several enzyme systems and metabolic pathways compete. Ornithine may be converted to L-glutamic acid, it may be channeled into polyamine biosynthesis, or it may enter into the urea cycle. All three fates of ornithine are of relevance to neoplasia for a number of reasons. The significance of ornithine metabolism to neoplasia is

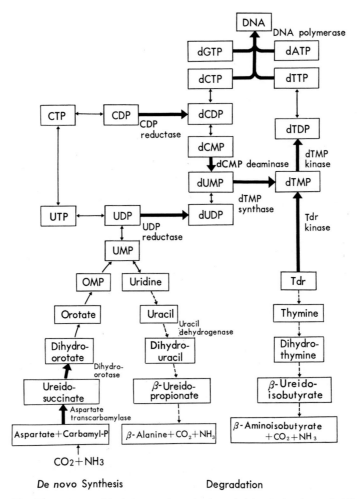

FIG. 13. The pattern of imbalance of synthesis and degradation in pyrimidine and nucleic acid metabolism in hepatomas

Heavy arrows indicate the enzymes which are increased in liver neoplasia. The dotted arrows represent enzymes which are missing or greatly reduced in the hepatomas. Other enzyme activities in the figure are either normal or no information is available for their behavior. The resulting metabolic imbalance shown in this scheme favors synthesis of DNA and prevents recycling of metabolites into degradative pathways.

Arrows: Thin=normal rate; thick=increased; dotted=decreased.

particularly seen in the fact that carbamyl phosphate and aspartate could be siphoned off into the urea cycle or alternately be utilized as precursors for nucleic acid synthesis. For these reasons, the fate of ornithine metabolism was examined from the point of view of its channeling to polyamine biosynthesis and into urea production. The metabolic pattern of the utilization of ornithine for polyamine synthesis was studied by Dr. Guy Williams-Ashman and his associates and will be reported elsewhere (52). At present, reference is made to studies carried out in my laboratories, which concentrated on the first step in ornithine

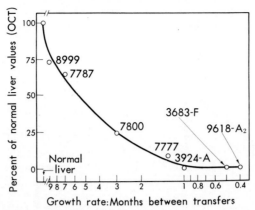

Fig. 14. Correlation of ornithine carbamoyltransferase (OCT) activity with growth rate in the hepatoma spectrum

utilization that channels this precursor into the urea biosynthesis by the enzyme ornithine carbamoyltransferase (EC 2.1.3.3).

The kinetic conditions for OCT activity were worked out in my laboratories, and the resulting enzyme assay provides a system which is limited only by the amount of OCT activity present. With this assay, which yields enzyme activity proportionate to the amount of enzyme and the time elapsed, the activity of OCT was examined in the spectrum of hepatomas of different growth rates (46a).

Figure 14 shows that with the increase in tumor growth rate there was a steady decline in OCT activity, and the rapidly growing liver tumors contained only traces or no measurable activity. There was no significant change in the regenerating liver for OCT activity as compared to the activity observed in the sham-operated controls.

The gradual decrease in this key ornithine-utilizing activity should result in a concurrent progressive decline in the activity of the urea cycle. This in turn should have a sparing effect on the availability of carbamylphosphate and aspartate for the nucleic acid synthetic processes. This interpretation is supported by the following experimental results and considerations. The activity of the enzymes that utilize carbamylphosphate, aspartate transcarbamylase, and the subsequent enzyme, dihydroorotase, increase parallel with the increase in hepatoma growth rate. Thus, the aspartate transcarbamylase/ornithine carbamoyltransferase ratios should progressively increase in parallel with tumor growth rate. This metabolic imbalance, which is apparently linked with tumor growth rate, should provide a selective advantage to the hepatoma cells in their metabolic pattern. Although previous to the present work the complete hepatoma spectrum was not systematically examined for the behavior of OCT activity, earlier data indicated that the enzyme decreases in some of the Morris hepatomas (10, 20), and the present study establishes its correlation with growth rate (46a). Observations which are in line with these results are those of Dickens and Weil-Malherbe (6), who observed very low urea production in primary liver tumors, and those of McLean et al. (16) who reported a progressive decrease in urea cycle enzyme activities in the liver during chemical carcinogenesis.

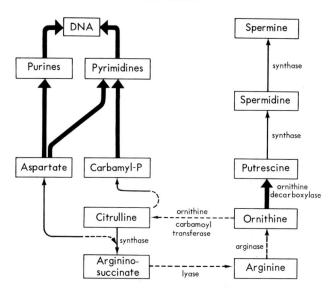

Fig. 15. Pattern of metabolic imbalance in ornithine metabolism in hepatomas
Biological advantage that an imbalance in ornithine decarboxylase/or ornithine carbamoyltransferase may confer to cancer cells.
Arrows: Thin=nomal rate; thick=increased; dotted=decreased.

Suggestions have been made that the decrease in urea cycle enzyme activities might be simply a sign of dedifferentiation (16). We examined the activity of OCT during differentiation, in newborn, young, and adult rats. The results show that the activities of OCT as expressed per μmoles of substrate metabolized per hr at 37° per cell in the newborn, in 50-g rat, and in the adult rat were 269, 435, and 673, respectively. In terms of μmoles/g/hr at 37°, the activities were 16,156, 17,511, and 15,962, respectively, for the three age groups. Thus, it can be seen that there is good activity present in the newborn rat liver, and this fact does not support the suggestion that in rapidly growing hepatomas the decrease of OCT activity is due to dedifferentiation.

The argument has also been proposed that, since urea cycle is a specialized function of liver cells, the loss of this function is merely a sign of loss of liver marker enzymes and of specialized functions. We consider this argument a difficult one to accept, since it simply gives a label for the alterations in the biochemical pattern and does not provide any experimental design or suggestions for the selective advantage of the loss of specialized function. A strong argument against the generalized nature of losing specialized functions in tumors is the fact, cited earlier in Table VI, that various tumors, instead of losing, gain new functions. It seems more productive to think in terms of a metabolic imbalance that represents an alteration in gene expression which apparently is linked with the proliferation rate of the tumors. The metabolic imbalance arising in ornithine metabolism is summarized in a preliminary fashion in Fig. 15. From this diagram the selective biological advantage that this metabolic imbalance confers to cancer cells can be perceived.

Specificity of Metabolic Imbalance to Neoplasia; Contrast of Hepatoma Metabolic Pattern with That of the Regenerating and Differentiating Liver

The question may be raised whether the alterations that occur in the hepatomas might represent only the type of biochemical change that occurs in rapidly growing tissues such as the regenerating liver. I think it would help the reader if I provide the information that our view rules out any such suggestion, because the evidence strongly favors the specificity of the metabolic pattern to neoplasia.

For instance, in regenerating liver the characteristic neoplastic alterations in carbohydrate metabolism, such as decrease in gluconeogenic enzyme activities, increase in glycolytic enzyme activities, increase in glycolysis, decrease in glycogen content, etc., do not occur. In brief, none of the carbohydrate metabolic changes that occur in the hepatomas in terms of metabolic imbalance appear to have been reported in the regenerating liver.

The lack of alterations in carbohydrate metabolism alone could serve to distinguish the metabolic pattern of hepatomas and regenerating liver. However, other characteristic changes that occur in hepatomas are also missing in the regenerating liver. For instance, there is a very marked decrease or absence of the degradative pathway of thymidine to CO_2, and the rate-limiting enzyme of the sequence, dihydrouracil dehydrogenase, is practically absent in the rapidly growing liver tumors. In the regenerating liver, which grows at approximately the same rate as the rapidly growing hepatoma, there is only a minor decrease in these parameters. In short, it is very easy to make a biochemical differential diagnosis between the extensive metabolic imbalance and enzyme deletions that occur in the hepatomas and the minor changes or lack of difference from normal liver that is the situation in the many parameters in the regenerating liver.

A somewhat frequent suggestion that the hepatomas might merely reflect the metabolism of a dedifferentiated tissue also has very little foundation in terms of biochemical imbalance and enzymology. As has been emphasized repeatedly, the embryonic rat liver is chiefly a hemopoietic organ and not liver. Therefore, most conclusions drawn from biochemical studies carried out in embryonic liver which have failed to take into consideration the extensive dilution of fetal liver cells with the different developmental forms of red blood cells, white blood cells, macrophages, etc., can have little validity. As our histological pictures showed (*38*), hemopoietic cells disappear soon after birth and thus postnatal differentiation can be analyzed with more confidence in the liver of the developing rats. The results obtained in postnatal differentiation do not provide support for claims that the metabolic imbalance and enzymology in the hepatomas represent dedifferentiation or blocked ontogeny. The facts show that no metabolic imbalance similar to that described in the hepatoma spectrum occurs at any stage of differentiation. For instance, as we mentioned, ornithine carbamoyltransferase is nearly absent in the rapidly growing hepatomas, but at no stage of postnatal differentiation is there such a lack of activity. The activity of this enzyme in the newborn rat in the average hepatic cell is about 40% of that found in the adult rat liver. As is well known, in carbohydrate metabolism in the rapidly growing hepatomas the key gluconeogenic enzymes are absent, whereas the key glycolytic ones are very highly increased. In contrast, the key gluconeogenic enzymes are high in the early postnatal development and they decline to normal values in the adult. The key glycolytic enzymes are low at birth and they slowly rise to normal values. If we consider thymidine metabolism, as we showed in the rapidly growing hepatomas, there is a very high rate of incorporation of Tdr into DNA in the presence of a completely vanished degradative pathway of Tdr. In the newborn rat liver there is high activity of the incorporation of Tdr to DNA. However, there is also an active degradative pathway which is about 50 or 60% of the highly active pathway observed in the adult rat liver. Thus, at no time during postnatal differentiation do we find any evidence for a metabolic imbalance that would resemble even remotely the metabolic picture observed in the hepatomas. These examples should suffice for this occasion to support the conclusion that the metabolic imbalance I outlined for the hepatoma spectrum is specific to neoplasia, and no similar one has been reported to occur either in the regenerating liver or at any stage of the postnatal differentiation in the rat liver.

Are the Approaches of the Molecular Correlation Concept Applicable to Tumors Other Than Hepatomas?

It seems appropriate to point out that publications during the past three years have provided evidence that the correlation of metabolic alterations and tumor growth rate also applies to kidney tumors (*14*), mammary cancer (*12*), and other neoplasms of different growth rates (*11*). Therefore, the approaches of the Molecular Correlation Concept appear to be valid not only for hepatomas, but also for various other types of neoplasms.

Acknowledgments

The research work outlined in this paper was supported by grants from the United States Public Health Service, National Cancer Institute, Grant No. CA-05034-12, The American Cancer Society, and the Damon Runyon Memorial Fund for Cancer Research, Inc.

REFERENCES

1. Allen, D. O., Munshower, J., Morris, H. P., and Weber, G. Regulation of adenyl cyclase in hepatomas of different growth rates. *Cancer Res.*, **31**, 557–560 (1971).
2. Ashmore, J., Weber, G., and Landau, B. R. Isotope studies on the pathways of glucose-6-phosphate metabolism in the Novikoff hepatoma. *Cancer Res.*, **18**, 974–979 (1958).
3. Burk, D., Woods, M., and Hunter, J. On the significance of glucolysis for cancer growth, with special reference to Morris rat hepatomas. *J. Natl. Cancer Inst.*, **38**, 839–863 (1967).
4. Canellakis, E. S. Pyrimidine metabolism. I. Enzymatic pathways of uracil and thymine degradation. *J. Biol. Chem.*, **221**, 315–322 (1956).
5. Criss, W. E., Litwack, G., Morris, H. P., and Weinhouse, S. Adenosine tri-

phosphate: adenosine monophosphate phosphotransferase isozymes in rat liver and hepatomas. *Cancer Res.*, **30**, 370–375 (1970).

6. Dickens, F. and Weil-Malherbe, H. The metabolism of normal and neoplastic tissue. A comparison of the metabolism of tumors of liver and skin with that of the tissues of origin. *Cancer Res.*, **3**, 73–87 (1943).
7. Elford, H. L., Freese, M., Passamani, E., and Morris, H. P. Ribonucleotide reductase and cell proliferation. I. Variations of ribonucleotide reductase activity with tumor growth rate in a series of rat hepatomas. *J. Biol. Chem.*, **245**, 5228–5233 (1970).
8. Ferdinandus, J. A., Morris, H. P., and Weber, G. Behavior of opposing pathways of thymidine utilization in differentiating, regenerating and neoplastic liver. *Cancer Res.*, **31**, 550–556 (1971).
9. Fritzson, P. The relation between uracil-catabolizing enzymes and rate of rat liver regeneration. *J. Biol. Chem.*, **237**, 150–156 (1962).
10. Jones, M. E., Anderson, A. D., Anderson, C., and Hodes, S. Citrulline synthesis in the rat. *Arch. Biochem. Biophys.*, **95**, 499–507 (1961).
11. Knox, W. E., Horowitz, M. L., and Friedell, G. H. The proportionality of glutaminase content to growth rate and morphology of rat neoplasms. *Cancer Res.*, **29**, 669–680 (1969).
12. Knox, W. E., Linder, M., and Friedell, G. H. A series of transplantable rat mammary tumors with graded differentiation, growth rate, and glutaminase content. *Cancer Res.*, **30**, 283–287 (1970).
13. Lea, M. A., Morris, H. P., and Weber, G. Comparative biochemistry of hepatomas. VI. Thymidine incorporation into DNA as a measure of hepatoma growth rate. *Cancer Res.*, **26**, 465–469 (1966).
14. Lea, M. A., Morris, H. P., and Weber, G. DNA metabolism in liver and kidney tumors of different growth rates. *Cancer Res.*, **28**, 71–74 (1968).
15. Lo, C., Cristofalo, V. J., Morris, H. P., and Weinhouse, S. Studies on respiration and glycolysis in transplanted hepatic tumors of the rat. *Cancer Res.*, **28**, 1–10 (1968).
16. McLean, P., Reid, E., and Gurney, M. W. Effect of azo-dye carcinogenesis on enzymes concerned with urea synthesis in rat liver. *Biochem. J.*, **91**, 464–473 (1964).
17. Morris, H. P. Studies on the development, biochemistry, and biology of experimental hepatomas. *Adv. Cancer Res.*, **9**, 227–302 (1965).
18. Ohe, K., Morris, H. P., and Weinhouse, S. β-Hydroxybutyrate dehydrogenase activity in liver and liver tumor. *Cancer Res.*, **27**, 1360–1371 (1967).
19. Omenn, G. S. Ectopic polypeptide hormone production by tumors. *Ann. Int. Med.*, **72**, 136–138 (1970).
20. Ono, T., Blair, D. R. G., Potter, V. R., and Morris, H. P. The comparative enzymology and cell origin of rat hepatomas. IV. Pyrimidine metabolism in minimal-deviation tumors. *Cancer Res.*, **23**, 240–249 (1963).
21. Ove, P., Jenkins, M. D., and Laszlo, J. DNA polymerase patterns in developing rat liver. *Cancer Res.*, **30**, 535–539 (1970).
22. Queener, S. F., Morris, H. P., and Weber, G. Dihydrouracil dehydrogenase activity in normal, differentiating and regenerating liver and in hepatomas. *Cancer Res.*, **31**, 1004–1009 (1971).
23. Queener, S. F., Morris, H. P., and Weber, G. To be published.
24. Queener, S. F., Morris, H. P., and Weber, G. Behavior of phosphorylases in normal and neoplastic tissues. To be published.

25. Sneider, T. W., Potter, V. R., and Morris, H. P. Enzymes of thymidine triphosphate synthesis in selected Morris hepatomas. *Cancer Res.*, **29**, 40–54 (1969).
26. Sweeney, M. J., Ashmore, J., Morris, H. P., and Weber, G. Comparative biochemistry of hepatomas. IV. Isotope studies of glucose and fructose metabolism in liver tumors of different growth rates. *Cancer Res.*, **23**, 995–1002 (1963).
27. Sweeney, M. J., Hoffman, D. H., and Poore, G. A. Enzymes in pyrimidine biosynthesis in Morris hepatoma. *Proc. Am. Assoc. Cancer Res.*, **8**, 66 (1967).
28. Sweeney, M. J., Hoffman, D. H., and Poore, G. A. Enzymes in pyrimidine biosynthesis. *Adv. Enzyme Regulation*, **9**, 51–61 (1971).
29. Wagle, S. R., Morris, H. P., and Weber, G. Comparative biochemistry of hepatomas. V. Studies on amino acid incorporation in liver tumors of different growth rates. *Cancer Res.*, **23**, 1003–1007 (1963).
30. Wagle, S. R., Morris, H. P., and Weber, G. Phosphopyruvate carboxylase in liver tumors of different growth rates. *Biochim. Biophys. Acta*, **78**, 783–785 (1963).
31. Weber, G. Behavior of liver enzymes in hepatocarcinogenesis. *Adv. Cancer Res.*, **6**, 403–494 (1961).
32. Weber, G. Behavior and regulation of enzyme systems in normal liver and in hepatomas of different growth rates. *Adv. Enzyme Regulation*, **1**, 321–340 (1963).
33. Weber, G. The molecular correlation concept: studies on the metabolic pattern of hepatomas. *CANN Monograph*, **1**, 151–178 (1966).
34. Weber, G. Carbohydrate metabolism in cancer cells and the molecular correlation concept. *Naturwissenschaften*, **55**, 418–429 (1968).
35. Weber, G. Hormonal control of metabolism in normal and cancer cells. *In* "Twenty-Second Annual Symposium on Fundamental Cancer Research," ed. by. R. B. Hurlbert, University of Texas Press, pp. 521–560 (1969).
36. Weber, G. Gene expression: Its regulation in normal and cancer cells. Rapporteur's Report. *Proc. 10th Int. Cancer Congr.*, **1**, 837–867 (1971).
37. Weber, G. and Cantero, A. Glucose 6-phosphatase activity in normal, precancerous, and neoplastic tissues. *Cancer Res.*, **15**, 105–108 (1955).
38. Weber, G. and Cantero, A. Glucose 6-phosphatase activity in regenerating, embryonic, and newborn rat liver. *Cancer Res.*, **15**, 679–684 (1955).
39. Weber, G. and Cantero, A. Glucose-6-phosphate utilization in hepatoma regenerating, and newborn rat liver, and in the liver of fed and fasted normal rats. *Cancer Res.*, **17**, 995–1005 (1957).
40. Weber, G. and Cantero, A. Fructose 1,6-diphosphate and lactic dehydrogenase activity in hepatoma and in control human and animal tissues. *Cancer Res.*, **19**, 763–768 (1959).
41. Weber, G., Ferdinandus, J. A., and Queener, S. F. Role of metabolic imbalance in neoplasia. *Proc. 10th Int. Cancer Congr.*, **1**, 510–532 (1971).
42. Weber, G., Henry, M. C., Wagle, S. R., and Wagle, D. S. Correlation of enzyme activities and metabolic pathways with growth rate of hepatomas. *Adv. Enzyme Regulation*, **2**, 335–346 (1964).
43. Weber, G. and Lea, M. A. The Molecular Correlation Concept of neoplasia. *Adv. Enzyme Regulation*, **4**, 115–145 (1966).
44. Weber, G. and Lea, M. A. The molecular correlation concept: an experimental and conceptual method in cancer research. *Methods Cancer Res.*, **2**, 523–578 (1967).
45. Weber, G. and Morris, H. P. Comparative biochemistry of hepatomas. III. Carbohydrate enzymes in liver tumors of different growth rates. *Cancer Res.*, **23**, 987–994 (1963).

46. Weber, G., Queener, S. F., and Ferdinandus, J. A. Control of gene expression in carbohydrate, pyrimidine and DNA metabolism. *Adv. Enzyme Regulation*, **9**, 63–95 (1971).
46a. Weber, G., Queener, S. F., and Morris, H. P. Imbalance in ornithine metabolism in hepatomas of different growth rates as exprssed in behavior of L-ornithine carbamoyltransferase activity. *Cancer Res.*, **32**, 1933-1940 (1972).
47. Weber, G., Singhal, R. L., Stamm, N. B., and Srivastava, S. K. Hormonal induction and suppression of liver enzyme biosynthesis. *Fed. Proc.*, **24**, 745 (1965).
48. Weber, G., Singhal, R. L., and Srivastava, S. K. Action of glucocorticoid as inducer and insulin as suppressor of biosynthesis of hepatic gluconeogenic enzymes. *Adv. Enzyme Regulation*, **3**, 43–75 (1965).
49. Weber, G., Singhal, R. L., and Srivastava, S. K. Insulin: suppressor of biosynthesis of hepatic gluconeogenic enzymes. *Proc. Natl. Acad. Sci. U.S.*, **53**, 96–104 (1965).
50. Weber, G., Stamm, N. B., and Fisher, E. A. Insulin: inducer of pyruvate kinase. *Science*, **149**, 65–67 (1965).
51. Weber, G., Stubbs, M., and Morris, H. P. Metabolism of hepatomas of different growth rates *in situ* and during ischemia. *Cancer Res.*, **31**, 2177–2183 (1971).
52. Williams-Ashman, H. G., Coppoc, G. L., and weber, G. Imbalance in ornithine metabolism in hepatomas of different growth rates as expressed in formation of putrescine, spermidine and spermine. *Cancer Res.*, **32** 1924–1932 (1972).

DEDIFFERENTIATION OF ENZYMES IN THE LIVER OF TUMOR-BEARING ANIMALS

Masami SUDA,[*1] Takehiko TANAKA,[*2] Susumu YANAGI,[*2] Shin-ichi HAYASHI,[*2] Kiichi IMAMURA,[*2] and Koji TANIUCHI[*2]

*Institute for Protein Research,[*1] and Department of Nutrition and Physiological Chemistry, Osaka University[*2]*

High activities of hexokinase, phosphofructokinase, and pyruvate kinase M_2 were observed in liver of tumor-bearing animals. These deviations in enzyme pattern are similar to those in embryonic and regenerating livers.

Findings obtained by parabiotic experiments suggested that some humoral factor elevated the level of pyruvate kinase M_2 in liver of tumor-bearing animals. The presence of this chemical factor was demonstrated in blood of tumor-bearing animals by isolated liver perfusion experiments. The factor was also extracted from Ehrlich ascites tumor cells.

It is concluded that elevation of the level of pyruvate kinase M_2 in the liver of tumor-bearing animals was due to the increase in net synthesis of the enzyme.

In normal animals, sugar metabolism in liver has been found to fluctuate under the influence of hormonal and dietary conditions because this organ plays a central role not only in glycolysis, but also in gluconeogenesis to maintain the normal range of blood sugar. It is generally agreed that there are at least three irreversible steps between glucose and pyruvate; hexokinase, phosphofructokinase, and pyruvate kinase, as shown in Fig. 1. Multimolecular forms of hexokinase were originally reported in 1963 by Sols and his associates (12) and by Walker (13). A more detailed report was made by Katzen and Schimke (2) who detected at least four hexokinases in mammalian tissues. In 1964, it was reported by Tanaka et al. (8, 9) that there were at least two types of pynuvate in rat tissues, type L and type M. Later, type M was separated electrophoretically and kinetically into two types, M_1 and M_2(10). Recently, it was reported by our laboratory that the four types of phosphofructokinase were distributed in mammalian tissues (7). Details regarding multimolecular forms of pyruvate kinase and phosphofructokinase are presented by Tanaka et al. (10) in this volume.

In 1965, Suda et al. (4) reported that there was a remarkable deviation in key enzyme patterns of glycolysis in the liver and that these enzymes were not under the hormonal regulation of the tumor-bearing animal. In 1966, Suda et al. (5) further demonstrated that hexokinase and type M pyruvate kinase were

---

[*1] 5311 Yamada-kami, Suita, Osata 565, Japan (須田正己)
[*2] Joancho 33, Kita-ku, Osaka 530, Japan (田中武彦, 柳　進, 林　伸一, 今村喜一, 谷内孝次).

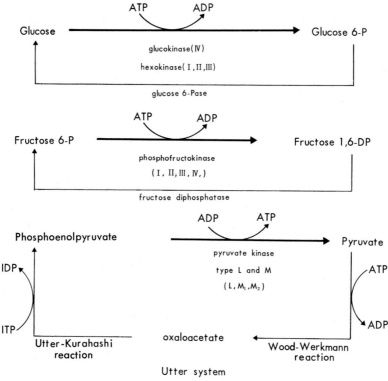

Fig. 1. Three irreversible steps of glycolytic and gluconeogenetic pathways and multimolecular forms of key glycolytic enzymes

synthesized rapidly when the isolated liver of normal rat was perfused with blood obtained from tumor-bearing rats. Furthermore, it was found that even in patients with gastric cancer, there was an increase of type M enzyme in the liver. We called this deviation in the enzyme pattern of tumor-bearing animals " dedifferentiation of the function in the host metabolism."

In addition to reviewing our experiments on metabolic dedifferentiation in tumor-bearing animals, this paper will demonstrate the recent observation that the increase in type M_2 pyruvate kinase in the liver of tumor-bearing animals was due to an increase in *de novo* synthesis of the enzyme, and that cell-free extracts of tumor cells caused a similar increase of pyruvate kinase M_2 in the liver of normal animals.

Animals: Male Sprague-Dawley albino rats were used throughout, except in the experiments shown in Tables I and II, in which ICR·JCL/T mice were used.

Tumors: The solid form of Walker carcinosarcoma *256* was transplanted into femoral muscle tissue of rat. Ehrlich ascites tumor cells were transplanted intraperitoneally into mice.

Enzyme assay: Assay procedures of glucokinase, hexokinase, phosphofructokinase, pyruvate kinase, and serine dehydratase employed in the experiments have been published in another paper (*4*). Assay of types M and L pyruvate kinase in the tissue extracts was carried out by the method published in a previous paper

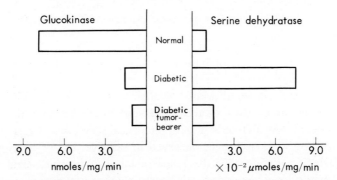

Fig. 2. Glucokinase and serine dehydratase in diabetic tumor-bearing rats (4)
A diabetic state was induced by intraperitoneal injection of alloxan monohydrate (6.0 mg/100 g of body weight).

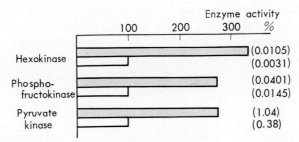

Fig. 3. Changes in levels of glycolytic enzymes in liver of tumor-bearing animals (4)
Numbers in parentheses show specific activity of enzymes in μmoles/mg/min. All the animals were fed laboratory chow *ad libitum*.
▧ tumor-bearing rat; ☐ normal rat.

(9). One unit of enzyme is expressed as the amount of enzyme which catalyzes the formation of one μmole of product per min under the assay conditions. The specific activity of the enzyme is expressed as units of enzyme per mg protein.
Starch block electrophoresis: The procedure of starch block zone electrophoresis of pyruvate kinase has been published elsewhere (9).
Polyacrylamide gel electrophoresis (10): A thin film (120 × 95 × 1 mm) of polyacrylamide gel (3.75%) was used as a supporting medium for an electrophoretic zymogram of pyruvate kinase. The buffer for electrophoresis was made up of 10 mM Tris-HCl, 5 mM $MgSO_4$, 0.5 mM dithiothreitol, and 0.5 mM fructose 1,6-diphosphate (pH 8.1). Before the electrophoretic run, the thin film was washed with the same buffer overnight in a cold room to remove ammonium persulfate, which was used as a catalyzer for polymerization of the amide. Electrophoresis was carried out for 3 hr at 22 V/cm. The enzyme activity was located by a slightly modified version of the method of Susor and Rutter (6).
Liver perfusion: Liver perfusion was carried out by a modified version of the method described by Miller (3), the details of which have been described already (5).

Deviation in Key Enzyme Patterns of Sugar Metabolism in the Liver of Tumor-bearing Animals

It is well known that glucokinase is induced by insulin, serine dehydratase by hydrocortisone, and that the levels of both enzymes are regulated reciprocally under various hormonal conditions. Walker carcinosarcoma-256 cells were implanted in the femoral muscle of albino rats and a diabetic state was induced by injection of alloxan. As shown in Fig. 2, reciprocal responses of serine dehydratase and glucokinase were no longer observed in diabetic tumor-bearing animals. From these facts, it was concluded that sugar metabolism in tumor-bearing animals deviated under hormonal regulation. Therefore, other key enzymes of glycolysis and gluconeogenesis were assayed to obtain more details about the change that occurred in host liver.

Figure 3 shows the levels of glycolytic key enzyme in the liver of rats with transplanted tumor cells. The three key enzymes of glycolysis, hexokinase, phosphofructokinase, and pyruvate kinase, increased in liver of tumor-bearing animals. In contrast, the levels of key enzymes of glyconeogenesis tested thus far did not show any significant change in liver of tumor-bearing animals, as shown in Fig. 4. From the results of two experiments in Figs. 3 and 4, it can be said that glucose metabolism tends toward glycolysis in liver of tumor-bearing

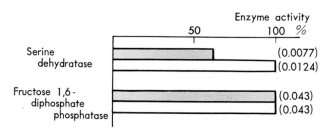

FIG. 4. Changes in levels of gluconeogenetic enzymes in liver of tumor-bearing animals (4)

Numbers in parentheses show specific activity of enzymes in μmoles/mg/min
All the animals were fed laboratory chow *ad libitum*.
▓ tumor-bearing rat; ☐ normal rat.

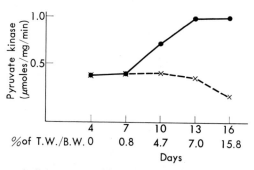

FIG. 5. Changes in liver pyruvate kinase activity after transplantation of Walker carcinosarcoma-256 into rats that underwent adrenalectomy (4)
Ratio of tumor weight to body weight is indicated at the bottom of the graph.
——— Adex, tumor-bearer; - - - Adex.

animals. Figure 5 shows the time course of the change in enzyme activity after transplantation of tumors. Rats were used in this experiment, although the same result was obtained when the tumor cells were transplanted to normal animals. Pyruvate kinase activity began to increase at the stage in which the tumor weight reached about 1.0 and continued to increase with the increase in tumor weight until the ratio reached about 10%.

Deviation in Patterns of the Multimolecular Forms of Pyruvate Kinase and Hexokinase

In 1967, it was reported by our laboratory that there were at least two pyruvate kinases in liver extract, type L and type M (*8, 9*). It was interesting to know whether or not the pattern of the multimolecular forms of pyruvate kinase in liver deviated after transplantation of tumor cells. Zone electrophoresis of pyruvate kinase in livers of normal and tumor-bearing animals was carried out on potato starch blocks. As shown in Fig. 6, muscle and hepatoma cells showed

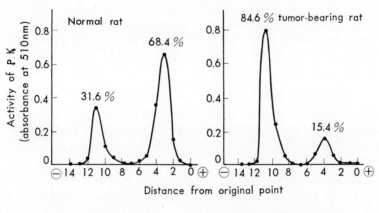

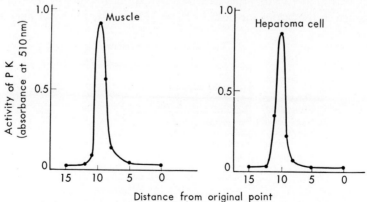

FIG. 6. Starch zone electrophoresis patterns of pyruvate kinase of muscle, ascites hepatoma cells, and livers of normal and tumor-bearing animals (*4, 9*)

Activity of pyruvate kinase is expressed as absorbance at 510 nm under the assay conditions described in experiments (*4, 9*).

Phosphate buffer: pH 8.0, EDTA 10^{-3} M, $\tau/2=0.31$, 350 V, 13 hr.

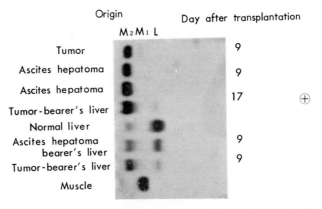

Fig. 7. Electrophoretic zymogram of pyruvate kinase of normal and tumor tissues and normal and tumor-bearing animals' livers

Walker carcinosarcoma-256 (tumor) and Yoshida ascites hepatoma AH-130 (ascites hepatoma) were used in the experiments. Nine or seventeen days after tumor transplantation, the rats were killed for the experiments. Polyacrylamide gel thin film (3.75%, 120×95×1 mm) was used as a supporting medium. The electrophoretic procedure is described in experiments.

Supporting media: 3.75% acrylamide gel. Buffer: 10 mM Tris-HCl, 5 mM $MgSO_4$, 0.5 mM dithiothreitol, 0.5 mM FDP, pH 8.1. Voltage: 22 V/cm. Time: 3 hr.

a similar single peak of enzyme activity, whereas liver showed two peaks. The more positively charged peak is type M enzyme, the more negatively charged peak type L enzyme. As seen clearly here, type M peaks were predominant in liver of tumor-bearing animals, whereas type L was predominant in liver of normal animals.

Figure 7 shows electrophoretic zymograms of pyruvate kinase on acrylamide gel thin film. In 1968, Susor and Rutter (6) reported that type M pyruvate kinase could be electrophoretically separated into two spots if fructose 1,6-diphosphate was contained in the buffer medium. Type M pyruvate kinase of liver showed a

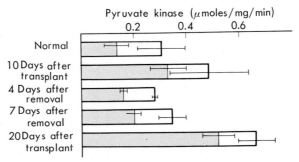

Fig. 8. Changes in patterns of pyruvate kinase type L and type M_2 after transplantation and removal of Walker carcinosarcoma 256

Ten days after transplantation of the tumor cells, the tumor tissues were removed by operation. The enzyme activities in liver were assayed 4 and 7 days after removal of the tumors. Shaded bars show type M_2 activity and white bars show type L activity.

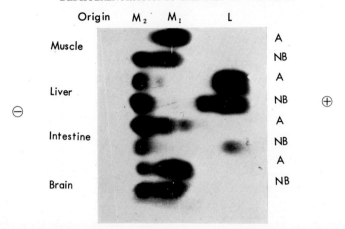

FIG. 9. Electrophoretic zymogram of pyruvate kinase of tissues of newborn and adult rats

Two-hour-old rats (NB) and about 6-month-old rats (A) were sacrificed for the experiment. Electrophoresis was carried out under the same conditions as in Fig. 7.

slightly different mobility from that of muscle enzyme. We call these two type M pyruvate kinases type M_1 and M_2. It can be clearly seen in Fig. 7 that there is a greater quantity of type M_2 in the liver of tumor-bearing animals.

The quantity of type M_2 pyruvate kinase in liver extract, which contains both type L and type M_2, is easily determined by using the antibody against M_1 enzyme. Both type M enzymes are not immunochemically differentiated from each other, and type L enzyme does not show any cross-reactivity to the antibody against M_1 enzyme. Type M_2 activity increased greatly with time in liver of rats after transplantation of tumor cells, whereas type L tended to decrease. When a solid tumor in the femoral muscle was removed ten days after transplantation, the level of type M_2 enzyme returned rapidly to the normal range.

Figure 9 shows an electrophoretic zymogram of pyruvate kinase of various tissues of newborn and adult rats. It should be noted here that the tissue of newborn rats (NB) showed quite a different pattern of pyruvate kinase (PK) from that of adult animals (A). Newborn rat muscle had M_1 and M_2 enzymes, whereas

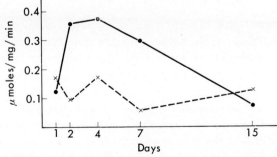

FIG. 10. Changes in pyruvate kinase type L and type M_2 activities after partial hepatectomy (9)

———— type M_2; - - - type L.

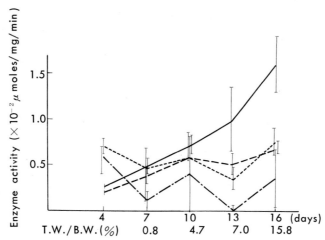

Fig. 11. Changes in liver hexokinase and glucokinase activities after transplantation of Walker carcinosarcoma-256 into animals that underwent adrenalectomy (4)
——— hexokinase, —·— glucokinase, tumor-bearer; ——— hexokinase, - - - glucokinase, normal.

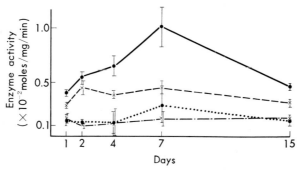

Fig. 12. Changes in liver hexokinase and glucokinase activities after partial hepatectomy (4)
——— hexokinase, ⋯ glucokinase, regenerating; ——— hexokinase, —·— glucokinase, normal.

adult tissue contained only one enzyme, M_1. Newborn rat brain and liver had more M_2 enzyme than adult tissue, and the intestines also showed quite a different pattern. Generally speaking, type M_2 was dominant enzyme in newborn rat tissues. Hepatoma cells and regenerating liver also contained large amounts of pyruvate kinase type M_2.

Figure 10 shows the time course of the level of pyruvate kinase in regenerating liver. Activity of type M_2 pyruvate kinase reached a maximum level after resection of two-thirds of the liver tissues, gradually decreased, and finally reached the normal range approximately two weeks after hepatectomy.

From the results shown in Figs. 9 and 10, it seems likely that type M_2 is a prototype of pyruvate kinase in rat tissues.

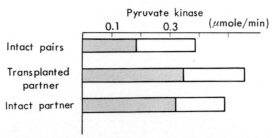

FIG. 13. Comparison of pyruvate kinases types L and M_2 between normal and tumor-transplanted parabiots (4)

Parabiosis was carried out according to the method of Jeffers (1). Other experimental conditions are described in the text. ■ type M_2; □ type L.

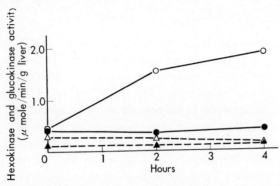

FIG. 14. Changes in pyruvate kinase types L and M_2 activity in the isolated liver perfused with normal and tumor-bearer's blood (5)

The liver was placed in the perfusion apparatus maintained at 37–39° and perfused via the portal vein with 90 ml of heparinized blood freshly oxygenated with a disc-rotating oxygenator. The details of experimental conditions have been described in another paper (5).

○ type M_2, △ type L, tumor-bearer's blood; ● type M_2, ▲ type L, normal blood.

The deviation in isozyme pattern was not only observed in pyruvate kinase. A similar pattern of deviation was also observed in hexokinase under the same conditions. Figure 11 shows the time course of hexokinase patterns in liver of tumor-bearing animals. Low K_m hexokinase in liver increased with the growth of tumor tissues, whereas the level of glucokinase did not change markedly. Glucokinase (hexokinase IV) tended to decrease rather than increase in liver of tumor-bearing animals. A similar increase in hexokinase was also observed in regenerating liver, as shown in Fig. 12.

Suggestive Evidence for the Presence of the Factor which Causes Deviation in the Enzyme Pattern

Usually, the tumor cells are transplanted in femoral muscle. However, deviation in isozyme pattern occurred in liver with growth of the tumor in the femoral muscle. This fact suggests the presence of a humoral factor which mediates some message from tumor tissues to liver and causes the deviation in the

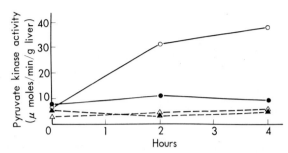

Fig. 15. Changes in hexokinase and glucokinase activities in isolated liver perfused with normal and tumor-bearer's blood (5)

The experimental conditions are the same as described in Fig. 14.

○ hexokinase, △ glucokinase, tumor-bearer's blood; ● hexokinase, ▲ glucokinase, normal blood.

pattern of multimolecular forms of pyruvate kinase and hexokinase in liver. Parabiotic experiments were carried out to clarify the above-mentioned problem (Fig. 13). A suspension of Walker carcinosarcoma-256 cells was injected into one of a parabiotic pair of animals. We called this one of the pair the "transplanted partner;" the other remained intact. To see whether parabiosis was satisfactory, phenolsulfophthalein was injected intravenously into one of the pair. If the dye was excreted in the urine of the uninjected partner, the operation was judged to be satisfactory. As shown in Fig. 13, the level of pyruvate kinase type M_2 increased not only in liver of the transplanted partner but also in that of the intact partner. This result strongly suggests the presence of a humoral messenger, originating in tumor tissues, which increases type M_2 enzyme in liver. To demonstrate more directly the presence of a humoral factor, two experiments were carried out, an isolated liver perfusion experiment and the direct injection of the extracts of the tumor cells into an intact animal.

The apparatus for liver perfusion described by Miller (3) was modified as follows. The exchange of gas was carried out by a disc-rotation oxygenator. The second heart was placed between liver and lung in order to prevent blood congestion occurring in liver during perfusion. Using a pH-stat, blood was automatically maintained at pH 7.4 by supplying 0.2 M $KHCO_3$ solution when the pH decreased and increasing partial pressure of CO_2 in gas phase of lung when the pH increased.

As shown in Fig. 14, the activity of type M_2 pyruvate kinase was greatly elevated within two hr after the start of the experiment, when isolated liver of normal rat was perfused with the blood obtained from tumor-bearing animals. In contrast, type L enzyme did not show any response. When the liver was perfused with normal rat blood, both types of pyruvate kinase showed no change, as seen in Fig. 14. A similar observation was also made in the case of hexokinase (Fig. 15). Low K_m hexokinase was greatly elevated only when the normal liver was perfused with blood from tumor-bearing rats; glucokinase showed no response. The results of these perfusion experiments suggest that the humoral factor originated from tumor cells being circulated in the body of the tumor-bearing host.

TABLE I. Increase in Activity of Type M_2 Pyruvate Kinase in Liver of Mice Transplanted with Ehrlich Ascites Tumor

No. of animals	PK Activity		
	Total	Type M	%
Normal			
1	0.668	0.052	7.7
2	0.630	0.062	10.4
3	0.713	0.078	10.9
4	0.583	0.054	9.2
5	0.537	0.084	15.6
6	0.560	0.076	13.5
Av.±SD	0.615±0.061	0.067±0.012	11.2±2.6
Tumor-bearing mice			
1	0.697	0.295	42.3
2	0.657	0.291	44.2
3	0.676	0.250	36.9
4	0.705	0.168	23.8
5	0.773	0.178	23.0
6	0.625	0.224	35.8
Av.±SD	0.688±0.045	0.234±0.049[a]	34.3±8.2[a]

Five-week-old mice (ICR · JCL/T) received intraperitoneal transplants of Ehrlich ascites tumor cells. Four days after transplantation, the enzyme activity in liver was assayed. Measurement of types L and M_2 pyruvate kinase (PK) in mouse liver was carried out by the same method as for rat liver enzymes, using the antibody against rat muscle enzyme (9). The enzyme activity was expressed as μmoles per min per mg of protein. Relative amounts of type M_2 have also been indicated as percent in the table.

[a] $P<0.001$

Table I simply shows that a similar deviation in the pattern of pyruvate kinase was observed in liver of mice transplanted with Ehrlich ascites tumor cells. Type M_2 pyruvate kinase showed immunochemical cross-reactivity to the antibody against the rat muscle enzyme. When Ehrlich ascites tumor cells were transplanted into five-week-old male mice (ICR·JCL/T), almost all the animals died six to seven days after transplantation. The enzyme activity in liver was assayed four days after transplantation. As can be seen in Table I, total pyruvate kinase did not increase remarkably. However, type M_2 pyruvate kinase of liver of tumor-bearing animals increased to more than three times that of normal control animals.

Ehrlich ascites tumor cells were collected and washed with saline solution. The tumor cells were suspended in an equal volume of saline solution, and the suspension was sonically disrupted for four min (20 kc). The sonicated suspension was centrifuged at 6,000 g for 20 min. The supernatant was collected, and the precipitate resuspended with 1.5-fold volume of saline solution. One ml of the supernatant or 0.5 ml of the suspension of the precipitate was injected intraperitoneally into each mouse, and the animals were killed for enzyme assay four days after injection. As shown in Table II, type M_2 pyruvate kinase of liver

TABLE II. Effect of Cell-free Preparation of Ehrlich Ascites Tumor Cells on Levels of Pyruvate Kinase (PK) Type M_2 in Mouse Liver

Treatment	No. of mice	PK Activity		%
		Total	Type M	
Normal	3	0.473±0.052	0.057±0.022	12.0
I.p. injection of 6,000 g supernatant	4	0.528±0.084	0.134±0.047[a]	25.4
I.p. injection of 6,000 g precipitate	5	0.495±0.093	0.149±0.014[b]	30.1
Tumor-bearing rats	2	0.546±0.030	0.262±0.0	48.0

[a] $P<0.1$. [b] $P<0.005$

of the mouse injected with the cell-free supernatant or the suspension of the cell-particulate fraction increased two to three times over that of control animals receiving intraperitoneal injections of saline solution. The possibility of contamination with intact tumor cells in both fractions was excluded for two reasons, microscopic examination showed no undisrupted tumor cells, and mice injected with either fraction, the 6,000 g supernatant or the precipitate, did not die until at least fourteen days after the injection. These experiments strongly suggest that the tumor cells produce the chemical messenger which modulates the level of the type M_2 pyruvate kinase in the host liver. The nature of the chemical factor is still unclear. Experiments recently carried out demonstrate that the supernatant obtained after centrifugation at 100,000 g still shows activity.

Increase in de novo Synthesis of Pyruvate Kinase Type M_2 in Liver of Tumor-bearing Animals

It should also be made clear whether or not the elevation in the level of pyruvate kinase M_2 in liver of tumor-bearing animals was the result of an increase

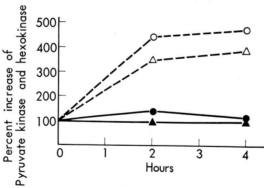

FIG. 16. Inhibitory effect of Actinomycin-S on pyruvate kinase type M_2 and hexokinase in isolated liver perfused with tumor-bearer's blood (5)

The experimental conditions are the same as described in Fig. 14. Three hundred μg Actinomycin was added to 90 ml of blood at the beginning of perfusion

▲ hexokinase, ● pyruvate kinase, Act S (+); △ hexokinase, ○ pyruvate kinase, Act S (—).

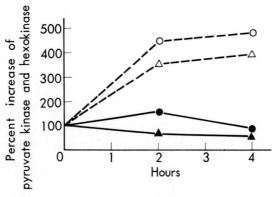

FIG. 17. Inhibitory effect of p-fluorophenylalanine (p-FPA) on pyruvate kinase type M_2 and hexokinase increase in isolated liver perfused with tumor-bearer's blood (5)

The experimental conditions are the same as described in Fig. 14. Three hundred mg p-fluorophenylalanine were added to 90 ml blood at the beginning of perfusion.

▲ hexocinase, ● pyruvate kinase, p-FPA (+); △ hexokinase, ○ pyruvate kinase, p-FPA (—).

TABLE III. Incorporation of ^{14}C-labeled amino acids into pyruvate kinase (PK) type M_2 of liver of tumor-bearing rats

Rats	PK Activity		Radioactivity of ppt. formed with		(a)−(b)	cpm/mg enzyme	cpm/mg protein
	Type L	Type M_2	Anti-PK serum (a)	Normal serum (b)			
	(nm/mg protein)						
Normal	347	59	67	8	59	9,900	2,240
	295	70	47	13	34	4,920	2,030
Tumor-bearing	302	202	559	5	554	31,200	3,370
	322	164	608	8	600	42,700	3,450

Normal or Walker carcinosarcoma-bearing (11 days) female rats, weighting 100 to 120 g, were injected with 40 μCi/100 body wt. of *Chlorella* protein hydrolysate-U-^{14}C, 2.4 mCi/mg atom C. The rats were sacrificed 2 hr later. Livers were homogenized with 4 volumes of 0.15 M NaCl-0.02 M potassium phosphate (pH 7.0), and centrifuged at 100,000 g for 60 min. The supernatant was allowed to stand at 37° for 30 min, then at 3° for 3 hr, then centrifuged. This pretreatment was carried out to minimize nonspecific precipitation.

Three ml each of the liver extracts were incubated at 30° for 30 min, then at 3° for 3 hr with 1 ml of anti-pyruvate kinase M_1 serum (a) and with 1 ml of normal serum (b). Twenty μg of muscle pyruvate kinase was added as a carrier to each incubation mixture. Precipitates formed were collected by centrifugation, washed 3 times with a small volume of cold NaCl-phosphate buffer, dissolved in 0.2 ml of 0.1 M NaOH, and assayed for radioactivity with a liquid scintillation counter.

in *de novo* synthesis of pyruvate kinase. To answer this question, two experiments were carried out; inhibition by Actinomycin-S and p-fluorophenylalanine, and incorporation of labeled amino acids into the enzyme protein.

As shown in Figs. 14 and 15, pyruvate kinase M_2 and low K_m hexokinase increased in isolated normal liver perfused with blood from tumor-bearing animals.

Actinomycin S inhibited the increase of both enzymes in isolated liver perfused with tumor-bearer's blood (Fig. 16). About 3.0 μg of the antibiotic per ml of blood was used in this experiment.

p-Fluorophenylalanine also inhibited the increase in pyruvate kinase M_2 and low K_m hexokinase in isolated liver perfused with blood of tumor-bearing animals (Fig. 17). About 3.0 μg of p-fluorophenylalanine was added to one ml of the blood perfusate.

These experiments using two kinds of inhibitors of protein synthesis, demonstrate that increases in type M_2 pyruvate kinase and in low K_m hexokinase are due to *de novo* synthesis of the enzymes. More direct evidence supporting this possibility was obtained by measurement of the rate of incorporation of ^{14}C-labeled amino acids into pyruvate kinase-type M_2.

Animals were sacrificed 2 hr after injection with a mixture of ^{14}C-labeled amino acids. Livers were removed and homogenized with four volumes of Tris buffer (0.1 M, pH 7.4). The homogenates were centrifuged at 100,000 g for 60 min. The supernatants were incubated with a sufficient amount of anti-pyruvate kinase M_1 serum or with normal serum for 30 min at 37° and then for 3 hr at 3°. The precipitates formed were collected by centrifugation (8,000 g for 20 min) and washed several times with the saline solution. Radioactivities were measured by a well-type scintillation counter. In Table III, (a) indicates radioactivity of precipitates formed by addition of anti-pyruvate kinase M_2 and (b) indicates that which was formed by the addition of normal serum. The difference between (a) and (b) means specific incorporation of ^{14}C-labeled amino acids into pyruvate kinase M_2 per unit liver weight. In liver of tumor-bearing animals, the rate of incorporation of ^{14}C-labeled animo acids into total pyruvate kinase type M_2 per unit of liver weight was about ten times that for normal animals. Furthermore, radioactivity per mg of pyruvate kinase type M_2 (calculated from the specific activity of type M_2 as 800 U/mg) increased threefold in liver of tumor-bearing animals. These results show that the rate of net synthesis of pyruvate kinase type M_2 is enhanced in liver of tumor-bearing animals.

Cellular Location of Type M_2 Pyruvate Kinase in Liver Tissue

Liver tissue is known to consist of several different kinds of cells. It is important to know in which kind of cells type M_2 increases in tumor-bearing animals. The indirect fluorescent antibody technique was employed to answer the above question.

Rabbit antibody against rat muscle pyruvate kinase was specifically purified by precipitation with its antigen. Chicken antiserum against rat γ-globulin was used as the secondary antibody and was labeled with fluorescein isothiocyanate. Preliminary experiments showed that type M_2 pyruvate kinase was located in liver parenchyma cells, and that the enzyme increased in parenchyma cells in liver of tumor-bearing animals.

REFERENCES

1. Jeffers, W. A. "The Rat in Laboratory Investigation," ed. by E. J. Farris, and J. Q. Griffith, J. B. Lippincott Co., Philadelphia, London, and Montreal, 451–452 (1949).
2. Katzen, H. M. and Schimke, R. T. Multiple forms of hexokinase in the rat: tissue distribution, age dependency, and properties. *Proc. Natl. Acad. Sci. U.S.*, **54**, 1218–1225 (1965).
3. Miller, L. L. Some direct actions of insulin, glucagon, and hydrocortisone on the isolated perfused rat liver. *Rec. Progr. Hormone Res.*, **17**, 539–568 (1961).
4. Suda, M., Tanaka, T., Sue, F., Harano, Y., and Morimura, H. Dedifferentiation of sugar metabolism in the liver of tumor-bearing rat. *GANN Monograph*, **1**, 127–141 (1966).
5. Suda, M., Tanaka, T., Sue, F., Kuroda, Y., and Morimura, H. Rapid increase of pyruvate kinase (M type) and hexokinase in normal rat liver by perfusion of the blood of tumor-bearing Rat. *GANN Monograph*, **4**, 103–112 (1968).
6. Susor, W. A. and Rutter, W. G. Some distinctive properties of pyruvate kinase purified from rat liver. *Biochem. Biophys. Res. Commun.*, **30**, 14–20 (1968).
7. Tanaka, T., An, T., and Sakaue, Y. Studies on multimolecular forms of phosphofructokinase in rat tissues. *J. Biochem. (Tokyo)*, **69**, 609–612 (1971).
8. Tanaka, T., Harano, Y., Morimura, H., and Mori, R. Evidence for the presence of two types of pyruvate kinase in rat liver. *Biochem. Biophys. Res. Commun.*, **21**, 55–60 (1965).
9. Tanaka, T., Harano, Y., Sue, F., and Morimura, H. Crystallization, characterization and metabolic regulation of two types of pyruvate kinase isolated from rat tissues. *J. Biochem. (Tokyo)*, **62**, 71–91 (1967).
10. Tanaka, T., Imamura, K., Ann, T., and Taniuchi, K. Multimolecular forms of pyruvate kinase and phosphofructokinase in normal and cancer tissues. *GANN Monograph*, **13**, 219–234 (1972).
11. Tanaka, T., Sue, F., and Morimura, H. Feed-forward activation and feed-back inhibition of pyruvate kinase type L of rat liver. *Biochem. Biophys. Res. Commun.*, **29**, 444–449 (1967).
12. Vinuela, E., Salas, M., and Sols, A. Glucokinase and hexokinase in liver in relation to glycogen synthesis. *J. Biol. Chem.*, **238**, PC1175–1177 (1963).
13. Walker, D. G. On the presence of two soluble glucose-phosphorylating enzymes in adult liver and the development of one of these after birth. *Biochim. Biophys. Acta*, **77**, 209–226 (1963).

ISOZYMES IN SELECTED HEPATOMAS AND SOME BIOLOGICAL CHARACTERISTICS OF A SPECTRUM OF TRANSPLANTABLE HEPATOMAS

Harold P. Morris

Department of Biochemistry, Cancer Research Unit, College of Medicine, Howard University*

Dyer *et al.* (*6*) observed a very high glutamic oxalacetic transaminase (GOT) activity in hepatoma 5123 and the blood of the host at the 18–19th transfer generations contrasted with the GOT of the poorly differentiated hepatoma 3683 and several other tumors. With additional transplantable hepatomas becoming available, Otani and Morris (*29*) extended these observations to other tumor lines and also investigated the isozymes by starch gel electrophoresis. Two isozymes were found. The relative quantative distribution of each isozyme varied in the five different hepatomas examined. The " anionic " to the " cationic " ratio was higher in the hepatomas than in the normaltissue.

Since the relative distribution of aspartate transaminase isozymes of hepatoma 5123B decreased on adrenalectomy, while that from normal livers was unchanged, the response to adrenal hormones of these isoenzymes of hepatoma 5123B were studied. Adrenalectomy resulted in a decreased specific activity, while hydrocortisone effected a slight increase as well as a change in isozyme distribution in favor of the cationic or mitochondrial component.

Studies by Otani and Morris (*31*) have recently been extended to the activities and relative isozyme distribution of three additional enzymes; glutamic-pyruvic transaminase (GPT), malic dehydrogenase (MDH), and lactic dehydrogenase (LDH) in hepatomas 3924A, 5123TC, 7793, and R7. The relative distribution of the isozymes of the different enzymes varied considerably among the tumors, but none resembled the distribution profile of the enzymes and isozymes of normal liver. The greatest deviation from the normal pattern was noted in hepatoma 3924A.

A spectrum of transplantable rat hepatoma tumor lines arranged according to growth rate is presented. These tumors grew satisfactorily only in the inbred rat strain in which the original tumor was induced. Ten different ingested carcinogens were used to induce this tumor spectrum. The rate of growth as measured by time between transfers and tumor size shows a 20-fold or more difference between the fastest and slowest growing lines. The histological characterizations of these tumors have been made according to the

---

* Washington, D.C. 20001, U.S.A.

following: Poorly differentiated, well-differentiated highly differentiated, or intermediate between any of these three classifications. Some changes in growth rate have been noted. Slower-acting carcinogens or low doses of rapidly acting carcinogens seem to induce well- to highly differentiated tumors possessing slower rates of growth.

The existence of two forms of glutamic-oxalacetic transaminase (GOT) in rat liver has been described by many investigators (2, 12, 20, 21). The isozymes of lactic dehydrogenase (LDH) (9, 14, 17, 37), malate dehydrogenase (MDH) (2, 4, 5, 19, 39), and glutamic-pyruvic transaminase (GPT) (5, 11, 40) have also been reported by many investigators. Our studies disclose how some of the patterns of these enzymes or their isozymes differ from the enzyme activities of normal liver in several transplanted hepatomas.

The development of a large spectrum of transplantable hepatomas that differ many fold in growth rate and degree of differentiation made possible studies of the relationship between growth rate and enzyme or isozyme activity in several of these hepatomas. Several other laboratories have also made use of the hepatomas of our spectrum, and some of their results are reported in this volume.

Glutamic-Oxalacetic Transaminase (GOT)

Ten years ago Dyer *et al.* (6) reported the GOT activity levels of the several

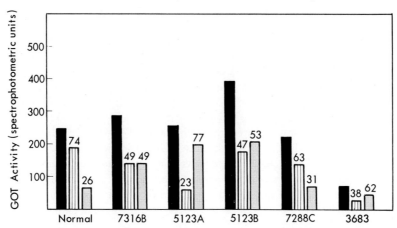

FIG. 1. Relative distribution of GOT isozymes in five transplantable hepatomas

Peak A represents the cationic and peak B the anionic component. The numbers above the bars represent the percent of total activity (29).

■ total; ▢ peak A; ▨ peak B.

Abbreviations used: GOT, glutamic-oxalacetic transaminase=L-asparate amino transferase (L-asparate : 2-oxoglutarate aminotransferase, EC 2.6.1.1); MDH, malic dehydrogenase (L-malate: NAD oxidoreductase, EC. 1.1.1.37); GPT, glutamic-pyruvic transaminase=alanine aminotransferase (L-alanine : 2-oxoglutarate aminotransferase, EC 2.6.1.2); LDH, lactate dehydrogenase (L-lactate : NAD oxidoreductase, EC1.1.1.27); NAD, nicotinamide-adenine-dinucleotide ; NADH, reduced form of NAD.

transplanted hepatomas then available (23). Hepatoma 5123 was found to have a GOT activity level 2.5 times that of normal liver. Because of the high GOT activity levels noted by Dyer et al. (6), Otani and Morris (29) examined several additional transplantable hepatomas to see if this was a characteristic finding for transplantable hepatomas. They also examined these hepatomas to determine the isozymic distribution of the two GOT isozymes and compared the GOT isozyme distribution in hepatomas with that found in the normal liver. The mobilities of the two isozymes of GOT after granular starch block electrophoresis of aqueous extracts of five hepatomas were found to be like that of normal or host liver. The relative quantitative distribution of each isozyme, however, varied both with the hepatoma and with control livers. With the five hepatomas studied, the anionic to cationic form (or soluble to mitochondrial form) (GOT) ratio was higher in the hepatomas than in the normal liver, as shown in Fig. 1 (29). One probable explanation could be that there are fewer mitochrondia in the hepatoma

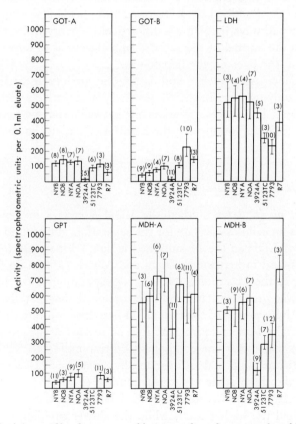

FIG. 2. Activity profile of enzyme and isozymes from four transplantable hepatomas

The figures in parentheses represent the number of determinations. The range of values is given by the verticle lines. NYB, normal young Buffalo-strain rats; NOB, normal old Buffalo-strain rats; NYA, normal young ACl-strain rats; NOA, normal old ACl-strain rats. Hepatoma 3294A in ACl-strain rats; hepatomas 5123 TC, 7793, and R7 in Buffalo-strain rats (31).

TABLE I. Some Biological Characteristics of Four Hepatoma Lines Examined for Glutamic-Oxalic Transaminase (GOT), Glutamic-Pyruvic Transaminase (GPT), Malic Dehydrogenase (MDH), and Lactic Dehydrogenase (LDH)

Average (months)	Tumor No.	Generation	Differentiation	Metastases
2.8	R7	19	WD, HCC, IHD & WD	+++
2.0	7793	24	WD, HD	++
1.1	5123TC	91	IWD, HD	+
1.0	3924A	266	Much N, Pd	

WD, well-differentiated; HCC, hepatocellular carcinoma; IHD, intermediate between high differentiation and well-differentiated; N, necrosis; Pd, poor differentiation. The histological characterization was by D. R. Meranze.

cells than in the hepatocyte of the normal liver from which the hepatoma was derived. The coupling reaction of mitochrondria-bound GOT and the tricarboxylic acid cycle described by Katunuma and Okada (15) suggests the role of two alternative pathways involving a difference in energy utilization noted in the deviation from the normal of the ratio of mitochondrial to supernatant distribution of GOT in the liver vs. the hepatomas.

In accordance with Weber and Lea's (42) molecular correlation concept, some enzyme activities are correlated positively or negatively with growth rate of hepatomas (23). The inverse relationship between enzyme activity on growth rate for GOT (31) was found with the high GOT activity in the slower growing tumors (Fig. 2, Table I).

Lactate Dehydrogenase (LDH)

Rosada et al. (6) have studied the LDH isozymes in hepatomas. Using electrophoresis on cellulose acetate for the separation, they found normal rat liver to have 92% of LDH in the M subunit fraction. Fine et al. (9) reported 95% as M subunits, and in hepatomas Otani and Morris (31) found 95% as isozyme corresponding to the M subunit and 5% corresponding to the H subunit. Our studies (31) using granular starch gel electrophoresis suggest that the occurence of the total activity profile of the two LDH isozymes indicate they are present in the tumors in the same relative amounts as found in the normal liver. The activity of the major isozyme component corresponding to the M subunit was essentially the same in normal rat liver and in hepatoma 3924A, but was considerably lower in the three other hepatomas studied (Fig. 2).

A direct relationship was obtained with LDH where the faster growing tumor, 3924A, exhibited the higher level, while the slower growing tumors had lower activity (31). This observation was in agreement with Rosada et al. (37).

Glutamic-Pyruvic Transaminase (GPT)

This enzyme exhibited a cationic peak which was coincident with the less cationic isozyme of GOT (31). Of the four hepatomas examined for GPT activity, none was found in the two faster growing hepatomas, 3924A and 5123TC (*31*) (Fig. 2). The GPT activity of the two slower growing tumors was approximately that found in the normal liver. No correlation of GPT activity with growth rate was noted.

Malic Dehydrogenase (MDH)

A successful separation of MDH isozymes was found (*31*). A sharp cationic peak A and a broad anionic peak complex B were obtained. This broad anionic complex could possibly consist of at least three components. The MDH level of 3924A hepatoma was significantly lower in both A and B components than that found in normals (Fig. 2). The MDH-A level was essentially in the range of the normal controls for the other three hepatomas. MDH-B levels of hepatomas 3924A, 5123TC, and 7793 were lower than the controls, but the MDH-B level was higher for hepatoma R1 (*31*) (Fig. 2).

Of these four tumor lines examined (*31*), it may be concluded that the relative distribution of the isozymes of the different enzymes varied considerably. None resembled the control liver distribution profile. These four hepatomas differed in at least one enzyme or isozyme.

It appears that, in the hepatomas examined for these four enzymes and isozymes, much diversity exists in neoplastic tissues. A given enzyme was found to be within the normal range in one tumor and abnormal (either high or low) in another. The level of an enzyme, therefore, is not necessarily altered in the same or opposite direction in the neoplastic state.

Adrenal Hormones

The comparative response of GOT of normal liver and of hepatoma 5123B (IWD-Pd)* to adrenal hormones was studied (*30*) in terms of activity and isozyme distribution. Single intraperitoneal injections of cortisone and hydrocortisone at 5 mg/kg were used. The GOT responded to a single dose of adrenal-corticoid hormone in approximately 5 hr, while the enzyme of normal liver showed no such response (Fig. 3) (*30*). No effect was observed in the specific activities or the relative isozyme distribution from normal female rat liver after either adrenalectomy or hydrocortisone treatment or both, although in these treatments there was some responsiveness to GOT induction in hepatoma 5123B. Adrenalectomy resulted in a decreased GOT specific activity in hepatoma 5123B, while hydrocortisone treatment resulted in a slight increase in specific activity as well as an alteration in the isozyme distribution in favor of the cationic or mitochondrial component. A combination of adrenalectomy and hydrocortisone treatment produced a normalization of the two effects on the specific activity and a change of the isozyme distribution favoring the cationic component.

Adrenalectomy of the tumor-bearing male rat liver resulted in a decrease and hydrocortisone a slight increase in the specific activity of the hepatoma 5123B

---

* IWD-Pd, intermediate between well-differentiated and poorly differentiated hepatocellular carcinoma by D. R. Meranze.

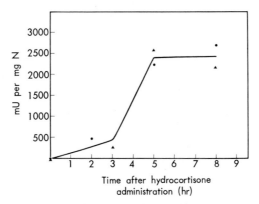

Fig. 3. Specific activity of hepatoma 5123B aspartate aminotransferase in adrenalectomized animals bearing hepatoma 5123B after intraperitoneally administrated hydrocortisone (0.05 mg/g body weight) (30)

The average of 4 determinations of different groups.

GOT. The GOT from normal male rat liver was unchanged after administration of cortisone. Cortisone or hydrocortisone administration to adrenalectomized male tumor-bearing rats neither normalized the effect on the tumor GOT activiy caused by adrenalectomy nor altered the isozyme profile. In these experiments the corticoid administration consisted of a single intraperitoneal injection. Sheid and Roth (38) injected corticoids over a five-day period at 2.5 times the dose we used, with the final injection 2.5 hr before sacrifice. They obtained a response in GOT to cortisone but not to hydrocortisone. Many investigators have shown that glucocortoids cause little change in GOT activity in intact rats, even with high doses.

We interpretated our results as the difference in the degree of sensitivity to glucocortoids after adrenalectomy as follows: (a) The tumor GOT showed responsiveness, while the normal liver GOT showed no sensitivity, and (b) the tumor in our experiments showed a change in the activity and an alteration in the isozyme profile in favor of the mitochondrial component. This perplexed us because the mitochondrial or cationic GOT component had been previously shown to be more refractory to change than the supernatant or anionic isozyme. Sheid and Roth (38), for example, found after prolonged administration of cortisone that the supernatant or anionic component was increased. The differences between our results and those reported by Sheid and Roth could be ascribed to measurements of different parameters, one determining the mitochondrial component completely or nearly completely released from the mitochondria and the other measuring only the portion loosely bound to the mitochondria. An increase in activity in the latter case could mean that, under the influence of some factor or factors such as a glucocorticoid, a larger portion of the bound mitochondrial enzyme was released.

Hepatoma Spectrum

A variety of carcinogenic chemicals have been administered in the diet to

TABLE II

	Carcinogenic-inducing agents	
1	N-2-Fluorenyldiacetamide	(N-2-FdiAA)
2	N-2-Fluorenylphthalamic acid	(N-2-FPA)
3	2-(4'-Methyl)benzoylaminofluorene	$(2[4'CH_3]BAF)$
4	2-Ethylnitrosaminofluorene	(2-ENAF)
5	N,N'-2,7-Fluorenylenebis (2,2,2-trifluoracetamide)	$(N, N'-2,7-FAA\ [F_6])$
6	Fluorenyl-2,7-N,N'-disuccinamic acid	(F-2,7-N,N'Dis)
6	2,4,6-Trimethylaniline	(2,4,6-TMA)
7	4'-Fluoro-4-acetylaminobiphenyl	(4'F-4-BAA)
9	N-Fluorenyl-2-nicotinamide	(N-F-2-NA)
10	Aflatoxin B_1	$(A-B_1)$

inbred strains of rats (Table II). These chemicals are not specific hepatocarcinogens but produce a wide variety of other types of neoplasms. Details of the induction and transplantation of the neoplasms of rat liver have been described (23). The details of these feeding experiments were varied. Some very strong carcinogens were used. For instance N-2-fluorenyldiacetamide, which induced 19 hepatomas, most of the transplantable hepatomas in our spectrum, was fed at several different levels from 0.05 to 0.015%. In some experiments the chemical was fed continuously, in others, at intervals; but in most experiments the carcinogen-fed animals were kept on the basal diet without the carcinogen for

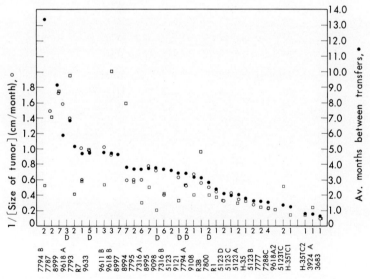

FIG. 4. Tumors are arranged from left to right in order of increasing rate of growth

Tumor identification or number is at the bottom. ● Av. time between transfers in months for the entire period of transplantation. □ Av. (months) between transfers for the last 10 transfers for each tumor. ○ size of tumors as 1/cm per month for the entire period of transfers. D, diploid number of chromosomes. The carcinogens are identified by number. cf. Table II for identification of abbreviation. NK, normal karyotype.

TABLE III. Histological and Biological Characteristics of Transplantable Hepatomas Arranged According to Increasing Growth Rate

Tumors	Inducing agent	Primary sex	Generations used	Months between transfers (Av.)	Histological type	Metastases (−/+)	Chromosome number
66	9	M	1–2	11.5	HD	2/0	42 NK
20	10	M	1–2	11.0	WD-HD	2/0	
16	10	M	1–4	10.5	W-HD	3/0	42–44
21	10	M	1–3	10.3	HD	3/0	
9618B	3	M	1–2	10.0	W-HD	4/0	42–44
9618A	3	M	1–5	8.0	HD	14/1	42 NK
47C	1	M	1–4	7.7	HD	1/3	
39A	1	M	1–5	7.2	HD-WD	2/4	
7787	2	F	7–14	6.3	HD-WD	8/0	44
28A	1	M	2–5	6.0	HD	2/2	82–85
44	1	M	1–4	5.8	WD	3/1	85
6	10	M	1–5	5.5	HD-WD	5/0	
8624[a]	4	M	1–5	5.0	WD-HD	5/0	
42A	1	M	1–4	4.7	WD	3/1	
9633	6	M	1–10	4.0	WD	3/2	42 NK
R3B	1	M	3–15	4.1	WD	5/5	43
9633F	6	M	5–9	4.0	WD	2/2	
7794B	2	M	10–17	3.1	HD	2/8	44
7794A	2	M	23–30	2.9	WD	6/1	42
R7	1	M	4–19	2.8	WD	4/6	44
38B	1	M	2–12	2.5	WD	8/2	
9611B	3	M	10–16	2.5	HD-WD	2/5	42–43
38A	1	M	1–2	2.3	WD	2/0	
8995	8	M	18–26	2.3	WD	0/10	43
9108	1	M	20–29	2.0	WD	4/6	42
5123C	2	F	71–77	2.0	WD	2/8	96
7793	2	F	21–28	1.8	WD	2/8	45
8994	8	M	29–38	1.8	Pd	9/1	72–82
7316B	7	F	32–41	1.8	WD	4/6	55
7795	2	M	29–39	1.8	WD	1/8	46–47
7800	2	M	42–48	1.7	WD	2/8	42
H35	1	M	55–64	1.7	WD	5/5	43–44
9121	1	M	20–29	1.7	WD	8/2	
7316A	7	F	33–42	1.6	WD	0/10	43
R1	1	M	22–31	1.6	WD	1/9	46
5123D	2	F	70–80	1.6	WD-Pd	0/10	45

Continued...

TABLE III. Continued.

5123A	2	F	80–89	1.5	WD	0/10	46
9121	1	M	24–33	1.5	WD	10/0	42
5123B	2	F	93–102	1.2	WD-Pd	4/6	46
9618A2	3	M	4–23	1.2	Pd	6/4	41
9098	1	M	32–41	1.2	WD-HD	7/3	42
5123TC	2	F	78–86	1.2	WD-Pd	1/9	47–49
7777	2	F	52–60	1.0	Pd	0/10	73
3924A	1	FC	264–272	1.0	Pd	3/7	65–73
7288C	5	M	93–102	1.2	Pd	8/2	43
H35TC2	1	M	95–104	0.7	Pd	7/3	52
7288CTC	5	M	50–59	0.7	Pd	10/0	67–69
H35TC1	1	M	120–129	0.7	Pd	2/8	49–50
3683F	1	M	261–270	0.6	Pd	5/5	39–40

Pd, poorly differentiated; WD-Pd, intermediate between well-differentiated and poorly differentiated; WD, well-differentiated; HD, highly differentiated; WD-HD, intermediate between well-differentiated and highly differentiated; NK, normal karyotype; FC, castrated female receiving testosterone.

[a] Tumor displaced by lymphoma cells. We thank Dr. D. R. Meranze et al. (13) and P. C. Nowell et al.(25, 26) for their help and personal communications.

lengths of time varying from a few weeks to several months before being killed, and the tumors present were inoculated into new hosts.

The second largest group of hepatomas were induced by N-2 fluorenylphthalamic acid and twelve tumors were in this group. The carcinogen was fed continuously for 10 months, and the animals were fed the basal diet without the carcinogen for 1 to 8 months before being killed and their tumor nodules transplanted. Of these 12 tumors, five were established as sublines from one hepatoma, 5123.

Four tumor lines were induced by 4-methyl-N-2-benzoylaminofluorene, and four lines were induced by aflatoxin B_1. Only one or two transplantable hepatomas are being maintained by the other carcinogens that have been studied (23).

Many characteristics of this large spectrum of hepatomas are listed in Table III. Included are the tumor designation, the inducing agent, the average months between transplants for the last ten transfers or, if less than ten were available, all of the transfers, the histological type, the extent of the lung metastases, and the chromosome number (13, 25, 26).

To gain information on the extent of regression or progression of growth rate in this spectrum of hepatomas, a comparison was made between the average time between transfers for all the transfers that had been made for any give tumor line and the average time between transfers for the last ten transfers of the same tumor. These data are presented in Fig. 4. Most of the rapidly growing tumors show little change in growth rate. Of the rapidly growing tumors, hepatoma 5123TC had a slightly slower growth rate. The growth rate of several of the slowly growing tumors had increased, although the growth rate of some of the slowly growing tumors was even slower for the last transfers, e.g., 9618A, 9618B, 8994, and R3B.

The growth rate data (Fig. 4) support the concept of diversity among a series of rat hepatomas as described by several other investigators (*8–44*).

If Potter and Watanabe's (*34*) concept of a tumor with a diploid number of chromosomes and a normal karyotype represents karyotypically the least-deviated stage of 42 normal-appearing chromosomes, then we have now induced a third transplantable hepatoma, No. 66, in addition to the two previous ones, 9618A and 9633. Biochemically, hepatoma 9618A has a near-normal glycogen content of 2–4%, whereas hepatoma 9633 is like the other hepatomas, with from 0 to 0.2% glycogen (*34*). Hepatoma 66 is a very slow-growing tumor and, for that reason, has not been studied extensively for its enzyme activities and has therefore not been examined for its regulation properties.

Each of the three transplantable hepatomas possessing diploid normal-appearing chromosomes was induced by a different carcinogen (Table III); 9618A was induced by 2-(4'CH_3)BAF, 9633 by F-2,7-N, N'Dis, and 66 by N-F-2NA.

No information is presently available concerning the mechanism of induction of hepatomas with 42 normal-appearing chromosomes. Additional studies are indicated first to see if more transplantable hepatomas with 42 normal-appearing chromosomes can be induced using any one or all three of these carcinogens. If more transplantable diploid normal-appearing chromosome tumors can be induced, additional neoplastic models would then be available to further examine Potter and Watanabe's (*34*) concept of biochemical similiarity of such " minimum deviation " hepatomas.

REFERENCES

1. Allen, D. O., Munshower, J., Morris, H. P., and Weber, G. Regulation of adenylcyclase in hepatomas of different growth rates. *Cancer Res.*, **31**, 557–560 (1971).
2. Berkes-Tomasevic, P. and Holzer, H. Zur Regulation der Synthese der Malatdehydrogenasen aus Cytosol und Mitochondrien von Ratten-Leber. *Eur. J. Biochem.*, **2**, 98–101 (1967).
3. Boyd, J. W. The intracellular distribution, latency, and electrophoretic mobility of L-glutamate-oxalacetate transaminase from rat liver. *Biochem. J.*, **81**, 434–441 (1961).
4. Christie, G. S. and Judah, J. D. Intracellular distribution of enzymes. *Proc. Royal Soc. (Biol.) (London)*, **141**, 420–433 (1953).
5. Delbruck, A., Schimassek, H., and Bartsch, K. Enzymverteilungsmuster in einigen Organen und Experimentellen Tumoren der Ratte und der Maus. *Biochem. Z.*, **331**, 297–311 (1959).
6. Dyer, H. M., Gullino, P. M., Ensfield, B. J., and Morris, H. P. Transaminase of liver tumors and serum. *Cancer Res.*, **21**, 1522–1531 (1961).
7. Elford, H. L. Mammalian ribonucleotide reductase and cell proliferation. *GANN Monograph*, **13**, 205–217 [1972].
8. Ferdinandus, J. A., Morris, Harold P., and Weber, G. Behavior of opposing pathways of thymidine utilization in differentiating, regenerating and neoplastic liver. *Cancer Res.*, **31**, 550–556 (1971).
9. Fine, I. H., Kaplan, N. O., and Kuftinec, D. Developmental changes of mammalian lactic dehydrogenase. *Biochemistry*, **2**, 116–121 (1963).
10. Fujii, S. Thymidine kinases of neoplastic tissues, regenerating and embryonic

liver, marrow cells, potato, and *Tetrahymena*. *GANN Monograph*, **13**, 107–119 (1972).

11. Hird, F. J. R. and Roswell, E. V. Additional transaminations by insoluble particle preparations of rat liver. *Nature*, **166**, 517–518 (1950).
12. Hook, R. H. and Vestling, C. S. Isozymes of glutamic-oxalacetic transaminase from rat liver. *Fed. Proc.*, **21**, 254 (1962).
13. Hruban, Z., Morris, H. P., Mochizuki, Y., Meranze, D. R., and Slesers, A. Light microscopic observations of Morris hepatomas. *Cancer Res.*, **31**, 752–762 (1971).
14. Johnson, H. L. and Kampschmidt, R. F. Lactic dehydrogenase isozymes in rat tissue tumors, and precancerous livers. *Proc. Soc. Exp. Biol. Med.*, **120**, 557–561 (1965).
15. Katunuma, N. and Okada, M. An alternative coupling reaction of mitochrondria-bound transaminase and tricarboxylic acid cycle and its metabolic role. *J. Vitaminol.*, **8**, 309–314 (1962).
16. Katunuma, N., Kuroda, Y., Matsuda, Y., and Kobayashi, K. Abnormal gene expression on the mode of amino-nitrogen excretion in rat hepatomas from phylogenic aspects. *GANN Monograph*, **13**, 135–141 (1972).
17. Kline, E. S. and Clayton, C. C. Lactic dehydrogenase isozymes during development of azo dye tumors. *Proc. Soc. Exp. Biol. Med.*, **117**, 891–894 (1964).
18. Linder, M. C., Moor, J. R., Munro, H. N., and Morris, H. P. Ferritin proteins in normal and malignant rat. *GANN Monograph*, **13**, 299–313 (1972).
19. Mann, K. G. and Vestling, C. S. Nature of the isozymes of rat liver mitochondrial malate dehydrogenase. *Biochim. Biophys. Acta*, **159**, 567–569 (1968).
20. Marino, Y., Itoh, H., and Wada, H. Crystallization of 2-oxoglutarate L-aspartate transaminases. *Biochem. Biophys. Res. Commun.*, **13**, 348–352 (1963).
21. Marino, Y., Kagemiyama, H., and Wada, H. Immunochemical distinction between glutamic-oxalacetic transaminases from the soluble and mitochondrial fractions of mammalian tissues. *J. Biol. Chem.*, **239**, PC 943–944 (1964).
22. Morris, H. P. and Wagner, B. P. The development of "minimal deviation" hepatomas. *Acta Unio. Int. Contra Cancrum*, **20**, 1364–1366 (1964).
23. Morris, H. P. and Wagner, B. P. Induction and transplantation of rat hepatomas with different growth rate (including "minimal deviation" hepatomas). *Methods Cancer Res.*, **4**, 125–152 (1968).
24. Murray, R. K., Hudgin, R. L., and Schachter, H. Studies of isozymes and heteroglycan metabolism of hepatomas. *GANN Monograph*, **13**, 167–180 (1972).
25. Nowell, P. C., Morris, H. P., and Potter, V. R. Chromosomes of "minimal devition" hepatomas and some other transplantable rat tumors. *Cancer Res.*, **27**, 1565–1579 (1967).
26. Nowell, P. C. and Morris, H. P. Chromosomes of "minimal deviation" hepatomas: a further report of diploid tumors. *Cancer Res.*, **29**, 969–970 (1969).
27. Ohashi, M. and Ono, T. Glucose 6-phosphate dehydrogenase isozymes in cultured Morris hepatoma cells. *GANN Monograph*, **13**, 267–278 (1972).
28. Ono, T., Wakabayashi, K., Uenoyama, K., and Koyama, H. Regulation of enzyme synthesis relating to differentiation of malignant cells. *GANN Monograph*, **13**, 19–29 (1972).
29. Otani, T. T. and Morris, H. P. Isozymes of glutamic-oxalacetic transaminase in some rat hepatomas. *Adv. Enzyme Regulation*, **3**, 325–334 (1965).
30. Otani, T. T. and Morris, H. P. The effects of glucocorticoids on the relative distribution of aspartate-aminotransferase isozymes of Morris hepatoma 5123B. *Cancer Res.*, **28**, 2092–2097 (1968).

31. Otani, T. T. and Morris, H. P. A comparison of enzyme patterns of four enzymes from hepatomas of different growth rate. *J. Natl Cancer Inst.*, **47**, 1247–1253 (1971).
32. Pedersen, P. L. The enzymology, ultrastructure, and energetics of mitochondria from three Morris hepatomas of widely different growth rates. *GANN Monograph*, **13**, 251–265 (1972).
33. Pitot, H. C., Iwasaki, Y., Inoue, H., Kasper, C., and Mohrenweiser, H. Regulation of the levels of multiple forms of serine dehydratase and tyrosine aminotransferase in rat tissues. *GANN Monograph*, **13**, 191–204 (1972).
34. Potter, V. R. and Watanabe, M. Some biochemical essentials of malignancy: the challenge of diversity. International Leukemia-Lymphoma Conf. Proc., Lea and Febiger Pub. Co., Philadelphia, pp. 34–45 (1968).
35. Potter, V. R., Watanabe, M., Pitot, H. C., and Morris, H. P. Systematic oscillations in metabolic activity in rat liver and hepatoma. Survey of normal diploid and other hepatoma lines. *Cancer Res.*, **29**, 55–78 (1969).
36. Potter, V. R., Walker, P. R., and Goodman, J. I. Survey of current studies on oncogeny as blocked ontogeny: isozyme changes in livers of rats fed 3′-methyl-4-dimethylaminoazobenzene with collateral studies on DNA stability. *GANN Monograph*, **13**, 121–134 (1972).
37. Rasada, A., Morris, H. P., and Weinhouse, S. Lactate dehydrogenase subunits in normal and neoplastic tissues of the rat. *Cancer Res.*, **29**, 1673–1680 (1969).
38. Sheid, B. and Roth, J. S. The distribution of L-aspartic aminotransferase in cell fractions of rat livers and some hepatomas. *Adv. Enzyme Regulation*, **3**, 335–350 (1965).
39. Sophianopoulos, A. J. and Vestling, V. S. Nature of two forms of malic dehydrogenase from rat liver. *Biochim. Biophys. Acta*, **45**, 400–402 (1960).
40. Swick, R. W., Barnstein, P. L., and Stange, J. L. The metabolism of mitochrondia proteins. I. Distribution and characterization of alanine aminotransferase in rat liver. *J. Biol. Chem.*, **200**, 3334–3340 (1965).
41. Watanabe, M., Potter, Van R., Reynolds, R. D., Pitot, C. H., and Morris, H. P. Enzyme patterns of Morris hepatoma 9618A under controlled feeding schedules. *Cancer Res.*, **29**, 1691–1698 (1969).
42. Weber, G. and Lea, M. A. The molecular correlation concept. An experimental and conceptual method in cancer research. *Methods Cancer Res.*, **2**, 523–578 (1967).
43. Weber, G. The molecular correlation concept: current status. *GANN Monograph*, **13**, 47–77 (1972).
44. Weinhouse, S., Shatton, J. B., Criss, W. E., Farina, F. A., and Morris, H. P. Isozymes in relation to differentiation in transplantable rat hepatomas. *GANN Monograph*, **13**, 1–17 (1972).

THYMIDINE KINASES OF NEOPLASTIC TISSUES, REGENERATING AND MEBRYONIC LIVER, MARROW CELLS, POTATO, AND *TETRAHYMENA*[*1]

Setsuro FUJII, Takaki HASHIMOTO, Takahiko SHIOSAKA,
Teruo ARIMA, Michiko MASAKA, and
Hiromichi OKUDA

Department of Enzyme Physiology, Institute for Enzyme Research, School of Medicine, Tokushima University[*2]

A simple method for separation of thymidine (Tdr) kinase was established and applied to the enzymes of ascitic tumor cells, such as Yoshida sarcoma, AH-130, Sarcoma-180, Ehrlich carcinoma, solid tumors such as Morris hepatomas 7793 and 7794A, rat regenerating liver, marrow cells and embryonic liver. All tumor tissues gave two peaks, while normal tissues, including regenerating liver and marrow cells, gave only one peak.

Two kinds of Tdr kinases were separated from a crude extract of Yoshida sarcoma by zone electrophoresis or DEAE-cellulose column chromatography and purified by streptomycin treatment, ammonium sulfate fractionation, hydroxylapatite and DEAE-cellulose column chromatography, and Sephadex G-200 gel filtration. Peak I was purified about 1,400-fold over the homogenate. The molecular weights of the enzymes in peaks I and II were estimated as 70,000 and 120,000, respectively. The K_m values for ATP of the enzymes in peaks I and II were 8×10^{-5} M and 7×10^{-5} M, and the K_m values for deoxythymidine were 3.3×10^{-5} M and 3.1×10^{-6} M, respectively. Both enzyme activities were inhibited by TTP.

Phosphorylation of Tdr in potato and *Tetrahymena pyriformis* was shown to be catalyzed by nucleoside phosphotransferase, not by Tdr kinase. The Tdr kinase-like activity of these tissues described in other reports appeared to be a combination of the activities of nucleoside phosphotransferase and of an ATP-hydrolyzing enzyme:

$$\begin{array}{ccc}
\text{ATP} & \longrightarrow & \text{ADP or AMP} \\
\uparrow & & \downarrow \\
\text{ATP-hydrolyzing enzyme} & & \\
\text{X} & \longrightarrow & \text{XMP} \\
\text{(nucleoside)} & \uparrow & \\
& \text{nucleoside phosphotransferase} &
\end{array}$$

[*1] This study was supported in part by a Grant-in-Aid for Cancer Research from the Ministry of Education.

[*2] Kuramoto-cho 3-18-1, Tokushima 770, Japan (藤井節郎, 橋本卓樹, 塩坂孝彦, 有馬暎雄, 真坂美智子, 奥田拓道).

The incorporation of Tdr, into DNA is proportional to the growth rate of various hepatomas (*12*). The first step in the incorporation of Tdr into DNA is the phosphorylation of this nucleoside to thymidine monophosphate (TMP), which is catalyzed by Tdr kinase. Tdr kinase activity is virtually undetectable in adult rat liver, but is high in regenerating liver (*2, 4, 20*), marrow cells (*5*), embryonic liver (*5, 10*), rapidly growing hepatomas (*5*), remaining kidney of animals after nephrectomy (*15*), and cultured mammalian cells after viral infection (*16*). This shows that Tdr kinase is subjected to induction and repression. Moreover, its activity is markedly inhibited by its end product, thymidine triphosphate (TTP) (*4, 11, 17*). The fact that Tdr kinase activity is regulated suggests that this enzyme is important in the control of DNA synthesis.

Okazaki and Kornberg (*17*) have reported the partial purification and properties of Tdr kinase from *Escherichia coli*, while Bresnick *et al.* (*6, 7*) have reported studies on the enzyme in various animal tumors and regenerating liver.

Shoup *et al.* (*18*) studied the phosphorylation of Tdr in *Tetrahymena pyriformis* and reported that TTP does not exert feedback control of Tdr kinase, suggesting that regulation of Tdr kinase activity in *T. pyriformis* may differ from that in mammalian tissues and *E. coli*.

Wanka *et al.* (*19*) reported that Tdr kinase in the roots of germinating corn and wheat seedlings could be fractionated with ammonium sulfate into two components, designated as components P and T. These were both enzymatically inactive, but high activity was restored by combining the two components. Component P was present in all plant materials examined, and Tdr kinase was found to be controlled by formation and degradation of component T.

Previously, our laboratory (*10*) reported that two fractions of Tdr kinase were separated from a crude extract of Yoshida sarcoma by zone electrophoresis or DEAE-cellulose column chromatography, while only one fraction was found in regenerating liver.

The present report describes the purification and characterization of Tdr kinase from tumor tissues, regenerating liver, *T. pyriformis*, and potato.

Enzyme Assay

Tdr kinase activity of mammalian tissues was measured by determining the conversion of labeled Tdr to TMP with ATP as phosphate donor, using DEAE-cellulose discs as described by Bresnick and Karzala (*8*). Enzyme activity of *E. coli* was determined by the method of Okazaki and Kornberg (*17*). Nucleoside phosphotransferase activity was measured by the method of using labeled Tdr as substrate and AMP as phosphate donor.

ATP-hydrolyzing activity was estimated as follows: A mixture of Tris-HCl (50 mM, 200 μl, pH 7.0), $MgCl_2$ (25 mM, 100 μl), ATP (20 mM, 100 μl), and 100 μl of enzyme solution was incubated at 37° for 30 min. The reaction was terminated by the addition of 2.0 ml of 5% trichloroacetic acid.

Inorganic orthophosphate was estimated colorimetrically by the method of Fiske and Subbarow (*9*) and protein concentration was measured by the method of Lowry *et al.* (*13*), with bovine albumin as a standard.

Simple Method for Separation of Tdr Kinase and Its Application to the Enzyme of Various Tumors and Tissues

Ascitic tumor cells (Yoshida sarcoma, AH-130, Sarcoma-180, and Ehrlich carcinoma), solid tumor tissues (Morris hepatomas 7793 and 7794A), regenerating liver, marrow cells, and embryonic liver were homogenized with about 4 volumes of 0.2 M Tris-HCl buffer, pH 8.0. The homogenate was centrifuged at $8,000\,g$ for 30 min, and the resulting supernatant was desalted by passage through a Sephadex G-25 column. The fraction containing enzyme activity was applied to a DEAE-cellulose column (1.5×10 cm) and equilibrated with 0.1 M Tris-HCl buffer, pH 8.0. The column was eluted with 80 ml of the same buffer and then with a linear gradient of 0 to 0.8 M NaCl in 0.4 M Tris-HCl buffer, pH 8.0. Fractions of 4 ml were collected at a flow rate of about 8.0 ml/min. Tumor tissues gave 2 peaks of Tdr kinase activity (Fig. 1) while regenerating rat liver, marrow cells, and embryonic liver gave only the first peak. A large difference was found in the stabilities of the Tdr-kinase enzymes in these tissues. The crude enzyme of Yoshida sarcoma was rather stable, while the enzyme activity of regenerating liver was extremely labile.

Therefore, the absence of the second enzyme peak in regenerating liver may be related to the instability of this fraction. However, this may not be the case, since the second peak did not become detectable in the presence of various stabilizers, such as ATP, TTP, and Tdr.

The existence of different enzyme peaks in tumor tissues was confirmed by using a different method as follows: Yoshida sarcoma cells were homogenized in 4 volumes of 5 mM Tris-HCl (pH 8.0), and the homogenate was centrifuged at $8,000\,g$ for 30 min. The resultant supernatant was desalted by passage through a Sephadex G-25 column and the eluate was applied to a DEAE-cellulose column equilibrated with 5 mM Tris-HCl buffer (pH 8.0). The column was washed with the same buffer and then successively eluted, stepwise, with 50 ml volumes of 10, 100, 150, and 500 mM Tris-HCl buffer (pH 8.0).

Three peaks of activity (peaks I, II, and III) were detected in the fractions eluted with 5, 100, and 500 mM Tris-HCl buffer (pH 8.0), respectively.

On chromatography of extracts of normal tissues, only two peaks (peaks I and II) were found. These results will be reported in detail elsewhere.

Purification of Tdr Kinase from Yoshida Sarcoma

To study the properties of the two Tdr kinase peaks from Yoshida sarcoma, purification of the enzyme was attempted. The temperature was maintained at 4° throughout the purification procedure.

Yoshida sarcoma cells were homogenized in 0.2 M Tris-HCl buffer (pH 8.0), centrifuged at $8,000\,g$ for 30 min, and 60% streptomycin solution was added to the resulting supernatant to a final concentration of 2%.

The mixture was stirred overnight and then centrifuged at $8,000\,g$ for 30 min. Solid $(NH_4)_2SO_4$ was added to the supernatant to give 40% saturation. The resulting precipitate was dissolved in 0.1 M potassium phosphate buffer (pH 8.0).

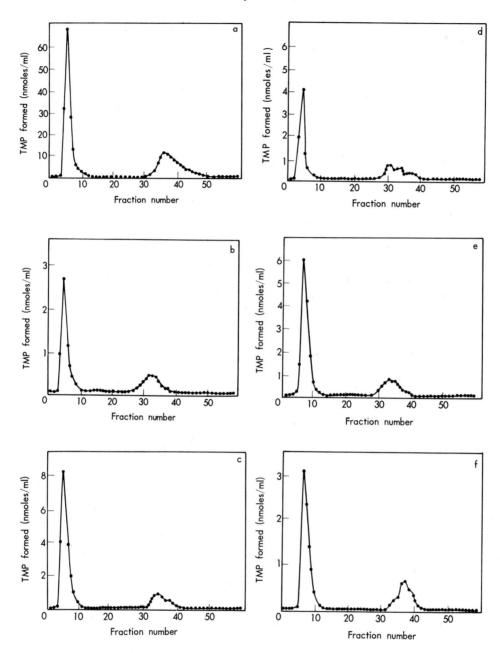

The solution was centrifuged and the supernatant applied to a column of Sephadex G-25 equilibrated with 0.005 M potassium phosphate buffer (pH 8.0). The column was eluted with the same buffer. The fraction containing enzyme activity was collected and applied to a hydroxylapatite column (3.2×15 cm) equilibrated with 0.005 M potassium phosphate buffer (pH 8.0). The column was washed with the same buffer and then with 0.015 M potassium phosphate buffer (pH 8.0). Tdr kinase was eluted with 0.075 M potassium phosphate buffer (pH 8.0). The

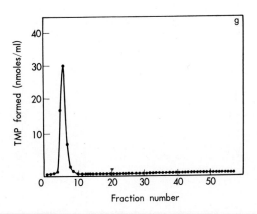
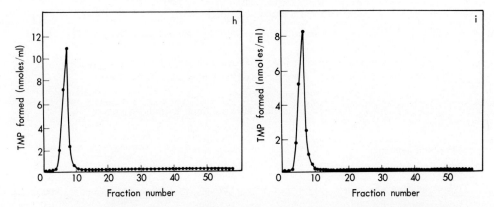

Fig. 1. Elution pattern of Tdr kinase of various tissues from a DEAE-cellulose column

a: Yoshida sarcoma; b: Ehrlich ascites tumor; c: ascites hepatoma AH-130; d: sarcoma 180; e: Morris hepatoma 7793; f: Morris hepatoma 7794A; g: rat embryonic liver; h: rat regenerating liver; i: rat marrow cells.

fractions containing the bulk of the activity were pooled, and solid $(NH_4)_2SO_4$ was added slowly with stirring to give 40% saturation. After standing for 12 hr, the solution was centrifuged at 8,000 g for 15 min. The precipitate was dissolved in 0.005 M Tris-HCl buffer (pH 8.0) containing 0.05 M NaCl and desalted by passage through a Sephadex G-25 column. The desalted enzyme solution was applied to a DEAE-cellulose column equilibrated with 0.005 M Tris-HCl buffer (pH 8.0), containing 0.05 M NaCl, and stepwise elution was carried out.

As shown in Fig. 2, peak I was eluted with 0.05 M NaCl in 0.005 M Tris-HCl buffer (pH 8.0) and peak II with 0.4 M NaCl in the same buffer.

The fractions in peak I were combined and applied to a Sephadex G-200 column equilibrated with 0.05 M Tris-HCl buffer (pH 8.0) containing 0.1 M KCl. The column was eluted with the same buffer. A summary of the purification procedure is shown in Table I.

The enzyme in peak I was purified about 1,400-fold by this procedure. The fractions in peaks I and II, respectively, were combined and rechromatographed

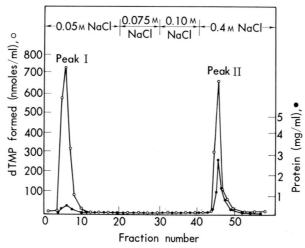

FIG. 2. DEAE-cellulose column chromatography of Tdr kinase from Yoshida sarcoma

Fractions of 6 ml were collected. The procedures used for chromatography are described in the text. 0.005 M Tris-HCl buffer, pH 8.0 was used.

TABLE I. Purification of Tdr Kinases from Yoshida Sarcoma

Fraction and step	Protein (mg)	Total activity (nmoles)	Specific activity (nmoles/mg protein)	Yield (%)
Whole homogenate	19,420	120,650	6.2	100
Crude extract	12,160	112,080	9.2	93
Streptomycin treatment 1st amm. sulf.	2,102	231,910	111.0	192
Hydroxylapatite	390	101,654	261.0	84
2nd amm. sulf.	126	69,824	554.0	58
DEAE-cellulose (peak I)	12	30,269	2,522.0	25
(peak II)	59	10,098	160.0	8.3
Sephadex G-200 (peak I)	24	21,880	9,117.0	18

on DEAE-cellulose column. As shown in Fig. 3, peak I was eluted with 0.05 M NaCl in 0.005 M Tris-HCl buffer (pH 8.0), while peak II was eluted with 0.4 M NaCl in the same buffer.

The molecular weights of the enzymes in peaks I and II were estimated by a modification of the method of Andrews (1) and found to be about 70,000 and 120,000, respectively. The K_m values of the enzymes in peaks I and II for ATP were 8×10^{-5} M and 7×10^{-5} M, and those for Tdr were 3.3×10^{-5} M and 3.1×10^{-6} M. The other properties of the enzymes in peaks I and II were very similar. The optimum pH values of both were around 8. As phosphate acceptors, Tdr and deoxyuridine were effective, while deoxycytidine, uridine, and cytidine were not. ATP and dATP were good phosphate donors while dCTP and dGTP were less effective. IMP, AMP, and phenyl phosphate did not act as a phosphate

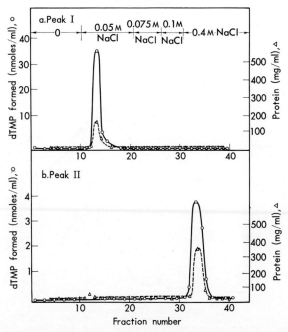

FIG. 3. Rechromatography of peaks I and II on DEAE-cellulose column
Fractions of 4 ml were collected. The procedure is described in the text.
0.005 M Tris-HCl buffer, pH 8.0 was used.

donor. TTP, sulfhydryl inhibitors (*i.e.*, iodoacetate, N-ethylmaleinamide, *p*-hydroxymercuribenzoate), and sodium dodecyl sulfate inhibited the activities of the enzymes in both peaks I and II.

Properties of the Aggregated and Disaggregated Forms of Tdr Kinase

Bresnick *et al.* (7) has reported the isolation of two forms of Tdr kinase from Walker carcinoma. In dilute solutions such as 0.05 M Tris-HCl buffer (pH 8.0), the enzyme was in an aggregated form which was eluted in void volume from a Sephadex G-200 column, suggesting that its molecular weight was above 6×10^5 M. Enzyme in a disaggregated form, with a molecular weight of approximately 110,000, was obtained from a Sephadex G-200 column equilibrated with 0.2 M KCl. As mentioned above, the Tdr kinase of tumors was separated into two peaks by DEAE-cellulose column chromatography. Therefore, it was interesting to see whether these two peaks of Tdr kinase in tumor tissues were in an aggregated form. Confirming the results of Bresnick *et al.* (7), a crude preparation of Yoshida sarcoma Tdr kinase was eluted from a Sephadex G-200 column with 0.005 M Tris-HCl buffer (pH 8.0) in void volume. When the ionic strength was increased by the addition of 0.05 M KCl, a greater volume of elution solvent was required, as shown in Fig. 4. To examine the conversion of the aggregated to the disaggregated form, Tdr kinase of Yoshida sarcoma was partially purified as follows. After streptomycin treatment as described above, the supernatant was applied to a Sephadex G-25 column equilibrated with 0.005 M

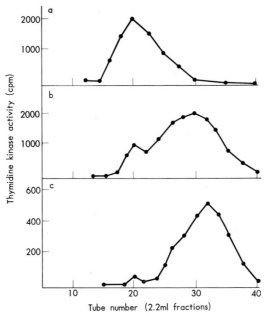

Fig. 4. Gel filtration of various Yoshida sarcoma Tdr kinase fractions on Sephadex G-200 columns (2.8×25 cm)
  a: Crude extract 5 mM Tris-HCl (pH 8.0); b: crude extract 5 mM Tris-HCl (pH 8.0)-0.05 M KCl; c: partially purified preparation 5 mM Tris-HCl (pH 8.0).

potassium phosphate buffer (pH 8.0). Elution was carried out with the same buffer. Calcium phosphate gel (120 mg/ml) was added to the eluate, and the suspension was centrifuged at 3,000 g for 5 min. Then the gel was suspended in 0.005 M potassium phosphate buffer (pH 8.0). After centrifugation at 3,000 g for 5 min, the gel was resuspended in 0.1 M potassium phosphate buffer (pH 8.0) and centrifuged at 3,000 g for 5 min. $(NH_4)_2SO_4$ was added to the supernatant to 30% saturation. The precipitate was collected and dissolved in 0.005 M potassium phosphate (pH 8.0). This solution was passed through a Sephadex G-25 column equilibrated with 0.005 M potassium phosphate buffer (pH 8.0) and eluted with the same buffer. The active fractions were combined and adsorbed on a column of hydroxylyapatite equilibrated with 0.005 M potassium phosphate buffer, and the enzyme was eluted with 0.05 M potassium phosphate buffer. Recovery of the enzyme was 21.5%, and its specific activity was 17 times that of the crude extract. The partially purified preparation thus obtained was subjected to gel filtration on a Sephadex G-200 column, using 0.005 M Tris-HCl buffer as developing solution.

As shown in Fig. 4, the Tdr kinase was eluted in more than void volume, indicating that purification caused retardation of the elution of this enzyme from a Sephadex G-200 column and that a certain modifier in the crude preparation was removed during the purification procedure. Similar retardation of the elution pattern was observed on treatment of the crude enzyme with RNase, but not with DNase. Partially purified Tdr kinase of Yoshida sarcoma or enzyme which had been treated with RNase was separated into two peaks by DEAE-cellulose column chromatography.

Phosphorylation of Tdr by Enzymes from T. pyriformis *and Potato*

The Tdr phosphorylating activities of enzymes from various sources were estimated by using ATP as the phosphate donor. The results are shown in Table II. Extracts of potato and *T. pyriformis* have higher specific activities for phosphorylation of pyrimidine than that of regenerating rat liver.

Studies on Tdr kinase in various animals and *E. coli* have generally indicated that TTP acts as an inhibitor of enzyme activity. As shown in Fig. 5, the Tdr phosphorylation activities of potato and *T. pyriformis* were not inhibited by TTP even at a concentration of 0.32 mM. Shoup *et al.* (*18*) obtained similar results on the enzyme in *Tetrahymena*. At this concentration, TTP almost completely inhibited the Tdr phosphorylation of regenerating liver and Yoshida sarcoma. These results suggest that the enzyme involved in Tdr phosphorylation in potato and *T. pyriformis* differs from that of animal tissues and *E. coli*.

The properties of the Tdr-phosphorylating enzyme of potato and *T. pyriformis* were studied by the following methods.

TABLE II. Tdr-phosphorylating Activities of Germinating Potato, *T. pyriformis*, Regenerating Rat Liver, and Yoshida Sarcoma

Enzyme source	Specific activity (nmoles TMP/mg protein)
Germinating potato	5.9
Tetrahymena pyriformis	3.6
Regenerating rat liver	1.2
Yoshida sarcoma cells	9.7

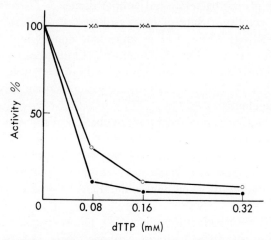

FIG. 5. Inhibitory effect of TTP on Tdr-phosphorylating activity
× *T. pyriformis*; △ potato; ○ regenerating rat liver; ● Yoshida sarcoma cells.

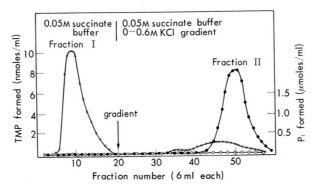

FIG. 6. DEAE-Sephadex A-50 column chromatogram of germinating potato extract
× nucleoside phosphotransferase activity; ● ATP-hydrolyzing enzyme activity; ○ Tdr kinase activity.

TABLE III. Restoration of Tdr Kinase Activity by Combination of Nucleoside Phosphotransferase and ATP-hydrolyzing Enzyme

Fraction	Tdr kinase-like activity (nmoles/mg protein)	Nucleoside phosphotransferase activity (nmoles/mg protein)	ATP-hydrolyzing enzyme activity (μmoles/mg protein)
I	0	5.56	0
II	0	0.56	8.50
I+II	2.80	5.00	7.50

An extract of potato was applied to a DEAE-Sephadex A-50 column. The elution pattern is shown in Fig. 6.

Brawerman and Chargaff (3) and Maley and Maley (14) demonstrated that nucleoside phosphotransferase, an enzyme catalyzing the synthesis of mononucleotides, was present in plants, bacteria, and animal tissues. Therefore, nucleoside phosphotransferase activity (using Tdr as substrate and AMP as phosphate donor), Tdr kinase activity (using ATP as phosphate donor), and ATP-hydrolyzing activity of fractions from a DEAE-Sephadex column were tested.

The first peak (fraction I) contained nucleoside phosphotransferase activity, and the later (fraction II) had the ATP-hydrolyzing enzyme. No Tdr kinase was observed in the eluate.

Tdr kinase-like activity appeared upon incubation of fraction I with fraction II, as shown in Table III.

The Tdr kinase-like activity and nucleoside phosphotransferase activity of the potato extract increased markedly upon storage of potatoes in the cold for 12 days before preparation of the extract, while the ATP-hydrolyzing activity did not change.

An extract of *T. pyriformis* was subjected to DEAE-Sephadex A-50 column chromatography, but the three enzyme activities were not separated, ATP-hydrolyzing, Tdr kinase, and nucleoside phosphotransferase activities appearing

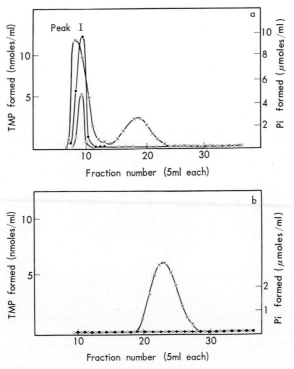

FIG. 7. a: Gel filtration pattern of *T. pyriformis* extract on Sephadex G-200 column
× nucleoside phosphotransferase activity; ○ Tdr kinase activity; ● ATP-hydrolyzing enzyme activity.
b: Gel filtration pattern on a Sephadex G-200 column of peak I after treatment with trypsin. × nucleoside phosphotransferase activity; ○ Tdr kinase activity; ● ATP-hydrolyzing enzyme activity.

in the same fraction. Then the extract was subjected to gel filtration on a Sephadex G-200 column. The elution profile obtained is shown in Fig. 7.

Tdr kinase, nucleoside phosphotransferase, and ATP-hydrolyzing activities were obtained in early fractions of the eluate, while only nucleoside phosphotransferase activity was found in later fractions. The early fractions of the eluate from Sephadex G-200 containing the three enzyme activities were collected in a Visking tube and concentrated by stirring the tube in 5% saturated polyethylene glycol at 4° overnight. The concentrated fraction was again subjected to gel filtration on a Sephadex G-200 column, and the three enzyme activities were again found in early fractions of the eluate. These fractions were combined, treated with trypsin at 37° for 30 min, and again subjected to gel filtration on a Sephadex G-200 column.

As shown in Fig. 7b, the Tdr kinase and ATP-hydrolyzing activities completely disappeared upon treatment with trypsin, while the peak of phosphotransferase activity was retarded.

The extracts of Yoshida sarcoma and regenerating liver were subjected to gel filtration on a Sephadex G-200 column. Although nucleoside phosphotrans-

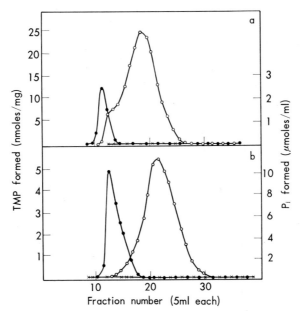

Fig. 8. Gel filtration patterns of extracts of (a) Yoshida sarcoma cells and (b) regenerating rat liver on a Sephadex G-200 column
● ATP-hydrolyzing enzyme activity; ○ Tdr kinase activity; × nucleoside phosphotransferase activity.

ferase activity was observed, Tdr kinase and ATP hydrolyzing activities were detected, as shown in Fig. 8.

REFERENCES

1. Andrews, P. The gel-filtration behaviour of proteins related to their molecular weights over a wide range. *Biochem. J.*, **96**, 595–606 (1965).
2. Bollum, F. J. and Potter, V. R. Nucleic acid metabolism in regenerating rat liver. VI. Soluble enzymes which convert thymidine to thymidine phosphates and DNA. *Cancer Res.*, **19**, 561–565 (1959).
3. Brawerman, G. and Chargaff, E. On the synthesis of nucleotides by nucleoside phosphotransferases. *Biochim. Biophys. Acta*, **15**, 549–559 (1954).
4. Breitman, T. R. The feedback inhibition of thymidine kinase. *Biochim. Biophys. Acta*, **67**, 153–155 (1963).
5. Bresnick, E., Thompson, U. B., Morris, H. P., and Liebelt, A. G. Inhibition of thymidine kinase activity in liver and hepatomas by TTP and d-CTP. *Biochem. Biophys. Res. Commun.*, **16**, 278–284 (1964).
6. Bresnick, E. and Thompson, U. B. Properties of deoxythymidine kinase partially purified from animal tumors. *J. Biol. Chem.*, **240**, 3967–3974 (1965).
7. Bresnick, E. Thompson, U. B., and Lyman, K. Aggregation of thymidine kinase in dilute solutions: properties of aggregated and disaggregated forms. *Arch. Biochem. Biophys.*, **114**, 352–359 (1966).
8. Bresnick, E. and Karzala, R. J. End-product inhibition of thymidine kinase activity in normal and leukemic human leukocytes. *Cancer Res.*, **24**, 841–846 (1964).

9. Fiske, C. H. and Subbarow, Y. The colorimetric determination of phosphorus. *J. Biol. Chem.*, **66**, 375–400 (1925).
10. Hashimoto, T., Shiosaka, T., Toide, H., Okuda, H., and Fujii, S. Thymidine kinases from normal and tumor tissues. *GANN*, **60**, 41–47 (1969).
11. Ives, D. E., Morse, P. A. Jr., and Potter, V. R. Feedback inhibition of thymidine kinase by thymidine triphosphate. *J. Biol. Chem.*, **238**, 1467–1474 (1963).
12. Lea, M. A., Morris, H. P., and Weber, G. Comparative biochemistry of hepatomas. VI. Thymidine incorporation into DNA as a measure of hepatoma growth rate. *Cancer Res.*, **26**, 456–469 (1966).
13. Lowry, O. H., Rosenbrough, N. J., Farr, A. L., and Randall, R. J. Protein measurement with the Folin phenol reagent. *J. Biol. Chem.*, **193**, 265–275 (1951).
14. Maley, G. F. and Maley, F. Nucleotide interconversions. X. Deoxyribo- and ribonucleoside 5'-phosphate synthesis *via* a phosphotransferase reaction in chick embryo extracts. *Arch. Biochem. Biophys.*, **101**, 342–349 (1963).
15. Mayfield, E. D., Jr., Liebelt, R. A., and Bresnick, E. Activities of enzymes of deoxyribonucleic acid synthesis after unilateral nephrectomy. *Cancer Res.*, **27**, 1652–1657 (1967).
16. McAuslan, B. R. and Joklik, W. K. Stimulation of the thymidine phosphorylating system in HeLa cells on infection with pox virus. *Biochem. Biophys. Res. Commun.*, **8**, 486–491 (1962).
17. Okazaki, R. and Kornberg, A. Deoxythymidine kinase of *Escherichia coil*. II. Kinetics and feedback control. *J. Biol. Chem.*, **239**, 275–284 (1964).
18. Shoup, G. D., Prescott, D. M., and Wykes, J. R. Thymidine triphosphate synthesis in *Tetrahymena*. I. Studies on thymidine kinase. *J. Cell. Biol.*, **31**, 295–300 (1966).
19. Wanka, F., Vasil, I. K., and Stern, H. Thymidine kinase: The dissociability and its bearing on the enzyme activity in plant materials. *Biochim. Biophys. Acta*, **85**, 50–59 (1965).
20. Weissman, S. M., Smellie, R. M. S., and Paul, J. Studies on the biosynthesis of deoxyribonucleic acid by extracts of mammalian cells. IV. The phosphorylation of thymidine. *Biochim. Biophys. Acta*, **45**, 101–110 (1960).

SURVEY OF CURRENT STUDIES ON ONCOGENY AS BLOCKED ONTOGENY: ISOZYME CHANGES IN LIVERS OF RATS FED 3'-METHYL-4-DIMETHYLAMINOAZOBENZENE WITH COLLATERAL STUDIES ON DNA STABILITY[*1]

V. R. POTTER, P. ROY WALKER,[*2] and Jay I. GOODMAN[*3]

McArdle Laboratory for Cancer Research, University of Wisconsin Medical School[*4]

Studies on pyruvate kinase, hexokinase, and aldolase isozymes were carried out on the earliest precancerous stages in livers of rats fed 3'-methyl-4-dimethylaminoazobenzene with reference to findings in Morris hepatomas and fetal and newborn liver. Parallel studies showed widespread DNA breakdown and new incorporation of ^{3}H-thymidine. The concept that "oncogeny is blocked ontogeny" is discussed in terms of "the Cohnheim and Durante alternatives," *i.e.*, as to whether the blocked ontogeny occurred in residual embryonic cells or in adult cells that reverted to an embryonic state. The experimental approaches to the blocked ontogeny hypothesis now seem widely applicable to a variety of systems in which the malignant transformation can be effected.

This conference marks a moment of historical significance in the long struggle to understand the nature of chemical carcinogenesis, and it is altogether fitting that a special meeting of Japanese and American investigators should be involved. Japanese studies on chemical carcinogenesis in liver go back to the 1930's, and recent striking advances in our knowledge of isozymes in hepatomas have also been the contribution of numerous Japanese investigators, as this monograph will record. From the American side, the development of the Morris hepatomas and their extensive use by nearly everyone at this conference has also been a noteworthy example of international scientific cooperation.

The confluence of the "isozyme" concept and the "minimal deviation" concept to reinforce the view that "oncogeny is blocked ontogeny" (*24, 25*) has occurred entirely during the decade of the 1960's and especially since 1967. The first use of the word isozyme was by Markert and Moller (*18*) in 1959, when they wrote as follows: "When subjected to a variety of physical, chemical, or serological

---

[*1] Financial support was provided by Grant CA-07175 from the National Cancer Institute.

[*2] Postdoctoral Fellow, 1970-1971, Damon Runyon Memorial Fund.

[*3] Recipient of USPHS postdoctoral fellowship 1-FO2-CA43985, awarded by the National Cancer Institute, NIH.

[*4] Madison, Wisconsin 53706, U.S.A.

tests enzymes from different organisms are commonly found to be different from each other even though catalyzing the same chemical reaction. In view of the demonstrated genetic control of protein synthesis it is not surprising that differences should exist in the structure of homologous enzymes or proteins synthesized by different species or even by animals of different genotype within the same species. Rather surprising, however, is the evidence demonstrating that several enzymes exist in multiple molecular forms not only within a single organism but within a single tissue. The existence of each of these enzymes as a family of closely related but distinguishable molecular types suggests the need for an extension of the classification of enzymes beyond that based on substrate specificity alone. We propose, therefore, to use the term isozyme to describe the different molecular forms in which proteins may exist with the same enzymatic specificity." [References omitted, *cf.* also Markert (*16*), Ikawa (*12*), and Uriel (*49*)].

Markert and Moller discussed the relation of isozymes to ontogeny (*cf.* their title (*18*)) and rejected the idea that changes in relative proportions of cells in tissues could account for changing isozyme patterns. Instead, they remarked, " More plausible is the hypothesis that the isozyme pattern of a tissue reflects the state of differentiation of its cells." In fact the two alternatives are not mutually exclusive and both suggestions are probably valid in particular circumstances. As recently as September 1967, Markert (*17*) discussed cancer as a disease of cell differentiation and examined the idea that carcinogenic stimuli are agents that " lead only to misprogramming of gene function." Incredible as it may seem, he did not mention the word isozyme, did not refer to his earlier publications on isozymes as the molecular clues to the normal process of differentiation (*16*, *18*), and did not suggest that isozymes might reveal the nature and extent of the "misprogramming" of gene function. Nor did any of the other participants in the 1967 Symposium on " The Developmental Biology of Neoplasia " refer to the Markert and Moller paper (*18*). Thus we may observe that the 1967 conference was on the verge of a consensus regarding the significance of fetal isozymes in hepatoma cells but the synthesis of ideas was never quite achieved. Although not explicitly relating isozymes and blocked differentiation, Farina *et al.* (*9*) nevertheless supplied highly significant data on carbohydrate isozymes and related them to glycolysis which was, in turn, related to stages of differentiation in connection with malignancy. At the same meeting, Potter (*23*) suggested that " the controls [for cell division] that act upon the differentiating cells at the appropriate time are unable to act upon their targets in cells that have undergone the malignant transformation." Referring to the " new " forms of isozymes found by the Weinhouse group (*9*) in hepatomas but not in adult liver, the question was raised by Potter (*23*) " whether the enzymes are indeed new or novel or whether they represent various levels of differentiation of the developing hepatocytes." There is a good possibility that a new and novel enzyme will be found eventually in a tumor line, but in every instance the question will have to be asked whether it is merely a derepressed part of a normal genome for the species or whether it is truly novel. At the time of the 1967 Symposium, I did not fully appreciate the possible role of isozymes in the biochemistry of cancer, although I emphasized

TABLE I. Ontogeny and Oncogeny

Oncology = the study of tumors	Ontology = the study of beings
Oncogeny = the formation of tumors	Ontogeny = the formation of beings
Oncogeny as a "locked-in" stage of ontogeny	Potter (24)
Cancer as "disdifferentiation"	Matsushima et al. (19)
Cancer as "unbalanced retrodifferentiation"	Uriel (50)
"Cancer tends to reproduce the fetal state"	Schapira et al. (33)

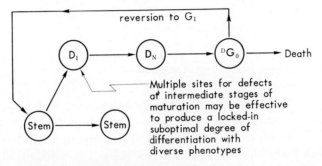

FIG. 1. Oncogency as a locked-in stage of ontogeny

We interpret Uriel's "retrodifferentiation" (50) as a reversion of differentiated G_0 cell ($^DG_0$) to a stem cell, followed by a failure in the differentiation process somewhere between the first stage (D_1) and the Nth stage (D_N), at which the ability of the cell to divide becomes responsive to organismic controls. Ontogeny is here defined as an unfolding of "environmental-response systems" for modulating gene expression according to a programmed time schedule. Oncogeny is here defined as an accumulation of locked-in gene configurations of availability that were scheduled to be unlocked at different times in normal ontogeny. In view of the findings by Katunuma et al. (this monograph) and Ichihara and Ogawa (this monograph), it must be asked whether genes that can be traced through phylogeny are unlocked at different times in ontogeny and sometimes randomly locked into an active state in oncogeny even though they are inactive in the normal adult.

the importance of studies on fetal and newborn liver as material for comparison with hepatoma tissues (23). Between the 1967 Symposium and the Canadian Cancer Conference of June 1968 (24), the highly significant paper by Matsushima et al. (19) appeared, and I realized that the Morris hepatomas should be compared with the whole gamut of changes that occur in fetal and newborn liver and that isozymes might be the key for distinguishing between fetal phenotypes and adult phenotypes, although quantitative changes in various enzyme activities had independently led to the conclusion that "oncogeny is blocked ontogeny" (24). Similar but less explicit views were reached almost simultaneously by Sugimura and co-workers (19, 42, 43, 44), Schapira et al. (33), and Uriel and co-workers (38, 49) (Table I, Fig. 1). The progress in cancer biochemistry has been especially rapid in view of the fact that the minimal deviation concept and the discovery of the first member of the series of Morris hepatomas dates only from 1960, when the first paper by Potter et al. (26) appeared. Moreover, it has only been since 1967 that the Morris hepatomas with 42 chromosomes have been identified (21).

TABLE II. Variants of the Blocked Ontogeny Hypothesis in Historical Perspective

I. "Cohnheim Alternative"–1880–Embryonic Rest Theory:
   1. Fetal liver cells in adult liver replace hepatocytes in presence of hepatocarcinogen and exhibit fetal liver isozymes.
   2. When carcinogen is withdrawn, normal ontogeny is recapitulated by most liver cells, but a few have been converted to hepatoma cells with diverse phenotypes, including some that exhibit fetal liver isozymes.

II. "Durante Alternative"–1874–Reversion of Mature Cells to Embryonic Stage:
   1. Many adult hepatocytes reactivate fetal genome in presence of hepatocarcinogen and exhibit fetal liver isozymes.
   2. Same as in "Cohnheim Alternative."

III. Alternatives may not be mutually exclusive.

Now it is clear that the "fetal isozyme" approach is converging with the "fetal antigen" studies (*38*, *48*) and that isozyme profiles of Morris hepatomas need to be compared not only with those of fetal and newborn liver, but also with precancerous liver, following the leads of Uriel and co-workers (*38*, *49*), and Endo et al. (*8*). The latter described the increase of a muscle-type liver aldolase in precancerous liver of rats fed 3′-methyl-4-dimethylaminoazobenzene (3′-methyl DAB). Although they referred to the enzyme as muscle-type aldolase it is clear that this is the form of the enzyme that predominates in fetal liver and, in the context of the blocked ontogeny hypothesis, it is also a fetal liver-type aldolase. It was called muscle aldolase for largely historical reasons (*24*) but is now referred to as aldolase A. The appearance of aldolase A in precancerous livers when no hepatomas are present is strong evidence that a reactivation of genes for this enzyme has occurred, and the occurrence of the enzyme in some hepatomas is evidence for the blocked ontogeny hypothesis. What is not established is whether certain cells of fetal type preexist in the liver and multiply to replace dying hepatic cells in the precancerous liver, becoming blocked at some stage as they recapitulate ontogeny, which we label the "Cohnheim alternative", or whether a substantial fraction of hepatocytes can revert to the fetal type under the influence of a carcinogen, which we propose to call the "Durante alternative" (Table II). In either case, blocked ontogeny hypothesis proposes that, while most of the cells recover to become normal hepatocyptes, a number of cells become blocked at various stages short of complete differentiation and develop into clones of hepatoma cells. The present report represents the beginning of an attempt to choose between these two alternatives. We hope later to look for noncarcinogenic stimuli that might produce reactivations of the fetal phenotype without cell replication. Such experiments might permit us to choose between the above alternatives.

Hepatoma Studies

Table III shows the enzyme activities of three Morris hepatomas compared with the activities found in normal adult liver and fetal liver (17 days). Hepatoma 9618A has a near normal pyruvate kinase activity and also has glucokinase activity.

TABLE III. Enzyme Activities in Normal Liver and Minimal Deviation Hepatomas

Tissue	Enzyme activity (μmoles/min/g tissue)				FDP/F1P ratio
	Pyruvate kinase	Hexokinase	Glucokinase	FDP-Aldolase	
Liver (adult)	54.6	0.12	0.75	3.60	1.00
Liver (fetal)	16.4	0.38	0	1.30	43.5
9618A	40.0	0.22	0.24	3.38	1.75
7800	10.0	0.27	0	3.14	4.10
5123C	8.5	0.17	0	3.02	2.50

Hepatoma tissue was obtained from several animals and pooled for the assays. Adult liver values are the mean values for animals maintained on a 30% protein diet and killed just before the controlled feeding period. Fetal liver was pooled from 17-day fetuses from a single litter.

Hepatomas 7800 and 5123C have much lower pyruvate kinase activities, and glucokinase could not be detected in these cells. All three hepatomas have slightly elevated hexokinase activities and near normal fructose diphosphate (FDP)-aldolase activities, although the FDP/F1P ratio is greater than one in all three cases, indicating increased production of aldolase A. Thus, in many respects the hepatomas resemble the fetal tissue rather than the adult liver. Similar results have been obtained by many workers (1, 9, 11, 13–15, 19, 32–34, 43, 51, 54).

Precancerous Liver

Experiments with precancerous livers using a new protocol were begun by Sneider and Potter (36) in 1968. Diets containing 3′-methyl-DAB at a level of 0.05% were provided by Drs. J. A. and E. C. Miller and were fed during only 8 hr of each day, using the controlled feeding schedule described as " 8+16 " (25, 26). Survival was very satisfactory, in contrast to the 100% deaths at 10 to 20 days when diets containing 0.06% of the dye were used (36). Animals were killed at weekly intervals from 0 to 11 weeks and somewhat more frequently in a second report (35). Using a variety of parameters, both studies indicated a crisis at 2 to 3 weeks followed by partial recovery in enzyme patterns (36) and distribution of cell types as judged by isolation of two types of nuclei (35).

The present experiments followed the same controlled feeding schedule and employed a basal diet and a diet containing 0.05% 3′-methyl-DAB, described earlier (36). Animals were killed at weekly intervals during the first 5 weeks only. Later, experiments over longer periods will be reported (see Addendum).

Isozyme Studies

Experiments were carried out for the first 5 weeks of exposure to 0.05% 3′-methyl-DAB. Twenty-seven male Sprague-Dawley rats were trained to the " 8+16 " feeding regimen on the basal diet for a period of 1 week (36). Fifteen rats were shifted to the dye-containing diet at day 0 (Fig. 2), and 3 rats were sacrificed for analysis. Groups of 3 rats were killed at weekly intervals from the

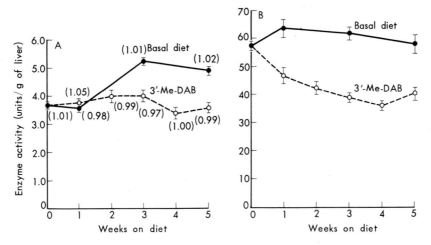

FIG. 2. Effect of feeding 3′-methyl-DAB on (A) rat liver aldolase and (B) rat liver total pyruvate kinase activity

Aldolase was assayed by the method of Blostein and Rutter (4) using FDP and F1P as substrates. Pyruvate kinase was assayed by the method of Bücher and Pfleiderer (2, 6) in the presence of 0.5 mM FDP. Each point is the mean of three determinations ±SE.

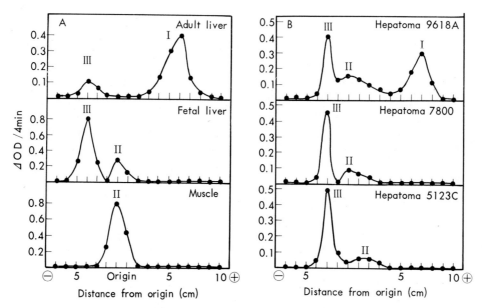

FIG. 3. Pyruvate kinase isozyme patterns of (A) normal liver, fetal liver, and muscle, and (B) Morris hepatomas 9618A, 7800, and 5123C

Starch block zonal electrophoresis was carried out as described by Tanaka et al. (47) in 0.25 M sucrose containing 75 mM Tris-HCl buffer (pH 8.4), 2.5 mM $MgSO_4$, 0.5 mM FDP, and 0.5 mM dithiothreitol at 250 V for 18 hr. The designations I, II, and III correspond to L, M_1, and M_2, as employed by Tanaka et al. (47).

dye-fed group and groups of 3 rats at 1, 3, and 5 weeks for the basal diet controls.

Figure 2A shows the results for FDP aldolase activity in the dye-fed and control groups, as well as the FDP/F1P activity ratios. Endo et al. (8) reported changes in the ratio of activities at 60 days but did not report data on animals killed at this early period. We will report later on animals killed at later times (see Addendum).

Figure 2B shows data on total pyruvate kinase activity in the same animals. A steady decline in activity occurred until the end of the fourth week, but then a return toward normal began. This slight rise is borne out by more detailed studies on pyruvate kinase isozymes.

Figure 3A shows electrophoretic patterns for pyruvate kinase isozymes using starch block separations in the presence of FDP. Findings by Suda et al. (41), Tanaka et al. (46, 47), and Susor and Rutter (45) for liver and muscle appear to be confirmed. No direct comparison can be made with the electrofocusing studies reported by Criss (7). In view of findings in fetal liver, it seems undesirable to designate isozymes according to their adult tissue distribution and we have therefore referred to the 3 peaks as I, II, and III. However, we should point out that these peaks correspond respectively to L, M_1 and M_2, as described by Tanaka. Figure 3A shows peaks I and III in normal adult liver, peaks II and III in fetal liver, and peak II in normal adult muscle. Peak III in liver is not identical with peak II in muscle.

Figure 3B shows pyruvate kinase patterns for the three Morris hepatomas. Hepatoma 9618A clearly has all three peaks, while hepatomas 7800 and 5123C have prominent activity in peak III, some activity in II, and none in I. It appears, therefore, that the enzyme isolated and characterized from hepatoma 3924A by

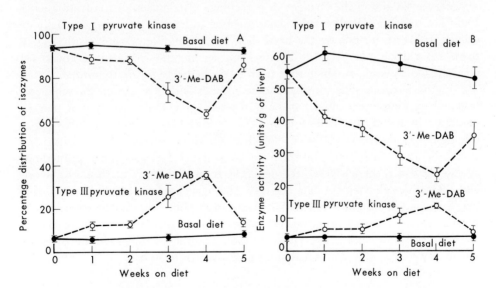

Fig. 4. Effect of feeding 3'-methyl-DAB on rat liver pyruvate kinase isozymes
A: Distribution of the isozymes as a percentage of the activity recovered from the block; B: contributions of the isozymes to the total pyruvate kinase activity. Each point is the mean of three determinations ±SE.

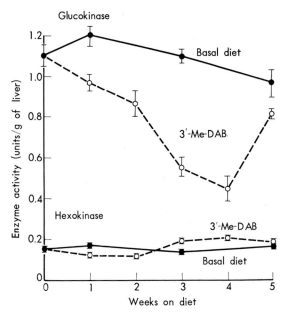

Fig. 5. Glucokinase and hexokinase in control and precancerous livers
Samples as in Figs. 2A and 2B. Enzyme activities determined as described by Walker and Parry (*52*).

Taylor *et al.* (*48*) is in fact the type III enzyme normally present in adult liver and not a new form of the enzyme. The same separation technique was applied to all of the samples for which the total pyruvate kinase activity was reported in Fig. 2B.

Figure 4A shows the changes in percentage distribution of pyruvate kinases I and III, and Fig. 4B shows the data in actual units. There was essentially no change in liver from rats on the basal diet, but in rats ingesting 3′-methyl-DAB there was a decline in pyruvate kinase I and an increase in pyruvate kinase III until the fourth week, at which time both forms began to return to normal.

Figure 5 reports data for glucokinase and hexokinase assays on the same animals. Hexokinase remained constant, whereas glucokinase showed a marked reduction in activity until the fourth week, followed by a significant upswing at this time. Comparison of Fig. 5 with Fig. 4B shows a marked parallelism between the glucokinase data and the pyruvate kinase I data and, when plotted one against the other, there is very little deviation from a straight line through the origin. Thus the glucokinase data reinforce the pyruvate kinase data in suggesting that a change in response to the dye occurred at about the fourth week. Subsequent studies referred to in the Addendum showed that this was a time of considerable variability and in experiments of longer duration the curves for pyruvate kinase and glucokinase continued the trend away from normal.

Studies on Livers with Prelabeled DNA

Changes in the isozymes pattern in precancerous liver and in hepatomas suggest that the availability of various parts of the genotype may be altered.

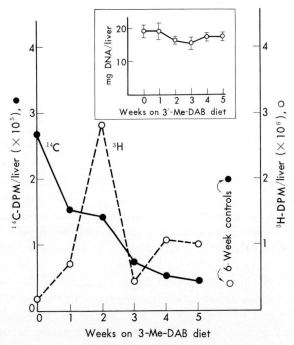

FIG. 6. Turnover of hepatic DNA in rats fed 3'-methyl-DAB

Immature (50 g, 25-day-old) male Sprague-Dawley rats were injected with thymidine-[methyl-^{14}C] (0.2 mCi, 8.1 μmoles/kg) to label their hepatic DNA. When animals reached 200 g (53 days old), they were placed on a diet containing 0.05% 3'-methyl-DAB. Groups of animals were killed at weekly intervals over a period of 5 weeks (one group of animals was maintained on a control diet for 6 weeks). Two hours prior to sacrifice, each animal was injected with thymidine-[methyl-^{3}H] (0.5 mCi, 0.07 μmole/kg). Liver homogenates were precipitated with 10% trichloroacetic acid (TCA) (final concentration), washed once with 10% TCA, and extracted with 95% ethanol containing 10% potassium acetate (39), 100% ethanol, chloroform-methanol (1:2), and ether. DNA was extracted from the lipid-extracted nucleic acid-protein precipitate as previously described for nuclei (37). The amount of DNA (see inset) was determined as previously described (3) while the radioactivity in the DNA was determined by liquid scintillation techniques. Each point represents the mean value obtained from 3 rats.

Studies paralleling those reported above were undertaken to determine the stability of prelabeled DNA in liver of rats maintained on a diet containing 0.05% of the hepatic carcinogen 3'-menthyl-DAB. Rat liver DNA was prelabeled with ^{14}C-thymidine and the rate of DNA synthesis was estimated by measuring the incorporation of ^{3}H-thymidine into DNA during a 2-hr period. The results of these experiments are shown in Fig. 6. There was a marked and progressive decrease in ^{14}C-DNA from the beginning, and after 5 weeks on the 3'-methyl-DAB diet the rats had lost 80% of their prelabeled DNA. The incorporation of ^{3}H-thymidine into hepatic DNA during a 2-hr pulse was always greater in the animals on the carcinogen diet than in the zero time and 6-week controls, indicating that the administration of 3'-methyl-DAB results in an increased rate of DNA synthesis, especially at the end of the second week. The fact that there was a loss in prelabeled DNA accompanied by an increase in the rate of DNA synthesis while total liver weight and mg DNA/liver remained constant over the course of the experiment (Fig. 6, inset) demonstrates that an initial effect of 3'-methyl-DAB is to cause a profound increase in turnover of hepatic DNA. From the data in Fig. 6, it can be calculated that the half-life of hepatic DNA in the animals on the 3'-methyl-DAB diet is 13 days. In this experiment, only a small number of cells contained ^{14}C-labeled DNA. Further experiments with more extensive labeling have now been done (see Addendum).

The degree of necrosis seen in histological sections appears insufficient to account for the loss of DNA observed. Therefore, it is suggested that this early increase in DNA turnover may be due in part to repair synthesis in response to covalent interaction of metabolites of 3'-methyl-DAB with DNA.

Radioactive metabolites of azo dye carcinogens have been found to bind covalently to rat liver DNA (53). Miller et al. (20) have shown that nonsemiconservative DNA synthesis (repair synthesis) occurs in cultured HeLa cells after treatment with mono- and difunctional alkylating agents. Recently, Stich et al. (40) have obtained a correlation between the oncogenicity of derivatives of 4-nitroquinoline 1-oxide and the capacity of the compounds to provoke DNA repair synthesis (as measured by autoradiographic demonstration of unscheduled incorporation of ^3H-thymidine) in Syrian hamster cells in tissue culture.

The induction of β-galactosidase in E. coli was impaired after in vivo alkylation of the cellular DNA by mono-and difunctional alkylating agents (5). It is possible that this is a result of the attempt of the cells to repair the damage rather than the alkylation per se. During the DNA repair process, single-strand breaks occur as the damaged segment is being excised (10), and it has been postulated (31) that genetic markers on pneumococcal DNA can be inactivated by a single-strand break.

DISCUSSION

Our data are consistent with the current literature on interaction between carcinogens and DNA, and suggest that repair synthesis of hepatic DNA is an early consequence of 3'-methyl-DAB ingestion. Since we might expect gene transcription to be altered during DNA repair, it is possible that this is one of the molecular mechanisms underlying the changes in gene expression observed in precancerous livers, but further experiments will be required to show whether DNA breakdown is an obligatory concomitant of the enzyme changes. Indeed, we might propose that DNA alteration in some cell or cells might be required for hepatoma production but not for the phenotypic changes (cf. Table II).

In future experiments we can examine the changes in the phenotypic expression of various isozymes in livers of animals exposed to carcinogens in the light of evidence for DNA damage and attempt to design experiments that will indicate whether the Cohnheim alternative or the Durante alternative or both may be applicable. If it proves possible to produce isozyme changes that result in phenotypic expressions seen in hepatomas and in fetal liver without DNA breakdown, repair, or new synthesis, and without permanent alteration (*i.e.*, without cancer), or if there are noncarcinogenic substances or procedures (*e.g.*, partial hepatectomy) that produce the phenotypic changes and DNA synthesis without DNA breakdown, it may be possible to separate the processes of "promotion" from the process of "initiation." In this context, the blocked ontogeny hypothesis can be studied not only in relation to the Cohnheim and Durante alternative, *i.e.*, a change in residual embryonic cells *vs.* a change in reverted cells, but also in relation to the somatic mutation theory *vs.* "misprogramming without mutation," a theory that might be implied from remarks by Markert, by Jacob and Monod, and by Pitot and Heidelberger. It should be noted that in using the term "misprogramming," Markert (*17*) did not say "without mutation" and that neither Jacob and Monod nor Pitot and Heidelberger predicted that carcinogenesis would be shown not to involve a somatic mutation (*21*). The experimental approaches to these problems now seem ripe for a concerted and cooperative attack on the fundamental requirements for the malignant transformation. The studies on chemical carcinogenesis herein reported can easily be extended to viral transformations and to cells in tissue culture, thus bringing about a coordinated attack on the problem instead of a series of isolated studies.

Addendum

Since this report was submitted additional experiments on the same overall problem have been carried out and reported by P. R. Walker and V. R. Potter in "Isozyme studies on adult, regenerating, precancerous, and developing liver in relation to findings in hepatomas" [*Adv. Enzyme Regulation*, **10**, 339-364 (1972)] and by Jay I. Goodman and V. R. Potter in "Evidence for DNA repair synthesis and turnover in rat liver following ingestion of 3'-methyl-4-dimethylaminoazobenzene" [*Cancer Res.*, **32**, 766–775 (1972)].

REFERENCES

1. Adelman, R. C., Morris, H. P., and Weinhouse, S. Fructokinase, triokinase, and aldolases in liver tumors of the rat. *Cancer Res.*, **27**, 2408–2413 (1967).
2. Bailey, E. and Walker, P. R. A comparison of the properties of the pyruvate kinases of the fat body and flight muscle of the adult male desert locust. *Biochem. J.*, **111**, 359–364 (1969).
3. Blobel, G. and Potter, V. R. Distribution of radioactivity between the acid soluble pool and the pools of RNA in the nuclear, nonsedimentable and ribosome fractions of rat liver after a single injection of labeled orotic acid. *Biochim. Biophys. Acta*, **166**, 48–57 (1968).
4. Blostein, R. and Rutter, W. J. Comparative studies on liver and muscle aldolase. *J. Biol. Chem.*, **238**, 3280–3285 (1963).
5. Brookes, P., Lawley, P. D., and Venitt, S. Nature of alkylation lesions and their repair: significance for ideas on mutagenesis. *Ciba Found. Symp. Mutation as Cellular Process*, 138–154 (1969).
6. Bücher, T. and Pfleiderer, G. Pyruvate kinase from muscle. *Methods Enzymol.* **1**, 435–440 (1955).

7. Criss, W. E. A new pyruvate kinase isozyme in hepatomas. *Biochem. Biophys. Res. Commun.*, **35**, 901–905 (1969).
8. Endo, H., Eguchi, M., and Yanagi, S. Irreversible fixation of increased level of muscle type aldolase activity appearing in rat liver in the early stage of hepatocarcinogenesis. *Cancer Res.*, **30**, 743–752 (1970).
9. Farina, F. A., Adelman, R. C., Lo, C. H., Morris, H. P., and Weinhouse, S. Metabolic regulation and enzyme alteration in the Morris hepatomas. *Cancer Res.*, **28**, 1897–1900 (1968).
10. Ginsberg, B., Yudelevich, A., Sadowski, P., Feiner, L., and Hurwitz, J. Biochemical mechanisms involved in breakage and repair of DNA. *In* " Exploitable Molecular Mechanisms and Neoplasia," Williams and Wilkins Co. Baltimore, pp. 373–394 (1969).
11. Haruno, K. Changes in the glutaminase activity of liver tissue from rats during the development of hepatic tumors by carcinogen feeding. *GANN*, **47**, 231–236 (1956).
12. Ikawa, Y. Isozyme histochemistry, especially in relation to malignant transformation and differentiation. I. Histochemical demonstration of lactic dehydrogenase isozymes. *GANN*, **56**, 201–217 (1965).
13. Knox, W. E., Jamdar, S. C., and Davis, P. A. Hexokinase, differentiation and growth rates of transplanted rat tumors. *Cancer Res.*, **30**, 2240–2244 (1970).
14. Linder-Horowitz, M. Changes in glutaminase activities of rat liver and kidney during pre- and post-natal development. *Biochem. J.*, **114**, 65–70 (1969).
15. Linder-Horowitz, M., Knox, W. E., and Morris, H. P. Glutaminase activities and growth rates of rat hepatomas. *Cancer Res.*, **29**, 1195–1199 (1969).
16. Markert, C. L. Epigenetic control of specific protein synthesis in differentiating cells. *In* " Cytodifferentiation and Macromolecular Synthesis," ed. by M. Locke, Academic Press Inc., New York, pp. 65–84 (1963).
17. Markert, C. L. Neoplasia: a disease of cell differentiation. *Cancer Res.*, **28**, 1908–1914 (1968).
18. Markert, C. L. and Moller, F. Multiple forms of enzymes: tissue, ontogenic, and species specific patterns. *Proc. Natl. Acad. Sci. U.S.*, **45**, 753–763 (1959).
19. Matsushima, T., Kawabe, S., Shibuya, M., and Sugimura, T. Aldolase isozymes in rat tumor cells. *Biochem. Biophys. Res. Commun.*, **30**, 565–570 (1968).
20. Miller, J. A., Lin, J. K., and Miller, E. C. N-(Guanosin-8-yl)- and N-(deoxyguanosin-8-yl)-N-methyl-4-aminoazobenzene: degradation products of hepatic RNA and DNA from rats administered N-methyl-4-aminoazobenzene. *Proc. Am. Assoc. Cancer Res.*, **11**, 56 (1970).
21. Nowell, P. C., Morris, H. P., and Potter, V. R. Chromosomes of " minimal deviation " hepatomas and some other transplantable rat tumors. *Cancer Res.*, **27**, 1565–1579 (1967).
22. Potter, V. R. Biochemical perspectives in cancer research. *Cancer Res.*, **24**, 1085–1098 (1964).
23. Potter, V. R. Summary of discussion on neoplasia. *Cancer Res.*, **28**, 1901–1907 (1968).
24. Potter, V. R. Recent trends in cancer biochemistry: the importance of studies on fetal tissue. *Can. Cancer Conf.*, **8**, 9–30 (1969).
25. Potter, V. R. Environmentally induced metabolic oscillations as a challenge to tumor autonomy. *Miami Winter Symp.*, **2**, 291–313 (1970).
26. Potter, V. R., Baril, E. F., Watanabe, M., and Whittle, E. D. Systematic oscil-

lations in metabolic functions in liver from rats adapted to controlled feeding schedules. *Fed. Proc.*, **27**, 1238–1245 (1968).
27. Potter, V. R., Pitot, H. C., Ono, T., and Morris, H. P. The comparative enzymology and cell origin of rat hepatomas. I. Deoxycytidylate deaminase and thymine degradation. *Cancer Res.*, **20**, 1255–1261 (1960).
28. Potter, V. R. and Watanabe, M. Some biochemical essentials of malignancy: the challenge of diversity. *In* "International Symposium on Leukemia—Lymphoma," ed. by C. J. D. Zarafonetis, Lea and Febiger, Philadelphia, pp. 33–46 (1968).
29. Reynolds, R. D. and Potter, V. R. Neonatal rise of rat liver tyrosine aminotransferase and serine dehydratase activity associated with lack of food. *Life Sci.*, **10** (Part II), 5–10 (1971).
30. Roberts, J. J., Crathorn, A. R., and Brent, T. P. Repair of alkylated DNA in mammalian cells. *Nature*, **218**, 970–972 (1968).
31. Rosenthal, P. N. and Fox, M. S. Effects of disintegration of incorporated ^{3}H and ^{32}P on the physical and biological properties of DNA. *J. Mol. Biol.*, **54**, 441–463 (1970).
32. Sato, S., Matsushima, T., and Sugimura, T. Hexokinase isozyme patterns of experimental hepatomas in rats. *Cancer Res.*, **29**, 1437–1446 (1969).
33. Schapira, F., Reuber, M. D., and Hatzfeld, A. Resurgence of two fetal-types of aldolases (A and C) in some fast-growing hepatomas. *Biochem. Biophys. Res. Commun.*, **40**, 321–327 (1970).
34. Shatton, J. B., Morris, H. P., and Weinhouse, S. Kinetic, electrophoretic, and chromatographic studies on glucose-ATP phosphotransferases in rat hepatomas. *Cancer Res.*, **29**, 1161–1172 (1969).
35. Sneider, T. W., Bushnell, D. E., and Potter, V. R. The distribution and synthesis of DNA in two classes of rat liver nuclei during azo dye-induced hepatocarcinogenesis. *Cancer Res.*, **30**, 1867–1873 (1970).
36. Sneider, T. W. and Potter, V. R. Deoxycytidylate deaminase and related enzymes of thymidine triphosphate metabolism in hepatomas and precancerous rat liver. *Adv. Enzyme Regulation*, **7**, 375–394 (1970).
37. Sneider, T. W. and Potter, V. R. Methylation of mammalian DNA: studies on Novikoff hepatoma cells in tissue culture. *J. Mol. Biol.*, **42**, 271–294 (1969).
38. Stanislawski-Birencwajg, M., Uriel, J., and Grabar, P. Association of embryonic antigens with experimentally induced hepatic lesions in the rat. *Cancer Res.*, **27**, 1990–1997 (1967).
39. Steel, W. J., Okamura, N., and Busch, H. Prevention of loss of RNA, DNA and protein into lipid solvents. *Biochim. Biophys. Acta*, **87**, 490–492 (1964).
40. Stich, H. F., San, R. H. C., and Kawazoe, Y. DNA repair synthesis in mammalian cells exposed to a series of oncogenic and non-oncogenic derivatives of 4-nitroquinoline 1-oxide. *Nature*, **229**, 416–419 (1971).
41. Suda, M., Tanaka, T., Sue, F., Harano, Y., and Morimura, H. Dedifferentiation of sugar metabolism in the liver of tumor-bearing rat. *GANN Monograph*, **1**, 127–141 (1966).
42. Sugimura, T. Decarcinogenesis, a newer concept arising from our understanding of the cancer phenotype. *In* "Chemical Tumor Problems," ed. by W. Nakahara, Japan Society for the Promotion of Science, Tokyo, pp. 269–284 (1970).
43. Sugimura, T., Matsushima, T., Kawachi, T., Hirata, Y., and Kawabe, S. Molecular species of aldolases and hexokinases in experimental hepatomas. *GANN Monograph*, **1**, 143–149 (1966).

44. Sugimura, T., Sato, S., and Kawabe, S. The presence of aldolase C in rat hepatoma. *Biochem. Biophys. Res. Commun.*, **39**, 626–630 (1970).
45. Susor, W. A. and Rutter, W. J. Some distinctive properties of pyruvate kinase purified from rat liver. *Biochem. Biophys. Res. Commun.*, **30**, 14–20 (1968).
46. Tanaka, T., Harano, Y., Morimura, H., and Mori, R. Evidence for presence of two types of pyruvate kinase in rat liver. *Biochem. Biophys. Res. Commun.*, **21**, 55–60 (1965).
47. Tanaka, T., Harano, Y., Sue F., and Morimura, H. Crystallization, characterization, and metabolic regulation of two types of pyruvate kinase isolated from rat tissues. *J. Biochem. (Tokyo)*, **62**, 71–91 (1967).
48. Taylor, C. B., Morris, H. P., and Weber, G. A comparison of the properties of pyruvate kinases from hepatoma 3924A, normal liver and muscle. *Life Sci.*, **8**, 635–644 (1969).
49. Uriel, J. Les isoenzymes. *Ann. Nutr. Aliment.*, **21**, B67–80 (1967).
50. Uriel, J. Transitory liver antigens and primary hepatoma in rat and man. *Pathol. Biol.*, **17**, 877–884 (1969).
51. Watanabe, M., Potter, V. R., and Morris, H. P. Benzpyrene hydroxylase activity and its induction by methylcholanthrene in Morris hepatomas, in host livers, in adult livers, and in rat liver during development. *Cancer Res.*, **30**, 263–273 (1970).
52. Walker, D. G. and Parry, M. J. Glucokinase in liver. *Methods Enzymol.*, **9**, 381–388 (1966).
53. Warwick, G. P. and Roberts, J. J. Persistent binding of butter yellow metabolites to rat liver DNA. *Nature*, **213**, 1206–1207 (1967).
54. Wu, C. and Morris, H. P. Responsiveness of gultamine-metabolizing enzymes in Morris hepatomas to metabolic modulations. *Cancer Res.*, **30**, 2675–2684 (1970).

ABNORMAL GENE EXPRESSION ON THE MODE OF AMINO NITROGEN EXCRETION IN RAT HEPATOMAS FROM PHYLOGENIC ASPECTS

Nobuhiko KATUNUMA, Yasuhiro KURODA, Yoshiko MATSUDA, and Keiko KOBAYASHI

*Department of Enzyme Chemistry, Institute for Enzyme Research, School of Medicine, Tokushima University**

There have been several valuable studies on abnormal gene expression in cancer cells from the ontogenic aspect, but no such studies have been published from the phylogenic aspect. It is well known that chickens and rats are respectively, uricoteric and ureotelic animals. Using this phenotypic difference as a marker, the authors carried out the following studies.

The incorporation rates of L-serine[U-^{14}C] into uric acid or allantoin in chicken liver, adult rat liver, regenerating rat liver, and fetal rat liver slices, and three kinds of hepatoma cells were compared. The radioactive carbon in L-serine[U-^{14}C] was efficiently incorporated into the carbons of uric acid at positions 4 and 5 by chicken liver slices, but little incorporation into uric acid and allantoin by rat liver was observed. On the other hand, two lines of rat ascites hepatoma, AH-130 and AH-7974, which have no urea cycle, showed the same degree of incorporation of L-serine[U-^{14}C] into uric acid as chicken liver slices. Morris 5123D, which has a partial urea cycle, showed an intermediate degree of incorporation.

Ornithine transcarbamylase, one of the most important enzymes of urea synthesis, was not detected in adult and newborn chick livers using L-ornithine[U-^{14}C], but considerable activity of this enzyme was detected at some stages of chick embryo liver development.

Some of the different phenotypes in phylogenetically different animal lines depend on the differences in gene expression of genotypes acquired by each animal line in early stages of evolution. Abnormal gene expressions may exist in some hepatomas arising not only from the ontogenic aspect, but also from the phylogenic aspect.

There have been several reports on abnormal gene expression from ontogenic aspects using organ-characteristic isozymes in hepatomas. We also reported the existence of abnormal gene expression in Morris hepatomas at different growing rates and in the AH-130 hepatoma using abnormal differentiation of organ-specific glutaminase isozymes (2-4). Kidney-type and brain-type glutaminase, which serve as indicators of abnormal differentiation of the organ-specific isozyme, were observed in some hepatomas, but their magnitudes had no direct relation-

* Kuramoto-cho 3-18-1, Tokushima 770, Japan (勝沼信彦, 黒田泰弘, 松田佳子, 小林圭子).

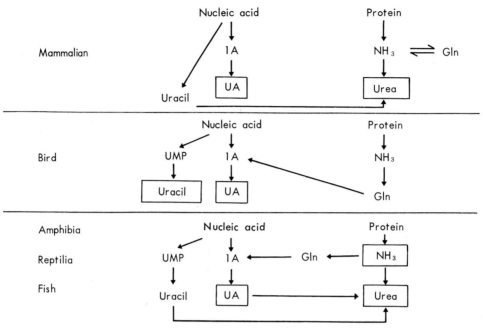

Fig. 1. Comparison of nitrogen excretion mechanisms from phylogenic aspects
IA, inosine 5-phosphate; Gln, glutamine; UA, uric acid.

ship to the growth rates of various hepatomas. That is, there have been several studies on abnormal gene expression in cancer cells from the ontogenic aspect which agree with the hypothesis for the explanation of the phenomenon by Potter (6), but no studies have been published on abnormal gene expression in cancer cells from the phylogenic aspect. It is well known that chickens and rats are uricotelic and ureotelic animals, respectively. Using this phylogenic phenotypic difference as a marker, we propose that abnormal gene expressions may exist in some hepatomas not only from the ontogenic aspect, but also from the phylogenic aspect. There are many kinds of metabolic regulation of purine biosynthesis that are different in mammals and birds. Excreted nitrogen from protein and nucleic acid catabolism was separated for mammals but not for birds as shown in Fig. 1.

A change from a low-protein (5% casein) to high-protein (50% casein) diet for both animals resulted in a remarkable increase in serum uric acid in chickens, but no increases were recognized in rats (Fig. 2).

Figure 3 shows that ^{14}C-radioactivity of L-serine[U-^{14}C] is strongly incorporated into uric acid in chicken liver slices and hepatoma AH-130 cell suspension while almost no incorporation into allantoin and uric acid was observed in adult rat liver slices. The same degree of incorporation was obtained using ^{14}C- glycine under the same experimental conditions. The 1/3 moiety of total radioactivity of uric acid formed was incorporated into C_4 and C_5 positions as shown in Fig. 4.

It is well known that carbon at positions 4 and 5, and nitrogen at position 7 of uric acid are derived from glycine. The incorporation rates of L-serine[U-^{14}C] into uric acid or allantoin in chicken liver, adult rat liver, regenerating liver, fetal

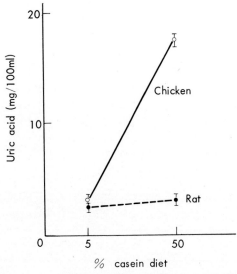

Fig. 2. Changes in serum uric acid concentration induced by a high protein diet

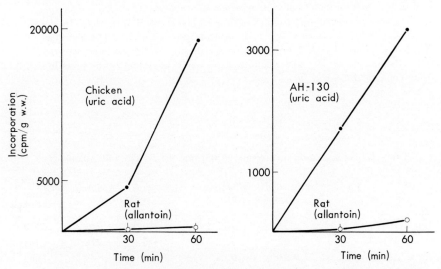

Fig. 3. Incorporation of L-serine[U-^{14}C] into uric acid and allantoin in chicken and rat liver slices and rat hepatoma cell suspension

liver slices, and three kinds of hepatoma were compared. The radioactivity of L-serine[U-^{14}C] was efficiently incorporated into uric acid in chicken liver slices, hepatomas AH-130 and AH-7974, and Morris hepatoma 5123D, but there was little incorporation into uric acid or allantoin in adult rat liver, regenerating rat liver, and fetal rat liver slices.

Table I shows the following important evidence. Two lines of rat ascites hepatoma, AH-130 and AH-7974, which have no urea cycle, showed the same degree

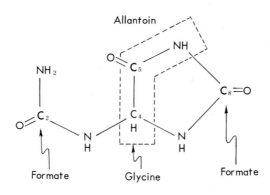

	Uric acid and allantoin (cpm)
Total	5640
$C_4 + C_5$	1950

FIG. 4. Distribution of radioactive carbon in uric acid and allantoin molecules
^{14}C-Glycine hydrolyzed from ^{14}C-uric acid, synthesized with conc. HCl at 160° for 16 hr was isolated and hydrolyzed to glyoxalate (8). The glyoxalate was precipitated by HCl-semicarbazide specifically. Uric acid was converted to allantoin by uricase, and allantoin was assayed according to the procedure of Young and Conway (9).

TABLE I. Incorporation of L-serine[U-^{14}C] into Uric Acid and Allantoin

	Uric acid	Allantoin	Uric acid	Allantoin	Specific count (cpm/μmole)
	(μmole/hr/g)		(cpm/hr/g)		
Chicken liver	2.72	0.25	19,400	740	6,770
Hepatoma AH-130	0.42	0.12	3,420	224	6,750
Hepatoma AH-7974	0.34	0.08	3,330	—	7,940
Morris hepatoma 5123D	0.18	1.43	788	888	1,040
Rat liver	0.27	1.81	0	200	96
Regenerating rat liver	0.47	2.16	0	260	99
Fetal rat liver	—	—	102	261	—

Two ml of Krebs-Ringer bicarbonate solution, 200–1,000 mg of liver slices or hepatoma cells, 0.5 μmole (0.5 μCi) of L-serine[U-^{14}C] and 10 μmoles of L-glutamine were incubated for 60 min at 37° with shaking. The reaction mixture was deproteinized with 25% trichloroacetic acid and ^{12}C-uric acid or ^{12}C-allantoin was added as a carrier to the extracts. The deproteinized supernatant was passed through a column of Dowex 50 form (H$^+$), the effluent was neutralized with 2N NaOH, then uric acid and allantoin were isolated from the extracts (5).
— undetectable ($\dotequal 0$).

of incorporation of L-serine [U-^{14}C] into uric acid as chicken liver slices. Morris hepatoma 5123D, which has a urea cycle, showed an intermediate degree of incorporation. Not only adult rat liver but also regenerating liver and fetal liver, both of which grow rapidly, did not incorporate any radioactivity from L-serine [U-^{14}C] into uric acid or allantoin.

On the problem of uricase, it is known that birds excrete nitrogen as uric acid and have no uricase. Remarkable decreases of uricase were also observed in hepatomas AH-130 and AH-7974, but Morris hepatoma 5123D contained an intermediate degree of uricase. When RNA synthesis in hepatoma AH-7974 was inhibited by Actinomycin-D, the question of how this incorporation would be changed arose. Table II illustrates the explanation that increases of incorpora-

TABLE II. Effects of Actinomycin-D on Incorporation of L-Serine[U-^{14}C] into RNA and Uric Acid by AH-7974 Cell Suspension

Condition	RNA	Uric acid
	(cpm/hr/g wet weight)	
None	904	1,430
	(ca. 113)	(3,300)
Actionmycin-D (2 μg/ml)	294	3,910
Actinomycin-D (4 μg/ml)	357	4,420

The numbers in parentheses represent the specific count.

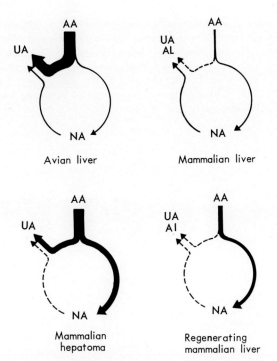

FIG. 5. Uric acid and allantoin formation in various tissues
AA, amino acid; Al, allantoin; UA, uric acid; NA, nucleic acid.

tion of radioactivity into uric acid in hepatomas are not due to RNA. Figure 5 shows these different modes of uric acid excretion.

Ornithine transcarbamylase, one of the most important enzymes of urea synthesis, was not detected in adult and newborn chick liver, but considerable activity of this enzyme was detected in some stages of chick embryo liver, as shown in Fig. 6.

It might be assumed that both genes for uric acid-synthesizing enzymes and urea cycle enzymes were acquired by rat and chicken in an early stage of evolution

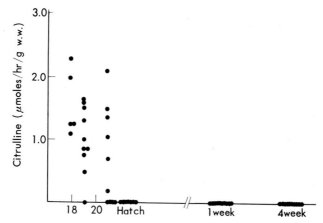

Fig. 6. Changes of ornithine-transcarbamylase activity in chicken liver during development

The liver (200 mg) was homogenized with 1 ml of 0.5% cetyltrimethylammonium bromide solution. The reaction mixture contained 0.4 ml of homogenate, 40 μmoles of L-ornithine, 5 moles of carbamylphosphate, 0.1 μmole of $MnCl_2$, and 0.5 ml of Tris-HCl buffer in a final volume of 1.5 ml. Reaction was carried out for 20 min at 37° and terminated by the addition of 0.5 ml of 10% trichloroacetic acid. Citrulline formed was estimated by the method of Hunninghake and Grisolia.

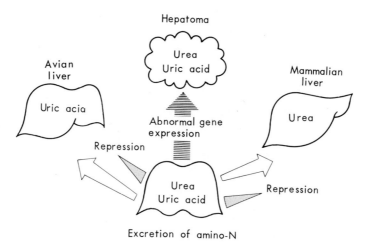

Fig. 7. Abnormal gene expression in hepatoma cells from phylogenic aspects

before the phylogenic differentiation into mammals and avian, and that in mammals generally the expression of the gene for uric acid-synthesizing enzymes was repressed in later evolutionary stages whereas in chickens that for urea-synthesizing enzymes was repressed. From these observations, the following hypothesis is offered for consideration. Different phenotypes in phylogenically different animal lines were acquired in early stages of evolution. Abnormal gene expressions may exist in some hepatomas arising not only from ontogenic but also from phylogenic aspects. Our hypothesis for the above is shown in Fig. 7.

REFERENCES

1. Hunninghake, D. and Grisolia, S. A sensitive and convenient micromethod for estimation of urea, citrulline and carbamyl-derivatives. *Anal. Biochem.*, **16**, 200–205 (1966).
2. Katunuma, N., Tomino, I., and Sanada, T. Differentiation of organ specific glutaminase isozyme during development. *Biochem. Biophys. Res. Commun.*, **32**, 426–432 (1968).
3. Katunuma, N., Kuroda, Y., Sanada, Y., Towatari, T., Tomino, I., and Morris, H. P. Anomalous distribution of glutaminase isozyme in various hepatomas. *Adv. Enzyme Regulation*, **8**, 281–287 (1970).
4. Katunuma, N., Kuroda, Y., Matsuda, Y., and Kobayashi, K. Abnormal gene expression on the mode of amino nitrogen excretion in rat hepatoma from phylogenic aspects. *Proc. 10th Int. Cancer Congr.* (1970) in press.
5. Korn, E. D. Techniques for isotope studies. *Methods Enzymol.*, **4**, 631–632 (1957).
6. Potter, V. R. Enzyme regulation in cancer cells. *Proc. 10th Int. Cancer Congr.*, (1970) in press.
7. Scherrer, K. and Darnell, J. E. Sedimentation characteristics of rapidly labelled RNA from Hela cells. *Biochem. Biophys. Res. Commun.*, **7**, 486–490 (1962).
8. Sonne, J. C., Buchanan, J. M., and Delluva, A. M. Biological precursors of uric acid. *J. Biol. Chem.*, **173**, 69–79 (1948).
9. Young, E. C. and Conway, C. F. On the estimation of allantoin by the Rimini-Schryver reaction. *J. Biol. Chem.*, **142**, 839–853 (1942).

RELATIONSHIP BETWEEN DEGREE OF DIFFERENTIATION AND GROWTH RATE OF MINIMAL DEVIATION HEPATOMAS AND KIDNEY CORTEX TUMORS STUDIED WITH GLUTAMINASE ISOZYMES[*1]

Nobuhiko KATUNUMA, Yasuhiro KURODA, Tasuku YOSHIDA, Yukihiro SANADA, and Harold P. MORRIS[*2]

*Department of Enzyme Chemistry, Institute for Enzyme Research, School of Medicine, Tokushima University,[*3] and the Nutrition and Carcinogenesis Section, Laboratory of Biochemistry, National Institutes of Health*

Glutaminase isozymes in three lines of Morris kidney cortex tumors, eight lines of Morris transplantable hepatomas, and a Yoshida ascites hepatoma growing at different rates were investigated to seek an answer to the question of whether there is any relationship between the growth rate and the isozyme ratio or the distribution of an organ-specific marker isozyme in hepatomas.

All the tumors tested contained both the phosphate-dependent glutaminase (PD) and the phosphate-independent glutaminase (PI) in various ratios. A proportionality was found between the growth rates of the hepatomas and the ratio of the isozyme contents (PD/PI). The specific activity of the glutaminase in all the kidney tumors tested was ten times higher than that in normal kidney, but no relationship was found between the growth rate and the isozyme ratio (PD/PI). The degree of differentiation of the organ-characteristic function in various hepatomas and kidney tumors was studied using organ-specific marker isozymes. In previous papers, kidney-type PI and product inhibition by glutamate are reported to have been observed in fetal liver, and the amount of kidney-type PI decreased gradually after birth (5, 6). Kidney-type PI and product inhibition were not found in adult liver or in any stage of regenerating liver. Kidney-type glutaminase and product inhibition, both of which serve as indicators of the degree of differentiation of the organ-specific isozyme, were observed in some hepatomas, but their magnitudes had no direct relationship to the growth rates of various hepatomas. On the other hand, only a small amount of kidney-type glutaminase and strong product inhibition were recognized in all the kidney tumors tested.

[*1] This work was supported by Grant DRG-959 from the Damon Runyun Memorial Fund.
[*2] Present address: Howard University, College of Medicine, Department of Biochemistry Washington, D.C. 20001, U.S.A.
[*3] Kuramoto-cho 3-18-1, Tokushima 770, Japan (勝沼信彦, 黒田泰弘, 吉田　翼, 真田幸弘).

The authors previously reported that there are two types of glutaminase (L-glutamine amidohydrolase; EC 3.5.1.2) in various organs (7, 8). One isozyme requires phosphate to show activity (phosphate-dependent, PD), while the other does not require phosphate either as a co-factor or as an activator (phosphate-independent, PI), but is activated by maleate or N-acetylglutamate (7). We reported that the PI isozymes in adult kidney and adult liver are entirely different proteins as shown by the purification of each enzyme to a homogeneous protein. The differentiation of these organ-characteristic isozymes poses an important problem. Therefore, the change in the PI isozyme patterns during development of rats has been studied. No kidney-type PI was observed in adult or regenerating liver, but kidney-type PI was found in fetal liver (5, 8). We now have the following hypothesis on the differentiation of organ-specific function; the differentiation of the organ-characteristic function is related to the disappearance of the isozyme which is not inherent to the particular organ. In the present work, these glutaminase isozymes in three lines of Morris kidney cortex tumors, eight lines of Morris hepatomas, and a Yoshida ascites hepatoma growing at different rates were investigated to resolve two problems: First, is there any relationship between the growth rates of tumors and the quantity or properties of these isozymes, and second, what is the degree of differentiation of organ-characteristic function using the organ-specific marker isozymes in these tumor lines? The results obtained on these liver tumors in the present study were compared with the conversion of the isozymes during development and regeneration of normal organs.

Metabolic Change of Glutaminase Activity in Relation to Growth Rate

1. Hepatomas

The relationship between glutaminase isozyme and fructose-6-phosphate (F-6-P) glutamine amidotransferase activities in rat liver is shown in Tables I and II. Since F-6-P amidotransferase and glutaminase are involved in the first step of glutamine anabolism and catabolism, respectively, some changes in both enzyme activities must be important in the regulation of glutamine metabolism. In the liver of partially hepatectomized rats, the specific activity of the glutaminase isozyme decreased markedly and, conversely, the F-6-P amidotransferase activity increased.

The PD/PI ratio increased in the regenerating liver; that is, the changes were parallel to the growth rate of the liver cells. Also, in the early stages of development, the specific activities of the glutaminase isozymes in liver were remarkably low and the PD/PI ratio was high, as in the regenerating liver.

Rapidly growing tissues, such as fetal liver and regenerating liver, generally have low glutaminase activity. Glutaminase activity is also generally low in rapidly growing hepatomas as in the case of fetal or regenerating liver. However, comparison of the specific activity of glutaminase in hepatoma tissues is technically difficult because these tissues are not only homogeneous but also contain necrotic, caseinous masses and blood. Therefore, the PD/PI ratio served as a better indicator since the change in parameters is parallel to the growth rate of the cells.

TABLE I. Changes of Glutaminase Isozyme and F-6-P Amidotransferase
in Regenerating Liver

Regenerating liver (period after partial hepatectomy)	Glutaminase			PD/PI ratio	F-6-P amidotransferase
	NH$_3$ formed (μmole/mg/hr)				Glucosamine-6-P formed (nmoles/mg/hr)
	PI		PD		
	Without maleate	With maleate (20 mM)	With PO$_4$ (100 mM)		
0 hr	0.738±0.19	1.230±0.20	0.642±0.02	0.87	8.1
6 hr	0.117	0.424	0.325	2.78	
1 day	0.106	0.296	0.310	2.92	31.5
2.5 days	0.043	0.230	0.406	9.45	23.4
3 days	0.094	0.340	0.457	4.87	12.5
3.5 days	0.104	0.352	0.616	5.92	
7 days	0.304	0.930	0.746	2.45	

Partial hepatectomy was carried out according to the method of Higgins (3) on male Donryu rats (150–200 g). No differences were observed between normal liver and the liver of sham-operated animals. Each value represents the average from 4 to 6 rats. These enzyme activities fluctuated only ±0.20 (SD) in animals on the control diet. Glucosamine synthetase activity was assayed by determining the amount of glucosamine produced during incubation. The reaction mixture contained 30 μmoles of L-glutamine, 3 μmoles of cysteine, 30 μmoles of F-6-P, and the enzyme preparation in 0.033 M Tris-HCl buffer (pH 8.3), in a total volume of 3 ml. The mixture was incubated for 30 min at 37°, and the reaction was terminated by boiling the mixture for 7 min. After removing the substrates by Dowex-50 column, the amount of glucosamine was determined by the method of Boas (1).

TABLE II. Changes of Glutaminase Isozymes during Development

Liver tissue	Glutaminase			PD/PI ratio
	NH$_3$ formed (μmole/mg/hr)			
	PI		PD	
	Without maleate	With maleate (20 mM)	With PO$_4$ (100 mM)	
Normal	0.940	1.460	0.980	1.10
1-Day neonatal	0.195	0.397	0.280	1.44
Fetal	0.139	0.360	0.361	2.60

The values are the averages from 10 to 20 rats. These enzyme activities fluctuated within ±0.1 (SD).

In contrast to the change in glutaminase activity, F-6-P amidotransferase activity increased in rapidly growing cells. The data on Morris hepatoma lines in Table III are arranged according to growth rate. A clear parallelism was found between the growth rates of these hepatomas and the PD/PI ratio. Knox et al. (11) also reported that the PD/PI ratio is higher in the Walker 256 carcinosarcoma, the Jensen D, and other tumors than in normal tissue. The activity of glutaminase, a catabolic enzyme of glutamine, decreased with the increased growth rates, while

TABLE III. Relationship between Growth Rates of Morris Hepatomas and Changes in Glutaminase Isozyme Patterns

Tissue	Glutaminase			PD/PI ratio
	NH$_3$ formed (μmole/mg/hr)			
	PI		PD	
	Without maleate	With maleate (20 mM)	With PO$_4$ (100 mM)	
Normal liver	0.738±0.190	1.230±0.206	0.642±0.20	0.87
AH-130	0.301±0.062	1.039±0.180	1.713±0.316	5.7
3924A	0.244±0.058	0.665±0.081	1.150±0.187	4.7
7777	0.240±0.049	0.429±0.068	0.551±0.055	2.7
5123C	0.578±0.071	1.560±0.441	0.792±0.272	1.4
7316A	0.122	0.286	0.400	3.3
7793	0.660±0.155	0.970±0.071	0.565±0.012	0.86
9618A	0.069	0.096	0.076	1.1

Each value represents the average of tumors from 2 to 4 tumor-bearing rats.

the activity of F-6-P amidotransferase, which is involved in the anabolic process of glutamine, increased.

2. *Kidney cortex tumors*

Glutaminase isozymes in three different kidney cortex tumors, 9789K, 9786K, and 8997K, with different growth rates, have also been investigated. These growth rates increased in the order of 9789K, 9786K, and 8997K. The specific activity of glutaminase in all three tumors was about ten times higher than in normal kidney (Table IV). There was no relationship between the growth rate and the PD/PI ratio, which showed the same value as in the normal kidney.

According to the method of Karnovsky (*4*), the glutamate formed by the glutaminases from added glutamine was detected using glutamic dehydrogenase

TABLE IV. Relationship between Growth Rate of Morris Kidney Tumor Lines and Changes in Glutaminase Isozyme Patterns

Tissue	Gultaminase			PD/PI ratio
	NH$_3$ formed (μmole/mg/hr)			
	PI		PD	
	Without maleate	With maleate (20 mM)	With PO$_4$ (100 mM)	
Adult kidney	0.5–0.6	3–4	4–5	9
Kidney tumors				
9789K	5.2	25	39	8
9786K	7.0	35	50	7
8997K	3.0	15	20	7

Each value represents the average of tumors from 2 tumor-bearing rats.

and tetrazolium. The location of glutaminase in tumors was determined by violet-colored staining. Histochemical examination revealed that the PI and PD in normal kidney were localized entirely in the straight tubules of the cortex. Furthermore, a high activity of aspartate transaminase in normal kidney was localized in tubules, but only a very small amount was present in the straight tubules (10). Since these tumors were stained strongly and homogeneously by the PI and PD staining method and stained very weakly by the aspartate transaminase staining method, it seems that these tumors must have been derived from the straight tubules of the cortex. This would explain why this series of tumors has such a high specific activity of glutaminase.

Degree of Differentiation of the Organ-specific Isozymes in the Tumor Lines

1. Hepatomas

Glutaminase isozymes were used in this study as markers for assessing the differentiation of the organ-specific function; therefore, the correlation between the glutaminase isozymes and their metabolic roles should be explained. We have found that two isozymes of glutaminase exist in various organs. One of these (PD) requires phosphate, and the other (PI) is not activated by phosphate but is strongly activated by maleate or N-acetylglutamate (8). The PI's in different organs are now considered to be isozymes. By purifying of the isozymes of PI in adult liver and kidney to homogeneous states, we found that they were different proteins (12). Thus they differed in molecular weight, reactivity with their respective antisera (immunochemical properties), polymerization by the substrate, and inducibility by a high-protein diet. These differences in organ-characteristic properties seem to be related to organ-specific function. The glutamine synthesized in various organs is transported in the blood, and most of its nitrogen is converted to urea in the liver, although some of it is excreted as ammonium salts in the urine. Since inorganic ammonia is the substrate for carbamyl phosphate synthetase, the amide of glutamine has to be hydrolyzed to ammonia by liver glutaminase prior to its incorporation into carbamyl phosphate. The most important regulatory mechanism for the control of glutamine metabolism in liver is the allosterism of glutaminase by its substrate, glutamine (9). At low concentrations of substrate the enzyme activity is rather low, but the presence of a large amount of the substrate causes polymerization of the enzyme molecules resulting in high activity. This may be considered to be one form of positive feedback control. Furthermore, the activators, N-acetylglutamate and compounds involved in the TCA cycle, facilitate the polymerization of the enzyme and cause a decrease in the K_m value for the substrate. This conformational change by the substrate is an important automatic form of regulation. N-Acetylglutamate and related compounds activate carbamyl phosphate synthetase, as shown by Grisolia and Cohen (2). These compounds also activate the liver-type PI, but not the kidney-type PI. Activation of the two successive reactions in the transfer of amide nitrogen of glutamine to carbamyl phosphate by common activators should provide a highly efficient regulatory mechanism. At the same time, a rich supply of adenosine triphosphate (ATP) and CO_2 is necessary for the synthesis of urea. The finding that the carboxylic acids of the TCA cycle activate both glutaminase and the TCA cycle provides another example of an intricate mechanism. Glutamine is also an important source of amide for the synthesis of purine bases and amino sugars.

This anabolic pathway of glutamine, via glutaminase, varies reciprocally with the activities of the enzymes in the anabolic pathway. In adult rats, no product inhibition by glutamate has been observed with liver and kidney type PI's, while the brain enzyme was strongly inhibited by the addition of glutamate (8). There is no physiological necessity for product inhibition in the liver or kidney because these two organs possess the urea cycle and an ammonium excretion system, respectively. There is no system for disposal of ammonia in brain tissue; instead, the glutaminase activity is controlled by the glutamate level in the tissue, so that the other product, ammonia, does not increase beyond a certain level. Strong product inhibition in which nitrogen is excreted by synthesis of uric acid directly from glutamine has been observed in all kinds of avian livers tested, and also in fetal rat liver which has not yet developed urea cycle enzymes.

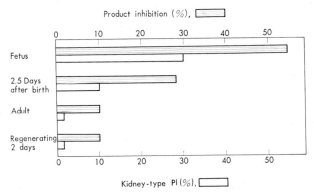

FIG. 1. Changes in the amount of kidney-type PI and product inhibition in liver during development

The values for the liver of fetal and adult animals are the averages for 10 to 20 rats; the averages for 4 to 6 rats is given for the neonatal and regenerating livers.

Changes in the PI isozyme pattern during the development and in regenerating liver are shown in Fig. 1. Fetal liver contains both the liver- and kidney-type enzymes. Kidney-type PI decreased gradually after birth, disappearing completely within one week. It should be noted that no kidney-type enzyme was found in the regenerating liver, although this tissue is rapidly growing. One might consider that regenerating livers have already completed differentiation of the isozyme. Evidence reported previously has demonstrated that the glutaminase activity not affected by p-chloromercuribenzoate (PCMB) and heat treatment is due to the kidney-type enzyme. The amount of kidney-type PI present in the liver may determine the remaining activity after PCMB and heat treatment under the constant conditions reported previously (6). As shown in Fig. 1, no product inhibition by glutamate was observed in adult liver PI, while fetal liver PI was

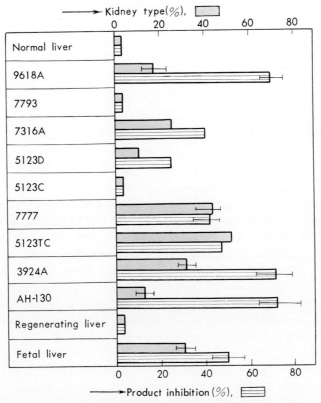

FIG. 2. Presence of kidney-type PI and product inhibition in Morris hepatomas
Each column represents the average number of tumors from 4 to 6 tumor-bearing rats.

inhibited 60 to 70%. This inhibition decreased gradually after birth, disappearing completely within one week. No product inhibition was observed at any stage in regenerating liver.

Figure 2 shows results obtained on the Morris hepatomas by two indicators of the degree of differentiation. The presence of kidney-type PI and of product inhibition was observed in some hepatomas, but the amount of kidney-type PI and the extent of inhibition showed no correlation with the growth rate. Thus, there is no direct relationship between the growth rate and the degree of differentiation of organ-specific isozymes. Hepatoma 7777 contained 50 to 60% kidney-type PI, and partially purified kidney-type enzyme from this tumor can be neutralized with antiserum from adult kidney type PI. Weinhouse and co-workers (13) also reported that the pyruvate kinase isozyme, which is not inherent in the particular organ, was observed in some Morris hepatomas.

As mentioned before, liver-type PI was inhibited by a low concentration of PCMB, and was not stained by the histochemical technique in the presence of PCMB, but hepatoma 7777, which contains an appreciable amount of kidney-type PI, was stained extensively and homogeneously even in the presence of a high concentration of PCMB.

TABLE V. Content of Kidney-type PI and Product Inhibition in Morris Kidney Tumors

		Content of kidney-type PI (%)	Product inhibition (%)
Kidney	Adult	100	0–10
	Neonatal (1–3 days)	50	27–44
	Fetus	50	
Kidney tumor	9789K	0–10	70
	9786K	0–10	50
	8997K	0–10	100

2. Kidney cortex tumors

As reported by Katunuma et al. (5), 50% of the PI activity in fetal kidney was that of liver-type PI, and the amount of liver-type PI decreased gradually after birth, disappearing completely within 4 to 5 days. We consider this phenomenon to be a functional differentiation in the kidney during development. More than 90% of the PI activity in the three kinds of kidney tumors tested was of the brain and liver type, judging from the PCMB and heat stabilities of the isozymes. No product inhibition was observed in PI of normal adult kidney, but the enzyme from these three kidney tumors was inhibited more than 50% by glutamate (Table V). No direct relationship was found between the growth rate and the degree of the product inhibition. Moreover, in these kidney tumors, no relationship was found between the growth rate and the degree of differentiation of the isozymes.

There have been no previous reports on the relationship between the degree of differentiation and the growth rate using enzymes such as the glutaminase isozymes which change irreversibly with the degree of differentiation. It should be pointed out that these isozymes do not change in regenerating liver or with various hormonal and dietary treatments.

There have been various opinions on this problem. Weinhouse indicated the dedifferentiational nature of cancer enzymes at the molecular level when muscle-type pyruvate kinase isozyme was detected in some hepatomas. Potter (15) also suggested recently that dedifferentiation of cancer enzymes is caused by a block of ontogeny. From the study of the distribution of organ-specific aldolase isozyme in hepatomas with different growth rates, Sugimura and his co-workers (14) suggested that the most suitable explanation for the peculiar pattern in these hepatomas was that of disdifferentiation. We did not observe any direct relationship between the growth rate and the degree of differentiation of the organ-specific glutaminase isozyme in various tumors whose isozyme pattern changed irreversibly during the course of development. Furthermore, it should be pointed out that from our data these marker isozymes did not appear in regenerating liver or after various hormonal and dietary treatments in contrast with the results on other isozymes by previous workers.

Acknowledgment

We are grateful to Prof. Sidney Weinhouse for his useful advice.

REFERENCES

1. Boas, N. F. Method for the determination of hexosamines in tissues. *J. Biol. Chem.*, **204**, 553–563 (1953).
2. Grisolia, S. and Cohen, P. P. Catalytic role of glutamate derivatives in citrulline biosynthesis. *J. Biol. Chem.*, **204**, 753–757 (1953).
3. Higgins, G. H. and Anderson, R. M. Experimental pathology of the liver. Restoration of the liver in the white rat following partial surgical removal. *Arch. Pathol.*, **12**, 186–202 (1931).
4. Karnovsky, M. J. and Himmelhoch, S. R. Histochemical localization of glutaminase I activity in kidney. *Am. J. Physiol.*, **201**, 786–890 (1961).
5. Katunuma, N., Katunuma, T., Tomino, T., and Matsuda, Y. Regulation of glutaminase activity and differentiation of the isozyme during development. *Adv. Enzyme Regulation*, **6**, 227–242 (1968).
6. Katunuma, N., Tomio, I., and Sanada, Y. Differentiation of organ specific glutaminase isozyme during development. *Biochem. Biophys. Res. Commun.*, **32**, 426–432 (1968).
7. Katunuma, N., Tomino, I., and Nishino, H. Glutaminase isozyme in rat kidney. *Biochem. Biophys. Res. Commun.*, **22**, 321–328 (1966).
8. Katunuma, N., Huzino, A., and Tomino, I. Organ specific control of glutamine metabolism. *Adv. Enzyme Regulation*, **5**, 55–69 (1967).
9. Katunuma, T., Temma, M., and Katunuma, N. Allosteric nature of a glutaminase isozyme in rat liver. *Biochem. Biophys. Res. Commun.*, **32**, 433–437 (1968).
10. Kishino, Y., Matsuzawa, T., and Katunuma, N. Histochemical study of aspartate transaminase in normal and pathological states using diazonium salt. *Symp. Pyridoxal Enzymes*, 223–229 (1968).
11. Knox, W. E., Tremblay, G. C., Spanier, B. B., and Friedell, G. H. Glutaminase activities in normal and neoplastic tissues of the rat. *Cancer Res.*, **27**, 1456–1458 (1967).
12. Krebs, H. A. Metabolism of amino acids. *Biochem. J.*, **29**, 1951–1969 (1935).
13. Lo, Chai-ho, Cristofalo, V. J., Morris, H. P., and Weinhouse, S. Studies on respiration and glycolysis in transplanted hepatic tumors of rat. *Cancer Res.*, **28**, 1–10 (1968).
14. Matsushima, T., Kawabe, S., Sibuya, M., and Sugimura, T. Aldolase isozymes in rat tumor cells. *Biochem. Biophys. Res. Commun.*, **30**, 565–570 (1968).
15. Potter, V. R. The recent advances in cancer biochemistry. The importance of studies on fetal tissue. *Can. Cancer Conf.*, **8**, 9–30 (1969).

ISOZYMES OF FRUCTOSE 1,6-DIPHOSPHATASE, GLYCOGEN SYNTHETASE, AND GLUTAMINE: FRUCTOSE 6-PHOSPHATE AMIDOTRANSFERASE[*1]

Shigeru TSUIKI, Kiyomi SATO, Taeko MIYAGI, and Hisako KIKUCHI

Biochemistry Division, Research Institute for Tuberculosis, Leprosy and Cancer, Tohoku University[*2]

In rat liver, the liver-type isozymes of the glycolytic pathway are replaced by other isozymes upon neoplastic transformation. Studies were conducted to see if this observation could be extended to other pathways of carbohydrate metabolism. The results obtained suggest that fructose 1,6-diphosphatase, glycogen synthetase, and glutamine: fructose 6-phosphate amidotransferase all undergo similar changes in their isozyme patterns.

Fructose 1,6-diphosphatases of rat liver and muscle differ in their sensitivity to AMP inhibition. Of the nine rat hepatomas studied, three slow-growing tumors retained the liver-type isozyme, four ascites hepatomas possessed the muscle-type isozyme exclusively, and the two most rapidly growing hepatomas were completely devoid of enzyme activity.

Studies on the glycogen synthetases of rat liver, muscle, and hepatomas in Tris-maleate buffer (pH 7.4) revealed that the liver enzyme is unique in that the D form exhibits an unusually high K_m value for UDP-glucose even in the presence of glucose 6-phosphate. No marked difference could be detected between muscle and hepatoma enzymes.

Glutamine: fructose 6-phosphate amidotransferase increases upon neoplastic transformation and is less stable, inhibited more severely by UDP-N-acetylglucosamine, and eluted more slowly from the DEAE-Sephadex column as compared with the liver enzyme.

During hepatocarcinogenesis, certain liver enzymes, including aldolase (*10*), hexokinase (*17*, *18*), and pyruvate kinase (*24*), undergo alterations in which the isozyme unique to liver is replaced by a nonhepatic counterpart. This alteration certainly provides a reasonable explanation for the powerful glycolytic capacity developed in hepatomas.

Enzyme alteration of this type, however, may not be restricted to the glycolytic enzymes. The reduction in gluconeogenesis and glycogen deposition observed in hepatomas has been attributed to reduction in the amount of key enzymes (*19*,

---

[*1] This work was supported by a scientific research fund from the Ministry of Education of Japan.
[*2] Hirosemachi 4-12, Sendai 980, Japan (立木 蔚, 佐藤清美, 宮城妙子, 菊池尚子).

20, 28–30), but would be explained equally well if certain key enzymes were replaced by less active counterparts. In this laboratory, studies were conducted to see if the observation made for glycolytic enzymes could be extended to other pathways of carbohydrate metabolism. The enzymes studied were fructose 1,6-diphosphatase (FDPase, EC 3.1.3.11), glycogen synthetase (EC 2.4.1.11), and glutamine: fructose 6-phosphate (Fru-6-P) amidotransferase (EC 2.6.1.16). FDPase is one of the key enzymes in gluconeogenesis. Glycogen synthetase plays an important regulatory role in glycogen synthesis, and the last enzyme catalyzes the conversion of Fru-6-P to glucosamine 6-phosphate (GlcN-6-P), the precursor of UDP-N-acetylglucosamine (UDP-GlcNAc) which is in turn the precursor of N-acetylglucosamine, N-acetylgalactosamine, and N-acetylneuraminic acid present in glycoproteins and mucopolysaccharides.

Being situated at key metabolic positions, these enzymes possess characteristic regulatory properties. FDPase and glutamine: Fru-6-P amidotransferase are regulated primarily by specific inhibition caused by AMP (23) and UDP-GlcNAc (9), respectively. The regulation of glycogen synthetase occurs primarily by the interconversion of two forms, *i.e.*, less active D and more active I. The interconversion is effected by a combined action of specific phosphatase and kinase. These properties were utilized in the present work to disclose any alteration that might occur in these enzymes during hepatocarcinogenesis.

Fructose 1,6-Diphosphatase

Since the data on FDPase have already been published (15, 16), only brief mention of this enzyme will be made here. Our interest in the FDPase of neoplastic tissues was aroused when appreciable activity was found in mouse Ehrlich ascites carcinoma previously reported to lack the enzyme. The FDPase level of the tumor was far below the level of liver, but well comparable to that of skeletal muscle.

Comparison of the FDPase enzymes purified from these mouse tissues revealed that one marked difference concerned the sensitivity to AMP inhibition, *i.e.*, a 50% inhibition of the liver enzyme required 130 μM AMP, whereas the muscle or Ehrlich enzyme required only 1 to 2 μM AMP. Other properties, such as the effect of substrate concentration, pH optimum, and the requirement for Mg^{2+} and EDTA, were all identical.

Observations were then extended to rat tissues and hepatomas. Table I summarizes the results of these experiments. Hepatomas displayed a wide diversity in FDPase levels. In confirmation of the report of previous workers (27), higher growth rates were associated with the lower FDPase levels, and the fastest growing AH-130 and AH-66F had no FDPase activity. Enzymes from two well-differentiated and slow-growing Morris hepatomas (7316A and 5123D) showed a high K_i value, comparable to that of the liver enzyme, whereas enzymes from four rapidly growing ascites hepatomas (3683, 3924A, 108A, and 173) were as highly sensitive to AMP inhibition as was the muscle enzyme. Liver-type FDPase was also present at a low level in the 35-tc_2, the slowest growing of all the ascites hepatomas examined. Since no muscle-type enzyme was detected in

TABLE I. Tissue Level and K_i for AMP of Fructose 1,6-Diphosphatase in Rat Tissues and Hepatomas (16)

Strain of rat and tissues or hepatomas	Growth rate (days)	FDPase level[a]	K_i for AMP[b]
Buffalo			
Liver		420	130
Kidney		240	175
Skeletal muscle		18	2.1
7316A	100	252	120
5123D	80	210	110
ACI/N			
AH-35-tc$_2$	30	3.0	110
AH-3683	10	0.9	2.3
AH-3924A		+	2.0
Donryu			
Liver		450	130
Skeletal muscle		5.7	2.0
AH-108A	17	3.7	2.0
AH-173	9	0.8	<5
AH-130	10	0	
AH-66F	8	0	

[a] μmoles of Fru-6-P formed/hr/g wet weight or ml of packed cells.
[b] μM concentration of AMP necessary for 50% inhibition.

this hepatoma, the possibility of liver having a small amount of muscle-type FDPase, which might be retained even after all the liver-type FDPase had been deleted, is unlikely.

Thus, not only did these studies confirm the previous finding that FDPase activity decreases with the increase in growth rate, but they further suggested that the decrease in activity might be associated with the transformation from liver-type to muscle-type enzyme.

Glycogen Synthetase

Studies carried out in this laboratory on the glycogen metabolism of rat hepatomas (*13, 14, 21, 22, 25*) demonstrated that the lack of glycogen deposition found in a number of hepatomas could not be explained simply in terms of the amount of glycogen synthetase. As shown in Table II, where the levels of glycogen synthetase and phosphorylase of two rat ascites hepatomas are compared with those of rat liver and skeletal muscle, even glycogen-deficient AH-130 retains a considerable glycogen synthetase activity. These results prompted us to investigate the glycogen synthetase of these hepatomas from the standpoint of qualitative alteration.

Recently, extensive studies were made on liver glycogen synthetase by Segal and co-workers (*3, 11*) and Hers *et al.* (*4*). The characteristic feature of the liver

TABLE II. Activities of Glycogen Synthetase and Phosphorylase in Liver,
Leg Muscle, and Hepatomas of the Rat

	Activities (nmoles/min/mg protein)[a]			
	Rat liver (well-fed)	Leg muscle	AH-130	AH-66F
Glycogen synthetase[b]				
−Glc-6-P	0.22±0.03	9.06±0.16	0.51±0.07	1.56±0.12
+Glc-6-P	11.00±0.55	41.9±1.0	2.56±0.18	6.38±0.46
Phosphorylase[c]				
−AMP	85.3±3.2	615±28.5	22.7±1.0	30.9±1.5
+AMP (1 mM)	92.5±1.8	3410±150	28.3±0.9	36.0±0.3

[a] The activities (mean±SE) were determined in the 5,000 g supernatant of tissue homogenates prepared in the presence of 40 mM NaF. The assay mixture also contained 40 mM NaF.
[b] Assay was made in Tris-maleate buffer, pH 7.4 (muscle and hepatomas) or pH 8.6 (liver) by a method similar to that described by Saheki et al. (13).
[c] Assayed at pH 6.1 (Tris-maleate buffer) in the direction of glycogen synthesis with 25 mM Glc-1-P as substrate.

enzyme appears to be that, in concentrated extracts (3) or in the presence of physiological concentrations of inorganic phosphate (2), glycogen synthetase D activity is low in the neutral pH range even at high levels of glucose 6-phosphate (Glc-6-P). Since maleate has almost the same effect as phosphate in depressing the D activity, comparison of the hepatoma enzymes with that of liver or muscle was made with the Tris-maleate buffer (pH 7.4) as assay medium.

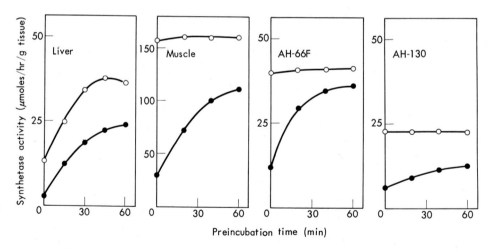

FIG. 1. Activation of glycogen synthetase in 5,000 g supernatants

Tissues were homogenized with 4 vol. of 0.5 M sucrose, 62.5 mM Tris-HCl (pH 7.4), 6.25 mM EDTA. After centrifugation at 5,000 g for 10 min, the supernatants were preincubated at 30° for various lengths of time. Glycogen synthetase was then assayed in 50 mM Tris-maleate buffer, pH 7.4, with (○) or without (●) 10 mM Glc-6-P. The concentration of UDP-glucose was 5 mM. The assay mixture also contained 40 mM NaF.

In the experiments shown in Fig. 1, the 5,000 g supernatants of rat tissue homogenates were incubated at 30° prior to the assay of glycogen synthetase. Due to D to I conversion, glycogen synthetase underwent an activation, the pattern of which was quite different between the liver and muscle enzymes. With the muscle enzyme, only the activity measured without Glc-6-P was increased, whereas with the liver enzyme, the activities measured both with and without Glc-6-P were increased almost concurrently. Figure 1 also shows that, under the same conditions, the glycogen synthetase of AH-66F and AH-130 behaved just like the muscle enzyme. The possibility therefore arose that the glycogen synthetase of these hepatomas might be of the muscle type.

In order to substantiate this possibility, glycogen synthetases D and I were prepared from liver, muscle, AH-66F, and AH-130. Liver glycogen synthetase D was prepared by a method similar to that of Hizukuri and Larner (5). About 50% of synthetase D activity was recovered on the final glycogen pellet, and the degree of purification was almost 280-fold.

Synthetase I was then prepared by incubating the purified glycogen-bound synthetase D with glycogen synthetase D phosphatase at 30°. The 105,000 g

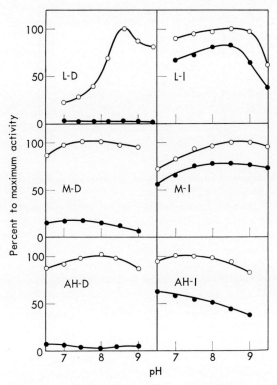

Fig. 2. pH-Activity relationship of partially purified glycogen synthetases D and I

The assay was made in 100 mM Tris-maleate buffer of indicated pH with (○) or without (●) 10 mM Glc-6-P. The concentration of UDP-glucose was 5 mM. L, liver; M, skeletal muscle; AH, AH-66F.

supernatant of rat liver homogenate prepared in the absence of NaF served as a source of the phosphatase.

Similar procedures were used for the preparation of glycogen synthetase from ascites hepatomas, but not for the synthetase I of AH-130, which was directly precipitated from crude extracts by ammonium sulfate. Synthetases D and I of muscle were prepared by the method of Villar-Palasi et al. (26).

Figure 2 shows the pH activity curves for synthetases D and I purified from various sources. Even in the presence of Glc-6-P, the activity of liver synthetase D was very low in the physiological pH range, but rose sharply above pH 8. In contrast, the activity of muscle synthetase D was almost independent of pH be-

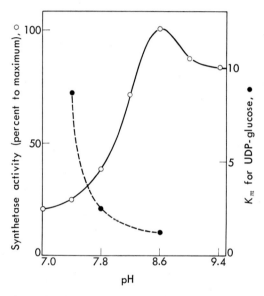

FIG. 3. Comparison of rat liver glycogen synthetase D activity and the K_m for UDP-glucose with varying pH

The assay of the partially purified D form was conducted in the presence of 10 mM Glc-6-P with 5 mM UDP-glucose as substrate.

TABLE III. K_m for UDP-glucose and Glucose 6-Phosphate of Glycogen Synthetase from Rat Tissues and Hepatomas in Tris-Maleate Buffer of pH 7.4

	Liver	Muscle	Hepatoma	
			AH-66F	AH-130
K_m for UDP-glucose[a]		(mM)		
D	8.0	0.50	0.22	0.54
I	0.4	0.50	0.23	0.44
K_m for Glc-6-P[b]				
D		0.60	0.23	0.13	0.29
I		0.08		0.009	0.017

[a] Determined at 10 mM Glc-6-P. [b] Determined at 1 mM UDP-glucose.

tween 7 and 9. The pH activity curve for synthetase D of AH-66F again resembled closely that of the muscle enzyme. The curves for the three I forms were similar.

An affinity for UDP-glucose of purified liver synthetase D was determined at several pH values in the presence of Glc-6-P and the K_m values derived from these data are plotted in Fig. 3. There was an inverse relationship between enzyme activity and K_m suggesting that the K_m for UDP-glucose was an important factor, accounting for the low activities of the synthetase D in the physiological pH range under our experimental conditions.

The K_m for UDP-glucose of synthetase D and I preparations from various sources was determined at pH 7.4 in the presence of Glc-6-P. As reported in Table III, the K_m values of muscle and hepatoma enzymes were almost identical for the D and I forms, but the synthetase D of liver exhibited a K_m value 20 times greater than that of synthetase I. With respect to the K_m for Glc-6-P (which is also reported in Table III), the hepatoma enzymes differ from the liver enzyme, resembling more closely the muscle enzyme.

The results presented above strongly suggest that the glycogen synthetase of liver differs from that of muscle. The most remarkable difference appears to be that in the presence of inorganic phosphate (3), sulfate (1), or maleate, the activity in the physiological pH range of liver glycogen synthetase D is very low, even at high levels of Glc-6-P, primarily because of the low affinity for UDP-glucose. The similarity of the kinetic properties of the AH-66F and AH-130 enzymes to those of the muscle enzyme is in agreement with the hypothesis that, in liver, glycogen synthetase unique to liver is replaced by the muscle-type enzyme upon neoplastic transformation.

Glutamine: Fructose 6-Phosphate Amidotransferase

The formation of GlcN-6-P from Fru-6-P and glutamine catalyzed by glutamine: Fru-6-P amidotransferase constitutes the initial step of the long metabolic road leading to the formation of UDP-GlcNAc and CMP-N-acetylneuraminic acid. Since growing cells require the continuous synthesis of membrane glycoproteins and glycolipids, the amidotransferase is essential for cell growth.

When the distribution of amidotransferase in rat tissues was studied, liver and brain were found to have much greater activity than any other parenchymal organ tested. The level of amidotransferase in rat and mouse tumors was determined after precipitation of the enzyme by ammonium sulfate. This treatment was necessary for the removal of the amidotransferase inhibitor found in tumor extracts (6). As shown in Table IV, all tumors examined were much greater in amidotransferase activity than was liver or brain.

The idea that the amidotransferase of tumors might differ from the liver enzyme occurred when the ammonium sulfate precipitates were examined for sensitivity to UDP-GlcNAc inhibition. Figure 4 shows that the enzymes from various tumors were inhibited by UDP-GlcNAc more profoundly than was the liver enzyme.

160 S. TSUIKI ET AL.

TABLE IV. Glutamine: Fructose 6-Phosphate Amidotransferase Activity of Ammonium Sulfate Precipitate Fraction (6)

	Activity (units/mg protein)[a]
Rat	
Liver	56.8
Brain	56.1
Yoshida sarcoma	185.5
AH-130	206.6
Mouse	
Liver	53.0
Ehrlich carcinoma	249.6

[a] One unit of enzyme was defined as the amount which catalyzed the formation of nmole of GlcN-6-P/hr.

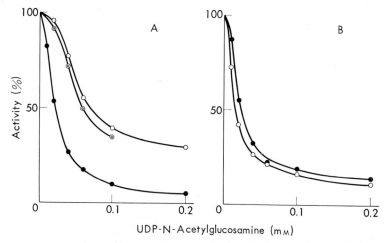

FIG. 4. Inhibition of amidotransferase activity of ammonium sulfate precipitate by UDP-GlcNAc (6)

A: ○ rat liver; ◎ liver from Yoshida sarcoma-bearing rat; ● Yoshida sarcoma. B: ○ AH-130; ● mouse Ehrlich carcinoma.

Since the amidotransferase was very unstable, its purification beyond the ammonium sulfate stage was difficult. In this laboratory, partial purification of the liver enzyme has been brought about by DEAE-Sephadex column chromatography (12). Further purification was possible by use of the hydroxylapatite column. Representative data of the progress of purification are shown in Table V.

DEAE-Sephadex chromatography as applied to the liver enzyme failed to accomplish a satisfactory purification of the tumor enzyme. Table VI shows that purification of Yoshida sarcoma enzyme requires the presence of GlcN-6-P plus glutamine in the medium. Since this was not a primary requisite for liver enzyme, the two enzymes differ in stability.

For both liver and tumor enzymes, however, chromatography on DEAE-Sephadex brought about a marked reduction in sensitivity to UDP-GlcNAc

TABLE V. Partial Purification of Amidotransferase from Rat Liver (12)

	Steps	Specific activity (units/mg)	Purity (X)	Recovery (%)
1.	105,000g supernatant	25.3	1	100
2.	40–60% $(NH_4)_2SO_4$ precipitate	58.8	2	80
3.	DEAE-Sephadex column	663.0	26	71
4.	Hydroxyapatite column	6,062.9	240	30

TABLE VI. Stability of Amidotransferase and Protecting Effect of Glucosamine 6-Phosphate

	+None		+GlcN-6-P, glutamine[a]	
	Specific activity (units/mg)	Recovery (%)	Specific activity (units/mg)	Recovery (%)
Liver enzyme				
$(NH_4)_2SO_4$ precipitate	70.0	100	50.2	100
After DEAE-Sephadex	772.0	85	565.5	90
Yoshida sarcoma enzyme				
$(NH_4)_2SO_4$ precipitate	122.0	100	178.5	100
After DEAE-Sephadex	279.0	43	813.0	78

[a] 0.05 mM GlcN-6-P plus 20 mM glutamine was present throughout the purification.

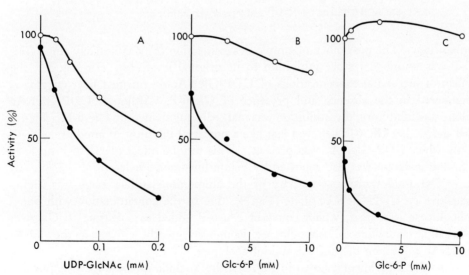

FIG. 5. Effect of Glc-6-P on UDP-GlcNAc inhibition of amidotransferase (12)
A: Enzyme was incubated with varying concentrations of UDP-GlcNAc in the absence (○) or presence (●) of 7 mM Glc-6-P. B and C: Enzyme was incubated with varying concentrations of Glc-6-P in the absence (○) or presence (●) of 0.1 mM UDP-GlcNAc. A and B: Rat liver enzyme purified by DEAE-Sephadex column. C: Yoshida sarcoma enzyme purified by DEAE-Sephadex column.

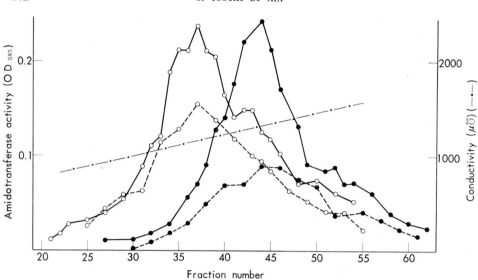

Fig. 6. Elution diagram of amidotransferase after DEAE-Sephadex chromatography

The ammonium sulfate precipitate was dissolved in 50 mM Na phosphate (pH 7.5) containing 50 mM KCl and 1 mM EDTA and after desalting by Sephadex G-25 was applied to a column (1.5 × 25 cm) of DEAE-Sephadex. After washing with the same buffer, a gradient of KCl from 50 to 500 mM was applied. Fractions of 2 ml were collected and assayed for amidotransferase in the absence (——) or presence (- - -) of 0.02 mM UDP-GlcNAc plus 7 mM Glc-6-P. ○ rat liver enzyme; ● Yoshida sarcoma enzyme. In this case, all the buffers used contained 0.05 mM GlcN-6-P and 20 mM glutamine.

inhibition. The possibility of simple desensitization (8) was unlikely since Glc-6-P was found to restore inhibition to its original level (12). In Fig. 5A, the effect of increasing concentrations of UDP-GlcNAc on purified liver enzyme is compared in the absence and presence of Glc-6-P. Although UDP-GlcNAc alone was inhibitory, the inhibition was markedly augmented by Glc-6-P. Figure 5B shows that Glc-6-P behaved just like a powerful inhibitor of amidotransferase only when UDP-GlcNAc was present. This unique effect of Glc-6-P on liver amidotransferase was also reported by Winterburn and Phelps (31).

The amidotransferase activity of the ammonium sulfate stage was highly sensitive to UDP-GlcNAc alone (Fig. 4), due to the contamination with phosphoglucose isomerase, which converts Fru-6-P added as substrate to Glc-6-P. Actually, crystalline phosphoglucose isomerase greatly enhanced the UDP-GlcNAc inhibition of purified amidotransferase.

In the experiments shown in Fig. 5C, the Yoshida sarcoma enzyme purified by DEAE-Sephadex column was studied under the same conditions as given in Fig. 5B. While the general pattern of inhibition was similar, the tumor enzyme was inhibited much more extensively both by UDP-GlcNAc alone and by UDP-GlcNAc plus Glc-6-P.

The elution diagrams of liver and Yoshida sarcoma enzymes after DEAE-Sephadex chromatography are compared in Fig. 6. The liver enzyme was eluted from the column by lower concentrations of KCl than was the tumor enzyme. Amidotransferase from other tumors such as Ehrlich ascites carcinoma was eluted at exactly the same position as was Yoshida sarcoma enzyme.

The amidotransferase of tumors thus differs from the liver enzyme in stability, sensitivity to feedback inhibition, and affinity for DEAE-Sephadex column. Although further investigation is necessary, the available data appear to suggest that the two enzymes represent a set of isozymes of amidotransferase. Preliminary immunochemical studies appear to support this conclusion.

DISCUSSION

The above results suggest that the changing pattern of isozymes previously observed in several key glycolytic enzymes during hepatocarcinogenesis also occurs in the key enzymes of other pathways of carbohydrate metabolism. For certain glycolytic enzymes, such as aldolase (10) and hexokinase (17, 18), the isozymes appearing anew in neoplastic liver are also found in fetal liver, and these findings provide a biochemical basis for the general idea of " dedifferentiation " as explaining the neoplastic state.

In neoplastic tissues as compared with control tissues, the activity of glutamine: Fru-6-P amidotransferase is high (Table IV), whereas the activity of UDP-GlcNAc 2′-epimerase is low (7). Thus the amidotransferase/epimerase activity ratio is 0.2 for livers, 5 for various hepatomas, and over 10 for Ehrlich carcinoma. Since fetal liver gives an activity ratio of almost 10 (7), it is very probable that the neoplastic change that occurs in the amidotransferase is closely related to dedifferentiation. The results of the present studies, therefore, suggest strongly that dedifferentiation affects not only the glycolytic but also almost all of the pathways of carbohydrate metabolism.

Another possible significance of isozyme studies of neoplastic tissues is that isozymes may provide explanations for the characteristic features of tumor metabolism. Thus the powerful glycolytic capacity of hepatomas is associated with the neoplastic change in glucose-phosphorylating enzyme. During hepatocarcinogenesis, glucokinase unique to liver tends to decrease and hexokinase activity appears high (17, 18).

Although Hers et al. (4) mentioned that hepatic glycogen synthesis is much more insensitive to inhibition by glycogen than synthesis in muscle, this alone would not be enough to explain the lack of glycogen deposition in hepatomas, since AH-66F and AH-130 possessing seemingly identical muscle-type glycogen synthetase differ markedly in glycogen content. We have reported previously that the rate of glycogen synthesis in AH-130 declines as glycogen accumulates, whereas in AH-66F, the rate is almost independent of cellular glycogen level (13). The site of glycogen inhibition of glycogen synthesis appears to be glycogen synthetase D phosphatase. To explain glycogen deficiency in hepatomas, therefore, we might expect multiple forms of the phosphatase, the pattern of which changes during hepatocarcinogenesis.

REFERENCES

1. De Wulf, H. and Hers, H. G. The interconversion of liver glycogen synthetase *a* and *b in vitro*. *Eur. J. Biochem.*, **6**, 552–557 (1968).
2. De Wulf, H., Stalmans, W., and Hers, H. G. The influence of inorganic phosphate, adenosine triphosphate and glucose 6-phosphate on the activity of liver glycogen synthetase. *Eur. J. Biochem.*, **6**, 545–551 (1968).
3. Gold, A. H. and Segal, H. L. Time-dependent increase in rat liver glycogen synthetase activity *in vitro*. *Arch. Biochem. Biophys.*, **120**, 359–364 (1967).
4. Hers, H. G., De Wulf, H., Stalmans, W., and Van Den Berghe, G. The control of glycogen synthesis in the liver. *Adv. Enzyme Regulation*, **8**, 171–190 (1970).
5. Hizukuri, S. and Larner, J. Studies on UDPG: α-1,4-glucan α-4-glucosyltransferase. VII. Conversion of the enzyme from glucose-6-phosphate-dependent to independent form in liver. *Biochemistry*, **3**, 1783–1788 (1964).
6. Kikuchi, H., Kobayashi, Y., and Tsuiki, S. L-Glutamine: D-fructose 6-phosphate amidotransferase in tumors and the liver of tumor-bearing animals. *Biochim. Biophys. Acta*, **237**, 412–421 (1971).
7. Kikuchi, K., Kikuchi, H., and Tsuiki, S. Activities of sialic acid-synthesizing enzymes in rat liver, and rat and mouse tumors. *Biochim. Biophys. Acta*, **252**, 357–368 (1971).
8. Kornfeld, R. Studies on L-glutamine D-fructose 6-phosphate amidotransferase. I. Feedback inhibition by uridine diphosphate-N-acetylglucosamine. *J. Biol. Chem.*, **242**, 3135–3141 (1967).
9. Kornfeld, S., Kornfeld, R., Neufeld, E. F., and O'Brien, P. J. The feedback control of sugar nucleotide biosynthesis in liver. *Proc. Natl. Acad. Sci. U.S.*, **52**, 371–379 (1964).
10. Matsushima, T., Kawabe, S., Shibuya, M., and Sugimura, T. Aldolase isozymes in rat tumor cells. *Biochem. Biophys. Res. Commun.*, **30**, 565–570 (1968).
11. Mersmann, H. J. and Segal, H. L. An on-off mechanism for liver glycogen synthetase activity. *Proc. Natl. Acad. Sci. U.S.*, **58**, 1688–1695 (1967).
12. Miyagi, T. and Tsuiki, S. Effect of phosphoglucose isomerase and glucose 6-phosphate on UDP-N-acetylglucosamine inhibition of L-glutamine: D-fructose 6-phosphate aminotransferase. *Biochim. Biophys. Acta*, **250**, 51–62 (1971).
13. Saheki, R., Sato, K., and Tsuiki, S. Glycogen synthesis and glycogen synthetase in rat ascites hepatomas of low and high glycogen content. *Biochim. Biophys. Acta*, **230**, 571–582 (1971).
14. Saheki, R. and Tsuiki, S. Glycogen synthesis and I to D conversion of glycogen synthetase in rat ascites hepatoma cells. *Biochem. Biophys. Res. Commun.*, **31**, 32–36 (1968).
15. Sato, K. and Tsuiki, S. Fructose 1,6-diphosphatase of mouse Ehrlich ascites tumor cells and its comparison with the enzymes of liver and skeletal muscle of the mouse. *Biochim. Biophys. Acta*, **159**, 130–140 (1968).
16. Sato, K. and Tsuiki, S. Fructose 1,6-diphosphatase of rat tissues and transplantable rat hepatomas. *Arch. Biochem. Biophys.*, **129**, 173–180 (1969).
17. Sato, S., Matsushima, T., and Sugimura, T. Hexokinase isozyme patterns of experimental hepatomas in rats. *Cancer Res.*, **29**, 1437–1446 (1969).
18. Shatton, J. B., Morris, H. P., and Weinhouse, S. Kinetic, electrophoretic, and chromatographic studies on glucose-ATP phosphotransferases in rat hepatomas. *Cancer Res.*, **29**, 1161–1172 (1969).

19. Shonk, C. E., Morris, H. P., and Boxer, G. E. Patterns of glycolytic enzymes in rat liver and hepatoma. *Cancer Res.*, **25**, 671–676 (1965).
20. Sweeney, M. J., Ashmore, J., Morris, H. P., and Weber, G. Comparative biochemistry of hepatomas. IV. Isotope studies of glucose and fructose metabolism in liver tumors of different growth rates. *Cancer Res.*, **23**, 995–1002 (1963).
21. Takeda, H., Sato, K., Saheki, R., Suzuki, R., and Tsuiki, S. Glucose-metabolizing enzymes in Yoshida sarcoma and its slow-growing variant subline. *GANN*, **58**, 459–466 (1967).
22. Takeda, H. and Tsuiki, S. Metabolism of glucose in Yoshida sarcoma and its slow-growing variant sublines. *GANN*, **58**, 221–227 (1967).
23. Taketa, K. and Pogell, B. M. Allosteric inhibition of rat liver fructose 1,6-diphosphatase by adenosine 5′-monophosphate. *J. Biol. Chem.*, **240**, 651–662 (1965).
24. Tanaka, T., Harano, Y., Sue, F., and Morimura, M. Crystallization, characterization and metabolic regulation of two types of pyruvate kinase isolated from rat tissues. *J. Biochem. (Tokyo)*, **62**, 71–91 (1967).
25. Tsuiki, S. and Saheki, R. Metabolism of glycogen in rat ascites hepatomas. *GANN Monograph*, **1**, 117–126 (1966).
26. Villar-Palasi, C., Rossel-Perez, M., Hizukuri, S., Huijing, F., and Larner, J. Muscle and liver UDP-glucose: α-1,4-glucan α-4-glucosyltransferase (glycogen synthetase). *Methods Enzymol.*, **8**, 374–384 (1966).
27. Weber, G. The molecular correlation concept: studies on the metabolic pattern of hepatomas. *GANN Monograph*, **1**, 151–178 (1966).
28. Weber, G., Banerjee, G., and Morris, H. P. Comparative biochemistry of hepatomas. I. Carbohydrate enzymes in Morris hepatoma 5123. *Cancer Res.*, **21**, 933–937 (1961).
29. Weber, G. and Morris, H. P. Comparative biochemistry of hepatomas. III. Carbohydrate enzymes in liver tumors of different growth rates. *Cancer Res.*, **23**, 987–994 (1963).
30. Weber, G., Morris, H. P., Love, W. C., and Ashmore, J. Comparative biochemistry of hepatomas. II. Isotopic studies of carbohydrate metabolism in Morris hepatoma 5123. *Cancer Res.*, **21**, 1406–1411 (1961).
31. Winterburn, P. J. and Phelps, C. F. Studies on the control of hexosamine biosynthesis by glucosamine synthetase. *Biochem. J.*, **121**, 711–720 (1971).

ISOZYMES AND HETEROGLYCAN METABOLISM OF HEPATOMAS

R. K. Murray, D. J. Bailey, R. L. Hudgin, and H. Schachter

*Departments of Biochemistry and Pathology,
University of Toronto*[*]

    The zymogram patterns of four cytoplasmic enzymes of a number of hepatomas were compared with those of normal rat liver by starch gel electrophoresis. With the exception of the Morris 7800 hepatoma, all of the hepatomas studied showed a lactate dehydrogenase pattern predominated by the M_4 isozyme. All of the tumors studied except the Novikoff hepatoma showed multiple molecular forms of sorbitol dehydrogenase identical to normal liver and complex patterns of ali-esterases. One principal zone of ADH activity was detected in both liver and all of the hepatomas except the Novikoff. This zone of ADH activity exhibited the same electrophoretic migration as the most basic azoprotein of rat liver, raising the possibility that ADH is a target protein for aminoazo carcinogens.

    Certain aspects of complex heteroglycan metabolism were investigated in Morris hepatomas 7777, 7800, and 5123D. Assays for CMP-sialic acid: glylcoprotein sialyltransferase and UDP-N-acetylglucosamine: glycoprotein N-acetylglucosaminyltransferase were performed on homogenates of these tumors. Whereas the sialyltransferase activity was decreased in the first two hepatomas relative to normal liver, the N-acetylglucosaminyltransferase activity was not altered. Since these glycosyltransferases have been previously shown to be located in the Golgi apparatus of normal rat liver, where they function in the biosynthesis of glycoproteins, the above results have been interpreted to indicate that a shift in the function of the Golgi apparatus has occurred in certain Morris hepatomas. An investigation of the ganglioside composition of these hepatomas revealed that the most complex gangliosides detectable in normal liver were missing in the tumors, but that overall the ganglioside patterns were more complex than those previously described for virally transformed cells or cultured hepatoma cells.

Our laboratory has previously examined certain aspects of the electrophoretic patterns of the cytoplasmic proteins and of several enzymes present in 4-(dimethylamino)azobenzene (DAB)-induced hepatomas and in the Morris hepatoma 5123tc (*9, 18*). We have been interested particularly in the relatively basic cytoplasmic proteins and enzymes of rat liver as it is in this fraction that the principal azoprotein of rat liver is located following the acute administration of aminoazo

---

[*] Toronto, Ontario, Canada.

carcinogens (*11, 35*). This fraction, purified relatively simply by elution through DEAE-cellulose at pH 8.0, has been shown to contain at least nine or ten cytoplasmic enzymes (*15*). Of special interest to us was the suggestion of Ketterer and Christodoulides (*10*) that the most basic azoprotein of rat liver may be a known sulfhydryl-containing enzyme. In the present contribution, we describe further electrophoretic studies on the soluble proteins and the multiple molecular forms of several cytoplasmic enzymes of rat liver and of a number of hepatomas. The enzymes studied were lactate dehydrogenase (LDH) (EC 1.1.1.27), sorbitol dehydrogenase (SDH) (EC 1.1.1.14), alcohol dehydrogenase (ADH) (EC 1.1.1.1), and the several ali-esterases (EC 3.1.1.1). Certain observations arising from these studies have raised the possibility that ADH may be the most basic azoprotein of rat liver.

More recently, we have also become interested in the study of complex heteroglycans in hepatomas. In particular, certain features of glycoprotein metabolism and of the gangliosides of hepatomas have been examined. Glycoproteins and gangliosides are important constituents of cytoplasmic membrane systems, and increasing attention is being paid to these compounds in view of the suggestion that membrane changes may be crucial phenotypic alterations in cancer cells (*32*). We report here a study of the activity of the Golgi apparatus (GA) of several Morris hepatomas, using the assay of two glycoprotein glycosyltransferases as indices of its function (*23*). This study was made possible by a collaborative project with Dr. H. Schachter and his colleagues. The pioneering work of Hakomori and Murakami (*7*) indicated that virally transformed cultured cells show significant alterations in their ganglioside composition. Thus it seemed of interest to analyze the ganglioside patterns of hepatomas growing *in vivo* in an attempt to evaluate the significance of altered ganglioside compositions in neoplastic cells.

Materials and Methods

1. Animals and cultured cells

Animals carrying Morris hepatomas were kindly supplied by Dr. H. P. Morris. Certain studies were also conducted on Morris hepatomas and on the Reuber H-35 hepatoma transplanted in the laboratory of Dr. H. C. Pitot. Rats carrying Novikoff hepatomas were kindly supplied by Dr. C. Heidelberger. Novikoff and Reuber H-35 hepatoma cells grown in culture were generously supplied by Miss Joyce Becker and Dr. V. R. Potter.

2. Preparation of liver and hepatoma extracts and electrophoresis

The 105,000g supernatant was prepared from livers and tumors homogenized in one volume of 0.2 M sucrose—0.05 M Tris-HCl (pH 7.4)—0.005 M $MgCl_2$. The methods for performance of starch gel electrophoresis (S.G.E.) and for the detection of zones of esterase and SDH activity have been previously described (*18, 17*). LDH isozymes were detected by the method of Fine and Costello (*6*). ADH activity was located by virtue of its "nothing dehydrogenase" activity (*17*). Enzyme activities were measured spectrophotometrically by methods described in Bergmeyer (1), and protein was measured by the method of Lowry et al. (14).

3. Detection of azoprotein

Adult male rats (approximately 200 g body weight) were injected intraperitoneally with 50 mg of 3′-methyl-DAB dissolved in corn oil. The animals were sacrificed 16 hr later and the 105,000g supernatant prepared. Aliquots of this were subjected to S.G.E. in buffers of various pH values, and the principal azoprotein detected by pouring 10% trichloroacetic acid (TCA) onto one-half of a sliced gel. Its electrophoretic position was compared with that of various soluble proteins or dehydrogenases detected on the other half of the sliced gel.

4. Assay of glycoprotein glycosyltransferases

The methods used in these studies have been fully described elsewhere (8, 23).

5. Extraction and analysis of gangliosides

The procedures used for ganglioside analysis have also been previously described (3, 36). The nomenclature used for description of the gangliosides is that of Svennerholm (30).

Electrophoretic Studies of Soluble Proteins and Enzymes of Liver and Hepatomas

An electrophoretic pattern of the soluble proteins of liver and various hepatomas is shown in Fig. 1. Some 12 proteins migrating toward the anode and approximately 8 proteins migrating toward the cathode ("basic" proteins) were seen in the extract of normal liver (slot 2). The latter group of proteins is of particular interest in relation to the present study. It is apparent that the faster-migrating basic proteins were absent in the Novikoff hepatoma (slots 5 and 8). These bands correspond to the h proteins of Sorof et al. (28). The positions of ADH and of the multiple molecular forms of SDH in this electrophoretic system are also indicated. ADH was seen to be the zone migrating second farthest toward the cathode. This is the position of one of the principal azoproteins of rat liver (Ref. 3 and *vide infra*).

Figure 2A shows the results of an experiment in which 3′-Me-DAB was injected into groups of normal rats and animals carrying the H-35 hepatoma. The soluble proteins were separated by electrophoresis, one half of the gel immersed in TCA, and the other half stained with Amido Black. The position of the principal azoprotein of normal liver is indicated in slots 3 and 5. Little or no protein-bound azo dye was detected in the hepatoma extracts. The pattern of soluble proteins detected on the other half of the gel is shown in Fig. 2B. You will note that the principal azoprotein corresponds in position to the protein migrating second farthest toward the cathode. As stated above, this is also the electrophoretic position of ADH. In other experiments in which electrophoresis was performed on cellulose acetate (sodium barbiturate buffer, 0.02 M, pH 8.6) or in other starch gel systems (either buffer of sodium glycine, 0.02 M, pH 10.0, or

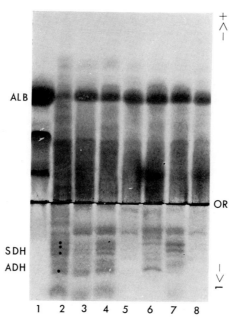

Fig. 1. Soluble proteins of rat liver and various hepatomas

Samples of the high-speed supernatants of liver and of the hepatomas were adjusted to a protein concentration of approximately 3 g/100 ml prior to application. The gel was stained with Amido Black. The positions of the multiple molecular forms of sorbitol dehydrogenase (SDH) and of the one zone of alcohol dehydrogenase (ADH) detected are indicated by the black dots (●). The buffer used for electrophoresis was 0.021 M sodium borate, pH 8.6. OR, origin; ALB, albumin.

Slot 1, rat serum; slot 2, liver; slot 3, Morris hepatoma 7794A; slot 4, Morris hepatoma 7800; slot 5, Novikoff hepatoma; slot 6, Reuber hepatoma H-35; slot 7, Morris hepatoma 5132C; slot 8, Novikoff hepatoma.

sodium phosphate, 0.02 M, pH 6.5), the azoprotein was also found to correspond in electrophoretic migration to that of ADH (as detected by the tetrazolium staining method).

The pattern of LDH isozymes in normal liver and in a number of hepatomas is shown in Fig. 3. The principal LDH isozyme of normal liver was seen to be the M4 component, with a small amount of the M3H band also present. The most striking feature was the presence in the 7800 hepatoma of significant amounts of all of the five LDH isozymes, the pattern generally resembling that of kidney. The other hepatomas grown *in vivo* showed a pattern qualitatively similar to that of liver. However, the two hepatomas grown in tissue culture showed a marked reduction of the M3H isozyme.

The pattern of multiple molecular forms of SDH in liver and various hepatomas is shown in Fig. 4. All of the multiple molecular forms of SDH found in liver were detected in the hepatomas, except for the case of the Novikoff tumor, which lacked SDH activity. The H-35 hepatoma grown in tissue culture retained all of the multiple molecular forms of SDH. The dehydrogenase zone migrating further toward the cathode than the SDH zones was ADH (" nothing dehy-

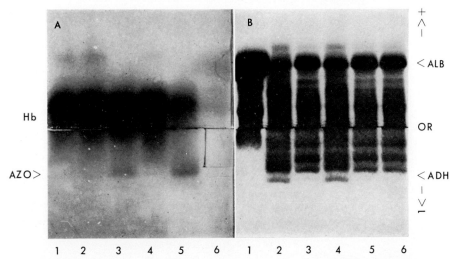

FIG. 2. Electrophoretic analysis of the azoproteins of rat liver and Reuber hepatoma H-35

Two control animals and three animals carrying the Reuber H-35 hepatoma were each injected with 150 mg of 3′-Me-DAB. The animals were sacrificed 16 hr later and samples of the 105,000g supernatant prepared for electrophoresis. After electrophoresis (0.021 M sodium borate, pH 8.6), one-half of the gel (A) was immersed in 10% trichloroacetic acid (to detect the azo dye-binding proteins) and one-half (B) was stained with Amido Black.

A. Slots 1, 2, and 4, Reuber hepatomas H-35; slots 3 and 5, livers; slot 6, plasma.

B. Slot 1, plasma; slots 2 and 4, livers; slots 3, 5, and 6, Reuber hepatomas H-35.

The position of the principal azoprotein is indicated. The other dense-staining area in A corresponds to hemoglobin (Hb). The positions of albumin (ALB) and alcohol dehydrogenase (ADH) are also indicated.

drogenase"). Only one ADH zone was detected in normal rat liver, and the tumors exhibited a zone of similar electrophoretic migration. The ADH zone was absent in the Novikoff hepatoma, which was quantitatively found to lack ADH activity.

The electrophoretic pattern of soluble ali-esterases is shown in Fig. 5. Apart from the Novikoff hepatoma, all of the tumors exhibited a complex pattern of soluble esterases, although quantitative and in some cases qualitative differences from normal liver were observable. Interestingly, the H-35 hepatoma exhibited one slow-migrating esterase not present in normal liver, and this zone was also observed in the H-35 cells grown in tissue culture (comparison of slots 4 and 8). Differences in the esterase pattern of the Novikoff tumor grown *in vivo* and in culture were also evident (comparison of slots 6 and 7).

Glycoprotein Glycosyltransferase Activities of Liver and Morris Hepatomas

Table I shows the sialyltransferase and N-acetylglucosaminyltransferase activities of homogenates of liver and of the 3 hepatomas studied. With regard to sialyltransferase activities, it is apparent that the enzyme activities were mark-

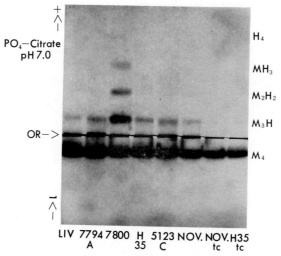

Fig. 3. LDH isozymes of liver and of various hepatomas

Samples of the high-speed supernatants of liver and of the 7794A, 7800, H-35, 5123C, and Novikoff hepatomas were applied. tc indicates samples of hepatomas grown in tissue culture. Approximately equal units of lactate dehydrogenase (LDH) activity were applied in each slot. The LDH isozymes are named in the right-hand margin. The electrophoretic buffer was sodium phosphate-sodium citrate, 0.02 M, pH 7.0.

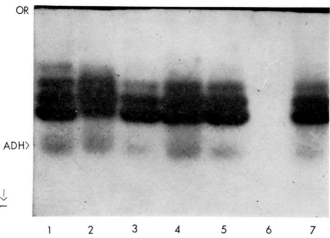

Fig. 4. Multiple molecular forms of sorbitol dehydrogenase of liver and of various hepatomas

Samples of the high-speed supernatants of liver and of the various hepatomas were applied at a protein concentration of approximately 3 g/100 ml. The electrophoretic positions of the four principal multiple molecular forms of SDH are evident, and the position of the only detectable zone of ADH activity is indicated in the margin. The slight variations in band patterns of SDH are not significant. The electrophoretic buffer was 0.021 M sodium borate, pH 8.6.

Slot 1, liver; slot 2, Morris hepatoma 7794A; slot 3, Morris hepatoma 7800; slot 4, Reuber hepatoma H-35; slot 5, Morris hepatoma 5123C; slot 6, Novikoff hepatoma; slot 7, Reuber hepatoma H-35tc.

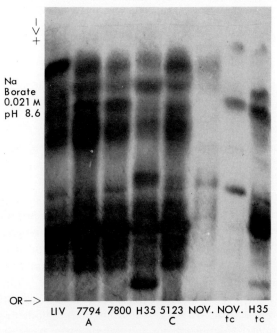

Fig. 5. Soluble esterases of liver and various hepatomas

Samples of the high-speed supernatants of liver and of the various hepatomas were applied at a protein concentration of approximately 2 g/100 ml. tc indicates the samples of hepatomas grown in tissue culture. β-Naphthyl acetate was the substrate, and Naphthanil Diazo Blue the indicator dye. The electrophoretic buffer was 0.021 M sodium borate, pH 8.6. No esterase activity was detected in the cathodal area of the gel.

Table I. Sialyltransferase (A) and N-Acetylglucosaminyltransferase (B) Activities of Liver and Hepatoma Homogenates

Tissue	No. of animals	A Mean±SD	B Mean±SD
Control liver	7	31.4±3.8	19.6±2.7
Hepatoma 7777	5	10.8±2.7	18.6±4.7
Host liver 7777	5	60.2±15.0	33.3±5.9
Hepatoma 7800	5	14.2±2.5	19.6±5.7
Host liver 7800	5	42.8±6.6	26.8±6.6
Hepatoma 5123D	5	26.8±3.3	30.4±10.7
Host liver 5123D	5	42.4±7.9	24.6±12.9

Values are expressed as nmoles/mg protein/hr.

edly lowered in hepatomas 7777 and 7800 but not significantly altered from the controls in hepatoma 5123D. The host livers showed significantly elevated enzyme activities for all 3 tumor types, the highest levels being in the livers of rats carrying hepatoma 7777. (The increases in activities of the glycosyltransferases in the host livers may relate to the well-established fact that the production of glycoproteins by the liver is stimulated in tumor-bearing animals.) In contrast

TABLE II. Ratio of Sialyltransferase Activity to N-Acetylglucosaminyltransferase Activities in Liver and Hepatomas

Hepatoma	Tumor	Host	Control
7777	0.58	1.81	
7800	0.72	1.60	
5123D	0.88	1.72	
			1.60

to the sialyltransferase activities, the activity of the N-acetylglucosaminyltransferase in the tumor homogenates was not significantly altered from the control values. Similarly, N-acetylglucosaminyltransferase activity in the host livers was not dramatically different from control values; only the hosts carrying hepatoma 7777 showed significant increase. It thus appears that sialyltransferase activity may be a very sensitive indicator of changes in GA function, whereas the N-acetylglucosaminyltransferase activity is more stable to changes in metabolic function. The ratio of sialyltransferase activity to N-acetylglucosaminyltransferase activity should therefore serve as a useful index of Golgi function. This ratio is shown in Table II. All three tumors can be seen to show decreased ratios as compared with control and host livers. In other experiments, methods previously used to prepare Golgi-enriched fractions from normal rat liver were applied to tumor homogenates. Four- to ninefold enrichments of the glycosyltransferases were found in membrane fractions from the tumors that, when examined by negative staining, showed the typical features of the GA of normal rat livers.

Ganglioside Patterns of Morris Hepatomas

A thin-layer chromatogram of the ganglioside patterns of human brain, hepatomas 5123D and 7777, and control liver is shown in Fig. 6. Control liver (slot 8) has a complex pattern, exhibiting bands corresponding in migration to most of the known gangliosides. The principal ganglioside bands seen in hepatoma 5123D extracts (slots 2–4) were those corresponding in migration to GM_3, GM_1, and GD_{1a}; traces of bands corresponding in migration to GD_{1b} and GT_{1b} were also observed. The principal bands exhibited by the ganglioside fraction of hepatoma 7777 (slots 5–7) were those corresponding in migration to GM_3, disialohematoside (21), and GD_{1a}; traces of bands corresponding in migration to GM_2, GM_1, and GT_{1b} were also visible. The ganglioside pattern of hepatoma 7800 is not shown but was similar to that of hepatoma 5123D. Thus, the major differences from control liver observed in the ganglioside fractions of all 3 hepatomas were the accumulation of GD_{1a} and diminutions of GD_{1b}, GT_{1b}, and GT_{1a}. Four samples of the ganglioside fraction of hepatomas 7777 and 5123D, and three samples of hepatoma 7800 were examined. The pattern of each of the hepatomas was constant. The ganglioside pattern of four host livers was also examined. Two of four samples showed a diminution of GT_{1b} and GT_{1a} but, otherwise the pattern was identical to that of control liver.

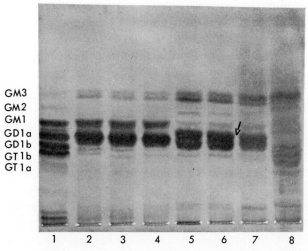

FIG. 6. Thin-layer chromatogram of various ganglioside preparations

Thin-layer chromatography for detection of gangliosides was performed as previously described (3, 36). The nomenclature for the gangliosides is that of Svennerholm (30). The band indicated by the arrow is tentatively designated as disialohematoside on the basis of its chromatographic migration (21).

Slot 1, human brain gangliosides; slots 2–4, ganglioside fraction from three individual 5123D hepatomas; slots 5–7, ganglioside fraction from three individual 7777 hepatomas; slot 8, ganglioside fraction from control rat liver.

DISCUSSION

1. Electrophoretic studies

Perhaps the most interesting finding to emerge from the electrophoretic studies is the observation that ADH and the most basic azoprotein of rat liver exhibit similar electrophoretic migrations under several conditions of electrophoresis. At the present time, no clear function of the azoproteins has been established. Arginase has been proposed as a target protein (27), and it has also been suggested (5, 10) that enzymes involved in the metabolism of carcinogens are theoretical candidates. Tasseron et al. (31) could find no arginase activity in a purified hydrocarbon-binding protein prepared from mouse skin. Sorof and his co-workers (27, 29) have discussed various factors—such as the possibility that interaction of a target protein with chemical carcinogens may inactivate it or cleave it and the known differences in the electrophoretic profiles of azoproteins following acute and chronic administration of azo dyes—that complicate the identification of azoproteins. However, on the basis of the following points, we tentatively propose that ADH may be a target protein for aminoazo dyes following their acute administration: (i) The previously described electrophoretic observations, (ii) fact that ADH is absent from the Novikoff hepatoma, (iii) suggestion of Ketterer and Christodoulides (10) that the principal azoprotein following acute administration may be a known sulfhydryl-containing enzyme, and (iv) ADH has a wide substrate specificity, several hydrophobic sites, and is present in relatively large amounts in liver. However, it is obvious that much work is required to support this hypothesis. In view of the fact that few, if any, studies have previously been made on the ADH activity of hepatomas, it is also of interest that the enzyme was detected in all of the tumors studied except for the Novikoff, and that its electrophoretic properties were similar in liver and the hepatomas.

Some of the other electrophoretic observations merit comment. LDH isozymes have been extensively studied in hepatomas and other tumors. Rosado et al. (22) have summarized their principal findings. Briefly, tumors in general showed a preponderance of the M form, but there appeared to be no clear correlation between either the total or subunit proportion of LDH activity and the growth rate or degree of differentiation of the tumor. The present observations would appear to agree with this statement. The finding that hepatoma 7800 exhibits a fetal type of LDH pattern confirms earlier observations by Ono (20).

It has previously been demonstrated (17) that the SDH of rat liver exhibits at least four multiple molecular forms. The nature of this polymorphism has not been established, but the multiple forms are thought to be very similar forms of the enzyme. It was also shown that hepatoma 5123tc preserved the same pattern of SDH enzymes as normal liver. In this study it has been shown that all of the hepatomas studied (except for the Novikoff) retained SDH activity and exhibited a pattern of SDH zones identical to that of liver. It is perhaps surprising that the activity of this enzyme should be so well retained by the hepatomas. However, it is relevant to note that Weinhouse (33) observed that the enzymes responsible for the subsequent metabolism of fructose were present at only slightly reduced activities in certain Morris hepatomas.

The esterase studies revealed that the Morris hepatomas retain a remarkably complex pattern of these enzymes. This is in marked contrast to the findings of Kreusser (13) in his study of the nonspecific esterases of normal and neoplastic tissues of the Syrian hamster. In this communication, Kreusser noted that the esterase patterns of the tumors studied did not resemble those of the parent tissues but did exhibit a marked tendency to assume a common pattern. It seems probable that the tumors studied were "dedifferentiated" and the present findings reemphasise the marked degree of biochemical differentiation of many of the Morris hepatomas.

2. Glycosyl transferases of liver and hepatomas

The possible role of the GA in plasma membrane biogenesis, the availability of suitable enzyme markers for the GA (23), and the fact that electron microscopic studies of the Morris hepatomas have indicated structural alterations in their GA (19) prompted the present study of glycosyltransferases in these tumors. It was found that the sialyltransferase activities in hepatomas 7777 and 7800 were significantly decreased relative to control values, whereas the N-acetylglucosaminyltransferase activities remained unchanged. These findings have been interpreted to indicate that the GA in hepatomas 7777 and 7800 functions in a different manner from the GA of normal or host liver cells. It is known that relatively fast-growing hepatoma cells are shifted from the major role in plasma glycoprotein synthesis and secretion carried out by normal liver to a nonsecreting, primarily reproductive mode (24); it appears reasonable to postulate that the GA in hepatomas 7777 and 7800 may therefore be shifted from a secretory mode to a predominantly membrane-generating mode, and this shift may be reflected in the altered sialyltransferase activities. It is apparent that hepatoma 5123D does not show significant changes in glycosyltransferase activities. Hepatoma 5123D is a relatively slow-growing tumor and hepatoma 7777 is a rapidly growing tumor; sialyltransferase activity may therefore depend, at least in part, on the growth rates of the hepatomas.

These studies have also shown that GA-enriched fractions can be obtained from all three hepatoma lines by methods previously applied to normal rat liver. It is therefore likely that methods can be developed for the production of highly purified hepatoma Golgi fractions. Such fractions can be further studied by both biochemical and electron microscopic methods in attempts to understand the functions of the GA in tumor cells.

3. *Ganglioside patterns of liver and hepatomas*

The observations on the ganglioside patterns of the Morris hepatomas are of interest in view of recent findings on the ganglioside patterns of cultured transformed cells. Hakomori and Murakami (7) found that, compared with control cells, polyoma-transformed BHK cells showed a marked decrease of hematoside (GM_3) and an accumulation of its precursor, ceramide lactoside. Mora et al. (16) reported that SV-40 transformed cells showed a loss of the more complex gangliosides found in control cells. Further observations have suggested that these changes can be explained by a deficiency of the galactosaminyltransferase converting GM_3 to GM_2 (4). Brady et al. (2) have described the ganglioside pattern of Morris hepatomas H5123 grown in tissue culture. The major findings were a marked decrease of GD_{1a} and an accumulation of GM_3 and GM_1 as compared with control hepatocytes. The present observations also show that chemically transformed hepatocytes growing *in vivo* demonstrate alterations of ganglioside patterns as compared with control liver. However, the changes are different from those described above for cultured neoplastic cells in that the hepatomas preserve a relatively complex pattern of gangliosides and appear to accumulate GD_{1a} but show reductions of GD_{1b}, GT_{1b}, and GT_{1a}. If alterations of gangliosides are important concomitants of malignancy, it would appear that the loss of the most complex ones would be of particular significance. Alterations of the activities of the various glycosidases degrading gangliosides or of the various glycosyltransferases involved in the biosynthesis of the more complex gangliosides could explain the differences of ganglioside patterns observed in the hepatomas. Investigation of the activities of these enzymes in liver and hepatomas would thus be of great interest. From observations in our laboratory, one might expect one difficulty to be incurred in such an investigation is that rat liver appears to possess a low capacity for synthesizing gangliosides (as evaluated by studying the incorporation of ^{14}C-glucosamine into individual gangliosides). Thus, presumably, the turnover of these compounds is relatively slow, and this may be reflected in low transferase activities. Also relevant to this discussion are the previously described low activities of the glycoprotein sialyltransferases in hepatomas 7777 and 7800. Siddiqui and Hakomori (26) have recently described findings very similar to ours in their investigation of the ganglioside patterns of several Morris hepatoma. These workers also demonstrated that fetal liver contained a pattern of gangliosides similar to that of adult liver and distinct from those of the hepatomas. Their findings are of particular interest in view of one of the principal themes of this symposium. Several workers have speculated that the alterations of gangliosides in surface membranes (2, 7) may be related to the abnormal biological properties of tumor cells. Studies on cultured cells indicate that gangliosides are enriched in surface membrane fractions (12, 25, 34). Recent studies in our laboratory (in collaboration with Dr. R. A. Hickie, University of Saskatchewan and Dr. V. Nigam, University of Sherbrooke) have demonstrated that all of the gangliosides present in whole rat liver are also represented in purified plasma membrane fractions from this organ. It appears safe to predict that there will be increasing attention paid to studies of ganglioside structure and metabolism in neoplastic cells. In particular, the series of hepatomas developed by Dr. Harold Morris should provide an excellent system for evaluating the significance of alterations of ganglioside composition in relation to the abnormal biological behavior of tumor cells.

Acknowledgments

This work was supported by grants from the Medical Research Council of Canada and the National Cancer Institute of Canada. Support, in part, from the Ontario Cancer Foundation is also appreciated. Finally, the invaluable cooperation of Dr. Harold P. Morris in making these studies possible is warmly acknowledged.

Addendum

Since this manuscript was submitted, two publications have appeared that are pertinent to the experiments described in this work relating to the possibility that ADH is one of the azoproteins of rat liver. Arslanian *et al.* (*1*) have described the purification and properties of rat liver ADH. Litwack *et al.* (*2*) have also described some of the properties of the principal azoprotein of rat liver. The properties of these two proteins do not appear particularly similar. It is thus possible that ADH may co-electrophorese with this protein in the electrophoretic conditions described herein. This comment does not affect the validity of the observations described but rather their possible significance.

1. Arslanian, M. J., Pascoe, E., and Reinhold, J. G. Rat liver alcohol dehydrogenase: purification and properties. *Biochem. J.*, **125**, 1039 (1971).
2. Litwack, G., Ketterer, B., and Arias, I. M. Ligandin: a hepatic protein which binds steroids, bilirubin, carcinogens and a number of exogenous organic anions. *Nature*, **234**, 466 (1971).

REFERENCES

1. Bergmeyer, H. U. " Methods of Enzymatic Analysis," Academic Press Inc., New York (1965).
2. Brady, R. O., Borek, C., and Bradley, R. M. Composition and synthesis of

gangliosides in rat hepatocyte and hepatoma cell lines. *J. Biol. Chem.*, **244**, 6552 (1969).

3. Cheema, P., Yogeeswaran, G., Morris, H. P., and Murray, R. K. Ganglioside patterns of three Morris minimal deviation hepatomas. *FEBS Letters*, **11**, 181 (1970).

4. Cumar, F. A., Brady, R. D., Kolodny, E. H., McFarland, V. W., and Mora, P. T. Enzymatic block in the synthesis of gangliosides in DNA virus-transformed tumorigenic mouse cell lines. *Proc. Natl. Acad. Sci. U.S.*, **67**, 757 (1970).

5. DeBaun, J. R., Rowley, J. Y., Miller, E. C., and Miller, J. A. Sulfotransferase activation of N-hydroxy-2-acetylaminofluorene in rodent livers susceptible and resistant to this carcinogen. *Proc. Soc. Exp. Biol. Med.*, **129**, 268 (1968).

6. Fine, I. H. and Costello, L. A. The use of starch electrophoresis in dehydrogenase studies. *Methods Enzymol.*, **6**, 958–972 (1963).

7. Hakomori, S. and Murakami, W. T. Glycolipids of hamster fibroblasts and derived malignant-transformed cell lines. *Proc. Natl. Acad. Sci. U.S.*, **59**, 254 (1968).

8. Hudgin, R. L., Murrary, R. K., Pinteric, L., Morris, H. P., and Schachter, H. The use of nucleotide-sugar: glycoprotein glycosyltransferases to assess Golgi apparatus function in Morris hepatomas. *Can. J. Biochem.*, **49**, 61 (1971).

9. Kalant, H., Murray, R. K., and Mons, W. Effect of EDTA on leakage of proteins from slices of normal rat liver and DAB-induced hepatoma. *Cancer Res.*, **24**, 570 (1964).

10. Ketterer, B. and Christodoulides, L. Two specific azo dye-carcinogen-binding proteins of the rat liver: the identity of amino acid residues which bind the azo dye. *Chem. Biol. Interactions*, **1**, 173 (1969).

11. Ketterer, B., Ross-Mansell, P., and Whitehead, J. K. The isolation of carcinogen-binding protein from livers of rats given 4-dimethylaminoazobenzene. *Biochem. J.*, **103**, 316 (1967).

12. Klenk, H. D. and Choppin, P. W. Glucosphingolipids of plasma membranes of cultured cells and an enveloped virus (SV 5) grown in these cells. *Proc. Natl. Acad. Sci. U.S.*, **66**, 57 (1970).

13. Kreusser, E. H. Nonspecific esterases in normal and neoplastic tissues of the Syrian hamster: a zymogram study. *Cancer Res.*, **26**, 2181 (1966).

14. Lowry, O. H., Rosebrough, N. J., Farr, A. L., and Randall, R. J. Protein measurements with the Folin phenol reagent. *J. Biol. Chem.*, **193**, 265 (1951).

15. Moore, B. W. and Lee, R. H. Chromatography of rat liver soluble enzymes and localization of enzyme activities. *J. Biol. Chem.*, **235**, 1359 (1960).

16. Mora, P. T., Brady, R. O., Bradley, R. M , and McFarland, V W. Gangliosides in DNA virus-transformed and spontaneously transformed tumorigenic mouse cell lines. *Proc. Natl. Acad. Sci. U.S.*, **63**, 1290 (1969).

17. Murray, R. K., Gadacz, I., Bach, M., Hardin, S., and Morris, H. P. Metabolic and electrophoretic studies of rat liver sorbitol dehydrogenase. *Can. J. Biochem.*, **47**, 587 (1969).

18. Murray, R. K., Kalant, H., Guttman, M., and Morris, H. P. Studies on composition and leakage of proteins and esterases of normal rat liver and Morris hepatoma 5123tc. *Cancer Res.*, **27**, 403 (1967).

19. Murray, R. K., Khairallah, L., Ragland, W., and Pitot, H. C. The biochemical morphology and morphogenesis of hepatomas. *Int. Rev. Exp. Pathol.*, **6**, 229 (1968).

20. Ono, T. Enzyme patterns and malignancy of experimental hepatomas. *GANN Monograph*, **1**, 189 (1966).

21. Penick, R. J., Meisler, M. H., and McLuer, R. H. Thin-layer chromatographic studies of human brain gangliosides. *Biochim. Biophys. Acta*, **116**, 279 (1966).
22. Rosado, A., Morris, H. P., and Weinhouse, S. Lactate dehydrogenase subunits in normal and neoplastic tissues of the rat. *Cancer Res.*, **29**, 1673 (1969).
23. Schachter, H., Jabbal, I., Hudgin, R. L., Pinteric, L., McGuire, E. J., and Roseman, S. Intracellular localization of liver sugar nucleotide glycoprotein glycosyltransferases in a Golgi-rich fraction. *J. Biol. Chem.*, **245**, 1090 (1970).
24. Schreiber, G., Boutwell, R. K., Potter, V. R., and Morris, H. P. Lack of secretion of serum protein by transplanted rat hepatomas. *Cancer Res.*, **26**, 2357 (1966).
25. Sheinin, R., Onodera, K., Yogeeswaran, G., and Murray, R. K. Studies of components of the surface of normal and virus-transformed mouse cells. *In* " The Biology of Oncogenic Viruses" (Proc. Second Lepetit Colloquium), North-Holland Publishing Co., Amsterdam, pp. 274–285 (1971).
26. Siddiqui, B. and Hakomori, S. Change of glycolipid pattern in Morris hepatomas 5123 and 7800. *Cancer Res.*, **30**, 2930 (1970).
27. Sorof, S. Carcinogen-protein conjugates in liver carcinogenesis. *In* "Physico-chemical Mechanisms of Carcinogenesis" (The Jerusalem Symposium on Quantum Chemistry), ed. by E. D. Bergmann and B. Pullman, The Israel Academy of Sciences and Humanities, Jerusalem, Vol. 1, pp. 208–217 (1969). (Distributed by Academic Press Inc., New York).
28. Sorof, S., Cohen, P. P., Miller, E. C., and Miller, J. A. Electrophoretic studies on the soluble proteins from livers of rats fed aminoazo dyes. *Cancer Res.*, **11**, 383 (1951).
29. Sorof, S., Young, E. M., McBride, R. A., and Coffey, C. B. On protein targets of chemical carcinogens: dissimilar molecular sizes of the principal protein conjugates. *Cancer Res.*, **30**, 2029 (1970).
30. Svennerholm, L. Ganglioside metabolism. *In* " Comprehensive Biochemistry," ed. by M. Florkin and E. H. Stotz, Academic Press Inc., New York, Vol. 18, pp. 201–227 (1970).
31. Tasseron, J. G., Diringer, H., Frohwirth, N., Mirvish, S. S., and Heidelberger, C. Partial purification of soluble protein from mouse skin to which carcinogenic hydrocarbons are specifically bound. *Biochemistry*, **9**, 1636 (1970).
32. Wallach, D. F. Cellular membranes and tumor behaviour: a new hypothesis. *Proc. Natl. Acad. Sci. U.S*, **61**, 868 (1968).
33. Weinhouse, S. Glycolysis, respiration and enzyme deletions in slow-growing hepatic tumors. *GANN Monograph*, **1**, 189 (1966).
34. Weinstein, D. B., Marsh, J. B., Glick, M. C., and Warren, L. Membranes of animal cells. VI. The glycolipids of the L cell and its surface membrane. *J. Biol. Chem.*, **245**, 3928 (1970).
35. Whitcutt, J. M., Sutton, D. A., and Nunn, J. R. Carcinogenesis: changes in the properties of some rat-liver proteins after administration of 4-dimethylamino-3′-methylazobenzene. *Biochem. J.*, **75**, 557 (1960).
36. Yogeeswaran, G., Wherrett, J. R., Chatterjee, S., and Murray, R. K. Gangliosides of cultured mouse cells: partial characterization and demonstration of ^{14}C-glucosamine incorporation. *J. Biol. Chem.*, **245**, 6718 (1970).

ISOZYMES OF BRANCHED CHAIN AMINO ACID TRANSAMINASE IN NORMAL RAT TISSUES AND HEPATOMAS

Akira Ichihara and Koichi Ogawa

*Institute for Enzyme Research, School of Medicine, Tokushima University**

Branched-chain amino acid transaminase (L-leucine: 2-oxoglutarate aminotransferase, EC 2.6.1.6) of rats has three types of isozymes, isozymes I, II, and III. Isozyme I is widely distributed in various tissues, while isozyme III is found only in the brain. These two isozymes have very similar enzymic characters and are both active with all three branched-chain amino acids, but they can be distinguished from one another by their chromatographic and immunochemical properties. Isozyme II has a different substrate specificity, specific for leucine only, and found only in the liver. This isozyme is readily inducible by various treatments, such as cortisol injection or administration of a high protein diet. Fetal rat liver contains only isozyme I, which decreases in activity during development, while isozyme II appears and increases after birth.

Various Yoshida ascites hepatomas contain both isozymes I and III, but not isozyme II. The isozyme III in hepatomas is indistinguishable from that in the brain in its enzymic properties, chromatographic behavior, and immunochemical cross-reaction. Different strains of Morris hepatoma showed a variety of isozyme patterns, *i.e.*, normal type, fetal type, or all three isozymes. There is no definite correlation, however, between the isozyme pattern and the growth rate of tumors. Among nine hepatomas induced by 3'-methyl-4-dimethylaminoazobenzene, seven showed the isozyme pattern of typical tumors with isozymes I and III, suggessting that the appearance of isozyme III is a characteristic feature of liver carcinogenesis.

Branched-chain amino acids (valine, leucine, and isoleucine) are all essential for animal growth, and their similarity in structure makes them compete nutritionally with each other. Their metabolic pathways are well known; valine is glycogenic, leucine is strongly ketogenic, and isoleucine has both properties (*20*). The first step in their metabolism was thought to be transamination rather than oxidative deamination, but transaminase was not characterized in detail until recently. In 1966, we (*8*) reported that there is a specific transaminase for these three amino acids in hog heart. These amino acids compete with each other as

---

* Kuramoto-cho 3-18-1, Tokushima 770, Japan (市原　明, 小川紘一).

substrates for this enzyme. Taylor and Jenkins (*28*) showed that the molecular weight of this transaminase is 75,000 and that it contained one pyridoxal phosphate moiety. Subsequently, we found another type of transaminase, in hog brain supernatant, which is very similar in substrate specificity to the enzyme in hog heart but differs in chromatographic and immunochemical characters (*3*). We tentatively named the heart enzyme isozyme I and the brain enzyme isozyme III. There are several reports on isozymes of other transaminases, but they are all isozymes with different subcellular localization (see Refs. in *2*). The isozymes of aspartate transaminase (L-aspartate:2-oxoglutarate aminotransferase, EC 2.6.1.1) have been studied in detail and differences in their amino acid sequences have been reported (*16*). Branched-chain amino acid transaminase was also localized in both the supernatant and mitochondrial fractions of beef heart, and it was found that the enzymes in the two fractions have different properties (*2*). The isozymes of aspartate transaminase were suggested to be physiologically important in the regulation of the gluconeogenic path, urea formation, and the tricarboxylic acid cycle (*4*, *13*). The physiological significance of the isozymes of branched-chain amino acid transaminase has not been studied in detail, but, since enzymes degrading branched-chain keto acids are localized in the mitochondrial fraction, it is likely that the transaminase in the supernatant is involved in control of the amino acid concentration in protein synthesis, and that the isozyme in the mitochondrial fraction is involved in degradation of branched-chain amino acids (*6*, *15*).

It has been shown that during cell growth the concentration of amino acids in the cell is controlled to meet the need for rapid protein synthesis and thus transaminase activity fluctuates (see Refs. in *9*). These findings prompted us to survey the isozyme patterns in various tissues of rats and to study their changes under various conditions, including hepatomas.

Isozyme Patterns in Various Rat Tissues

Figure 1 shows that when crude extracts of various rat tissues were subjected to DEAE-cellulose chromatography, most of the extracts were found to contain only one active fraction, which was eluted with 0.02 M phosphate buffer (isozyme I). However, rat brain contained isozyme III, eluted with 0.2 M phosphate buffer. It is interesting that liver contained another type of enzyme, isozyme II, which was eluted with 0.18 M phosphate buffer. Table I summarizes the properties of these three isozymes. Isozymes I and III were very similar in all of their properties except for their chromatographic and immunochemical characteristics (*18*), while isozyme II was quite distinct from the other two isozymes in that it was specific for leucine only and was found only in liver (*1*). It is interesting to note that among various living organisms examined only rodent livers contained isozyme II (Table II). The livers of other animals, including humans, contained either isozyme I or III, or both, but not isozyme II. It is also noteworthy that there are two transaminases (A and B) in *Escherichia coli* which have substrate specificities similar to the enzymes in rat liver with respect to branched-chain amino acids but a much broader substrate specificity (*21*). Recently,

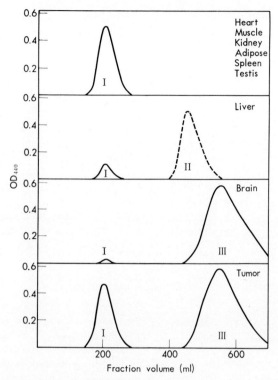

FIG. 1. Elution pattern of the branched-chain amino acid transaminase of various tissues of rats from DEAE-cellulose column (*18*)

The supernatant of tissue homogenates was applied to a DEAE-cellulose-column, and the column was eluted with a concentration gradient of 5×10^{-3} M to 3×10^{-1} M phosphate buffer. For tumor extract, Yoshida ascites hepatoma (AH-130) was used. ——, activity for all three branched chain amino acids; - - -, activity for leucine only.

TABLE I. Properties of Rat Isozymes of Branched-chain Amino Acid Transaminase (*1, 18*)

	Isozyme		
	I	II	III
Concentration of eluent buffer on DEAE-cellulose chromatography (M)	0.02	0.18	0.20
K_m for substrates (mM)			
Valine	4.3		2.5
Leucine	0.8	25.0	0.6
Isoleucine	0.8		0.5
α-Ketoglutarate	1.0	0.07	0.5
Localization in body	all tissues	liver	brain
Activation by 2-mercaptoethanol	+	−	+
Inhibition (%) by antiserum against			
Isozyme I	57	0	0
Isozyme III	0	0	94

TABLE II. Distribution of Isozymes of Branched-chain Amino Acid Transaminase in Various Animal Livers and Microorganisms

	Specific activity for leucine (units/mg protein)	Activity ratio[a]	Distribution of isozyme (%)		
			I	II	III
Human	38.3	1.1	92	0	8
Pig	9.3	1.0	20	0	80
Cow	2.8	1.2			
Dog	6.6	1.1	60	0	40
Cat	14.0	0.9	90	0	10
Rat	18.0	8.5	30	70	0
Rabbit	3.2	3.8	50	50	0
Guinea pig	16.2	10.8	60	40	0
Mouse	18.0	3.0	60	40	0
Chick	42.8	0.7	0	0	100
Frog	9.3	0.8	100	0	0
Fish	5.1	1.2			
Escherichia coli	88	2.0	0	30	70
Pseudomonas fluorescence	7700	1.2	100	0	0

[a] The activity ratio was calculated as the activity for leucine divided by that for isoleucine or valine.

Coleman *et al.* (*5*) crystallized an enzyme which corresponds to enzyme B in *E. coli* and to either isozyme I or III in our terminology. A survey of the isozyme patterns in different animals seems to provide information on molecular evolution and can be interpreted by gene expression along lines similar to the isozyme patterns in hepatomas discussed below.

Changes of Isozyme Activities under Various Physiological Conditions

The unique properties of isozyme II in rat liver suggest that this isozyme may play a role in leucine metabolism. It is induced rapidly, within 6 hr after cortisol injection, and this induction can be prevented by concomitant treatment with Actinomycin or Puromycin (*10*). It is interesting that a hypophyseal factor, probably growth hormone, inhibited the induction of isozyme II by cortisol (Table III). Hypophysectomy alone induced isozyme I in kidney, as did repeated injections of cortisol for several days (*24*). These results suggest that isozymes I and II are under the control of cortisol, but that the half-life of isozyme II is much shorter than that of isozyme I in kidney. Induction of isozyme II by a high-protein diet was not mediated by glucocorticoid but by increasing the amino acid concentration in liver cells. The effects of a high protein diet and cortisol were additive. Effects of amino acid concentration, cortisol, and growth hormone on the induction of tyrosine transaminase (L-tyrosine:2-oxoglutarate aminotransferase, EC 2.6.1.5) have been discussed in detail elsewhere (*24*).

TABLE III. Effect of Cortisol and Growth Hormone on Induction of Isozymes (24)

Treatment	Specific activity for leucine (units/mg protein)	
	Isozyme II in liver	Isozyme I in kidney
None	20.5±3.0	105±10
Cortisol	139.2±15.0	380±22
Growth hormone and cortisol	14.5±1.2	
Hypophysectomy	28.4±7.2	225±28
Adrenalectomy	10.5±3.0	66.4±4.0
Adrenalectomy and 75% protein diet	52.3±8.3	

For measurement of the activity of liver isozyme II, 10 mg/100 g body weight of cortisol was injected i.p. and the animals were killed 6 hr later. For measurement of isozyme I in kidney, 5 mg/100 g body weight of cortisol was injected daily for five days and 0.3 mg/100 g body weight of growth hormone was given daily for three days.
The values are mean±SE.

The physiological significance of isozyme II is not yet known, but leucine is not only a potent ketogenic amino acid but is also metabolized to β-hydroxy-β-methylglutaryl-CoA, which is a precursor of cholesterol (see Refs. in 20). These important metabolic pathways may account for the presence of the leucine-specific transaminase, isozyme II.

It has been shown that the transamination step is rate-limiting in the metabolism of branched-chain amino acids (6, 15) and, therefore, change of these isozyme activities under various conditions may be important in regulation of the metabolism of these amino acids.

Changes in the Isozyme Patterns in Developing, Regenerating, and Cancerous Liver

Fetal rat liver contains only isozyme I, the activity of which decreases during development, while isozyme II appears in the liver after birth and increases in activity to the adult level (Fig. 2). However, we (9) found no indication of the appearance of isozyme III during the developmental period as early as the 16th day after gestation.

Many functional enzymes, such as serine dehydratase (L-serine hydrolyase, EC 4.2.1.13), tryptophan pyrrolase (L-tryptophan-oxygen oxidoreductase, EC 1.13.1.12), tyrosine trnasaminase, and various glycolytic enzymes are not present in fetal liver but appear after birth (see Refs. in 9, 19). It seems probable that, among the isozymes of branched-chain amino acid transaminase, isozyme II, which is only found in liver and is readily inducible under various physiological conditions, is a functional enzyme.

In regenerating liver, isozyme II was markedly induced 6 hr after partial hepatectomy, while the activity of isozyme I remained constant. Enzyme III was not found in liver during regeneration.

In oncogenesis, many functional enzymes are deleted and the isozyme pattern changes to that found in fetal liver. This phenomenon has been called dis- (or de-)differentiation (25, 26) or blocked ontogeny (19). However, previous studies

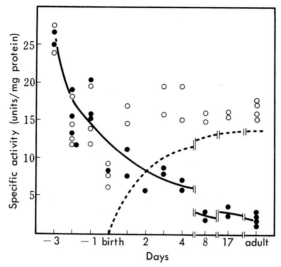

Fig. 2. Changes of isozyme activities during development (9)
● activity in units for isoleucine or valine; ○ activity for leucine.
Dotted line indicates difference of activities between valine and leucine, namely appearance of isozyme II.

on this phenomenon have been mostly on glycolytic enzymes (7, 22, 25, 26), which show a change to the isozyme pattern found in muscle. There are very few reports on changes in the enzymes involved in amino acid metabolism (11, 12, 14, 23).

Yoshida ascites hepatoma (AH-130) contains isozymes I and III in about equal amounts, but no isozyme II, as shown in Fig. 1 (18). Isozyme III from the hepatoma was purified and compared with the enzyme in rat brain. The two enzymes were indistinguishable in all characteristics examined, namely, their K_m values for substrates, pH optimum curves, electrophoretic mobilities, and immunochemical properties (Figs. 3 and 4). The appearance of isozyme III and disappearance of isozyme II were also observed in other strains of Yoshida ascites hepatoma (AH-60c, 66f, 143a, and 7974) and Yoshida sarcoma (17). The total transaminase activity, *i.e.*, the sum of the activities of isozymes I and III, was usually much higher than that in normal liver, which is the sum of the activities of isozymes I and II. Several Morris hepatomas were also examined and were found to have quite diverse isozyme patterns, namely, fetal type (7777, and 7795), normal type (7794A, and 7316A), and a mixture of all three isozymes (5123tc, 7793, and one strain of 7795). However, there was no correlation between the isozyme pattern and the growth rate of these tumors (17).

The above observations were made on established strains of hepatomas and so it seemed possible that the observed changes of the isozyme patterns could have occurred during transplantation rather than as a result of oncogenesis *per se*. Therefore, the isozyme patterns of primary hepatomas induced in rats by feeding 3′-methyl-4-dimethylaminoazobenzene were examined. The rats were fed the carcinogen from month 1 to month 5 and then laboratory chow was given for

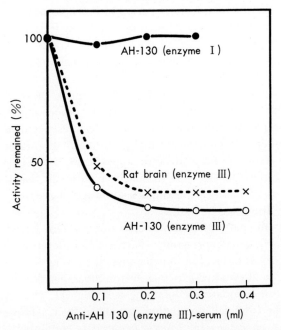

FIG. 3. Neutralization of activity of tumor isozyme III and brain isozyme III by antiserum against hepatoma isozyme III (*18*)

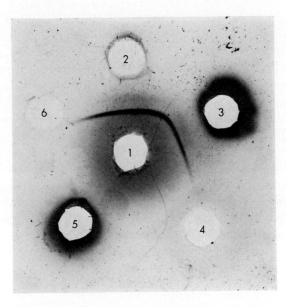

FIG. 4. Immunological cross-reactions of type III isozymes of rat brain and hepatoma (*18*)

1, antiserum against hepatoma isozyme III; 2, hepatoma isozyme III; 3, brain isozyme III; 4, hepatoma isozyme I; 5, rat liver isozymes I and II; 6, hog brain isozyme III.

several months to make the total feeding period of seven months. The incidence of tumors increased with duration of carcinogen administration. The histology and isozyme patterns of these tumors were examined. Cancer developed in nine rats and were classified as hepatomas, cholangiocellular carcinoma, and mixed type tumors. It was found that, with only one exception, isozyme III appeared in cancerous tissues. Noncancerous, hyperplastic tissues showed the normal pattern of isozymes. Seven of the nine cancerous tissues showed the typical isozyme pattern of a poorly differentiated tumor, the presence of isozymes I and III. One of the other two had the normal pattern and the other had all three types (*17*). These results indicate that the appearance of isozyme III is a very good indication of carcinogenesis. It should be mentioned that no isozyme III appeared in the livers of rats bearing tumors, although the activities of isozymes I and II tended to decrease.

Changes in isozyme patterns in tumor cells such as these have been explained as due to aberrations of gene expression. Isozyme II is derepressed and isozyme III is repressed in normal tissue, but in hepatomas the reverse situation may occur. There is, however, an alternative explanation for the appearance of isozyme III, *i.e.*, selection of a cell population containing isozyme III which is present as a small minority of the total cell population in normal liver. However, this possibility is unlikely because several Morris hepatomas contained all three types of isozyme, while some hyperplastic tissues contained no isozyme III. Recently we found that tissue culture cells isolated from rat liver, and cloned and grown from single cells contained only isozyme I, but produced isozyme III when grown in a medium containing 4-nitroquinoline 1-oxide (unpublished data).

We do not yet know whether a single cell can contain all of these isozymes. The finding of a hybrid enzyme and of isozymes composed of homologous subunits should effectively prove this possibility. Further studies are required on the isozymes of branched-chain amino acid transaminase. However, all three these types of isozymes were found in several hepatomas, so it seems very likely that all isozymes can be present in the same cells.

It is interesting that the mitochondrial fraction of normal liver also contains isozymes I and II and that the content of isozyme II is much higher than in the supernatant fraction (*1*), while the mitochondrial fraction of Yoshida ascites hepatoma AH-130 contains isozyme I, but not isozymes II and III (*17*). This suggests that aberrations of gene expression occur independently in the two fractions in hepatomas. This possibility is supported by the fact that in hog heart the type I isozymes in the supernatant and mitochondrial fractions have different properties, as discussed before (*2*). Thus the isozymes I and II in the two fractions of rat liver are probably different proteins which are controlled by different genes.

REFERENCES

1. Aki, K., Ogawa, K., and Ichihara, A. Transaminase of branched-chain amino acids. IV. *Biochim. Biophys. Acta*, **159**, 276–284 (1968).

2. Aki, K., Ogawa, K., Shirai, A., and Ichihara, A. Transaminase of branched-chain amino acids. III. *J. Biochem. (Tokyo)*, **62**, 610–617 (1967).
3. Aki, K., Yokojima, A., and Ichihara, A. Transaminase of branched chain amino acids. VII. *J. Biochem. (Tokyo)*, **65**, 539–544 (1969).
4. Charles, R., Tager, J. M., and Slater, E. C. Citrulline synthesis in rat-liver mitochondria. *Biochim. Biophys. Acta*, **131**, 29–41 (1967).
5. Coleman, M. S., Soucie, W. G., and Armstrong, F. B. Branched-chain acid aminotransferase of *Salmonella typhymurium J. Biol. Chem.*, **246**, 1310–1312 (1971).
6. Dawson, A. G., Hird, F. J. R., and Morton, D. J. Oxidation of leucine by rat liver and kidney. *Arch. Biochem. Biophys.*, **122**, 426–433 (1967).
7. Farina, F. A., Adelman, R. C., Lo, C. H., Morris, H. P., and Weinhouse, S. Metabolic regulation and enzyme alterations in the Morris hepatomas. *Cancer Res.*, **28**, 1897–1900 (1968).
8. Ichihara, A. and Koyama, E. Transaminase of branched chain amino acids. I. *J. Biochem. (Tokyo)*, **59**, 160–169 (1966).
9. Ichihara, A. and Takahashi, H. Tansaminase of branched chain amino acids. V. *Biochim. Biophys. Acta*, **167**, 274–279 (1968).
10. Ichihara, A., Takahashi, H., Aki, K., and Shirai, A. Transaminase of branched chain amino acids. II. *Biochem. Biophys. Res. Commun.*, **26**, 674–678 (1967).
11. Katunuma, N., Kuroda, Y., Sanada, Y., Towatari, T., Tomino, I., and Morris, H. P. Anomalous distribution of glutaminase isozyme in various hepatomas. *Adv. Enzyme Regulation*, **8**, 281–287 (1970).
12. Knox, W. E., Tremblay, G. C., Spanier, B. B., and Friedell, G. H. Glutaminase activities in normal and neoplastic tissues of the rat. *Cancer Res.*, **27**, 1456–1458 (1967).
13. Lardy, H. A., Paetkau, V., and Walter, P. Paths of carbon in gluconeogenesis and lipogenesis: the role of mitochondria in supplying precursors of phosphoenolpyruvate. *Proc. Natl. Acad. Sci. U.S.*, **53**, 1410–1415 (1965).
14. Linder-Horowitz, M., Knox, W. E., and Morris, H. P. Glutaminase activities and growth rates of rat hepatomas. *Cancer Res.*, **29**, 1195–1199 (1969).
15. McFarlane, I. G. and von Holt, C. Metabolism of leucine in protein-calorie deficient rats. *Biochem. J.*, **111**, 565–571 (1969).
16. Morino, Y. and Watanabe, T. Primary structure of pyridoxal phosphate binding site in the mitochondrial and extramitochondrial aspartate aminotransferases from pig heart muscle. Chymotryptic peptides. *Biochemistry*, **8**, 3412–3417 (1969).
17. Ogawa, K. and Ichihara, A. Isozyme patterns of branched chain amino acid transaminase in various rat hepatomas. *Cancer Res.*, **32**, 1257–1263 (1972).
18. Ogawa, K., Yokojima, A., and Ichihara, A. Transaminase of branched chain amino acids. VII. *J. Biochem. (Tokyo)*, **68**, 901–911 (1970).
19. Potter, V. R. Summary of discussion on neoplasia. *Cancer Res.*, **28**, 1901–1907 (1968).
20. Rodwell, V. R. Carbon catabolism of amino acids. *In* " Metabolic Pathways," ed. by D. M. Greenberg, Academic Press Inc., New York, Vol. 3, pp. 191–235 (1969).
21. Rudman, D. and Meister, A. Transamination in *Escherichia coli*. *J. Biol. Chem.*, **200**, 591–604 (1953).
22. Sato, K. and Tsuiki, S. Fructose 1,6-diphosphatase of rat tissues and transplantable rat hepatomas. *Arch. Biochem. Biophys.*, **129**, 173–180 (1969).

23. Sheid, B., Morris, H. P., and Roth, J. S. Distribution and activity of aspartate aminotransferase in some rapidly proliferating tissues. *J. Biol. Chem.*, **240**, 3016–3022 (1965).
24. Shirai, A. and Ichihara, A. Transaminase of branched chain amino acids. VIII. *J. Biochem. (Tokyo)*, **70**, 741–748 (1971).
25. Suda, M., Tanaka, T., Sue, F., Harano, Y., and Morimura, H. Dedifferentiation of sugar metabolism in the liver of tumor-bearing rat. *GANN Monograph*, **1**, 127–141 (1966).
26. Sugimura, T., Matsushima, T., Kawachi, T., Hirata, Y., and Kawabe, S. Molecular species of aldolases and hexokinases in experimental hepatomas. *GANN Monograph* **1**, 143–149 (1966).
27. Sugimura, T., Sato, S., and Kawabe, S. The presence of aldolase C in rat hepatoma. *Biochem. Biophys. Res. Commun.*, **39**, 626–630 (1970).
28. Taylor, R. T. and Jenkins, W. T. Leucine aminotransferase. II. *J. Biol. Chem.*, **241**, 4396–4405 (1966).

REGULATION OF THE LEVELS OF MULTIPLE FORMS OF SERINE DEHYDRATASE AND TYROSINE AMINOTRANSFERASE IN RAT TISSUES[*1]

Henry C. Pitot, Yoshifumi Iwasaki,[*2] Hideo Inoue,[*3]
Charles Kasper, and Harvey Mohrenweiser[*4]

*The Departments fo Oncology and Pathology, McArdle Laboratory,
University of Wisconsin Medical School*

The enzymes serine dehydratase and tyrosine aminotransferase have been extensively studied as models of the environmental regulation of enzyme levels in liver, hepatomas, and other tissues. Studies from this laboratory have demonstrated that both crude and purified forms of each of these enzymes are actually a mixture of very closely related molecules having the same catalytic activity.

Serine dehydratase, which is unique to the liver in the rodent, occurs in two forms separable by DEAE-cellulose ion exchange chromatography and acrylamide gel electrophoresis. The two forms of the enzyme have been crystallized and are extremely similar, having the same molecular weight, subunit composition, kinetic constants for substrates, heat inactivation characteristics, phosphate content, and spectral characteristics. The amino acid composition of the two forms shows slight changes in the proline and lysine content, the significance of which is not certain at present. Preliminary studies of fingerprints of the peptides of the two forms show a maximum of 1 or 2 altered peptides between the two forms. Studies on the regulation of the two forms demonstrate clearly that the synthesis of the more electropositive form is regulated by the hormone glucagon, while that of the other form is regulated by cortisone. Isotopic studies indicate no precursor-product relationship between the two forms.

Tyrosine aminotransferase has been separated into four forms by the use of hydroxylapatite chromatography. The form having the least affinity for the column appears to be present in all tissues thus far studied, heart and kidney as well as liver. The level of this form does not change under various hormonal and dietary conditions and is most easily lost during the purification of the enzyme. Of the other three quite similar forms, that having the greatest affinity for the column appears to be regulated by insulin levels in the organism.

[*1] The work reported in this paper was supported in part by grants from the National Cancer Institute (CA-07175) and the American Cancer Society (E-588).

[*2] Present Address: 1st Division, Department of Medicine, Medical School, Kyoto University, Kyoto 606, Japan (岩崎良文).

[*3] Present address: Department of Biochemistry, The Dental School, Osaka University, Osaka 530, Japan (井上秀夫).

[*4] National Cancer Institute, Wisconsin 53706, U.S.A.

The exact regulation of the two intermediate forms is not certain as yet, but appears to involve steroids as well as cyclic nucleotides. The two intermediate forms are not present before birth, but appear on the day of birth and are present thereafter. The insulin-regulated form does not appear until the animal is weaned. Four Morris hepatomas were studied and found to contain all but the insulin-regulated form.

These studies indicate that, for these two enzymes, structurally quite similar forms of each exist, each form under the regulation of a different hormone or environmental condition. A proposed nomenclature to take into account the possibility of multiple forms of an enzyme resulting from post-transcriptional events is presented.

The presence of multiple molecular forms of proteins having identical catalytic activity has been known for some years. In rat liver, the demonstration of four forms of glucose phosphotransferase (ATP : D-hexose 6-phosphotransferase, EC 2.7.1.1 and 2.7.1.2) by Sols and his associates (17), Weinhouse and his associates (18), and others (11, 15) serves to show that certain isozymic forms of an enzyme are characteristically present in certain organs. The various forms of lactate dehydrogenase (L-lactate: NAD oxidoreductase, EC 1.1.1.27) have now been well documented and, in fact, the levels of the various isozymic forms of this enzyme in different organs appear to be the result of differential rates of degradation of each of the two separable subunits (5).

Relatively few studies have been carried out on the effects of the environment on the level of various isozymes in a specific organ. The extreme sensitivity of glucokinase synthesis to the presence of insulin (17, 18) is one such example. Since glucokinase is found only in liver, it would appear that in part this is a function of the unique differentiation of this organ. Previous studies from this laboratory have demonstrated the unique characteristics of the induction of the enzyme, serine dehydratase (L-serine hydro-lyase (deaminating), EC 4.2.1.13) in rat liver (9, 10). The experiments described in this paper demonstrate the isolation and characterization of two forms of serine dehydratase from rat liver as well as the isolation of four forms of the enzyme tyrosine aminotransferase (L-tyrosine:2-oxoglutarate aminotransferase, EC 2.6.1.5). Tomkins and his associates (1) have recently presented evidence that tyrosine aminotransferase induced by a synthetic corticoid is composed of four similar or identical subunits. However, the subunits seem to be enzymatically inactive. Earlier studies by Kenney (12) and Oliver and associates (8, 16) indicated that rat liver tyrosine aminotransferase occurred in several separable forms. In addition, Holt and Oliver (8) have suggested that the multiple molecular forms were under the regulation of several different hormones.

Multiple Forms of Serine Dehydratase

As has been described earlier (9), two forms of serine dehydratase have been characterized from rat liver. Serine dehydratase was crystallized from rat liver by a modification of the procedure of Nakagawa and Kimura (14). The enzyme

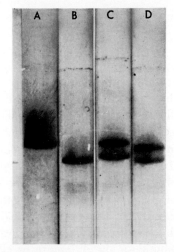

FIG. 1. Polyacrylamide gel electrophoretic pattern of crystalline serine dehydratase (SDH) isolated from rat liver

Gels A and B are crystalline forms I and II. Gel C represents a mixture of the two crystalline isozymes, and gel D shows the crystalline enzyme prior to separation of the two forms.

TABLE I. Amino Acid Composition of the Two Forms of Serine Dehydratase (SDH)

Amino acid	Serine dehydratase	
	I	II
Lysine	22.5	21.5
Histidine	5.8	5.9
Arginine	8.1	8.4
Aspartic acid	16.4	16.8
Threonine[a]	17.2	17.3
Serine[a]	22.5	21.9
Glutamic acid	34.8	35.4
Proline	14.8	16.4
Glycine	30.3	29.9
Alanine	38.7	37.9
Half-cystine[b]	5.8	6.3
Valine[c]	31.4	31.1
Methionine	5.5	5.5
Isoleucine[c]	15.8	16.0
Leucine	37.6	37.0
Tyrosine	3.6	3.8
Phenylalanine	8.6	8.8
Tryptophan	3.3	2.9

The values are expressed as moles of amino acid per 34,000 g of protein and represent the average recoveries after 24, 48, and 72 hr of hydrolysis.

[a] Values obtained by extrapolation to zero time. [b] Determined as cysteic acid.
[c] Average value for 72-hr hydrolysis period.

was prepared from the livers of animals maintained on a 90% protein diet for 1–2 weeks. Electrophoresis of the crystalline enzyme on polyacrylamide gel shows the staining pattern seen in Fig. 1. The two forms of the enzyme have been isolated separately (*10*) and shown to be essentially identical immunochemically and to have the same heat stability and kinetic constants for the substrates of the enzyme. The spectral characteristics of the two forms, both in the visible and ultraviolet, were essentially identical. Their molecular weights were also identical, and it was apparent that they each consisted of two subunits, probably identical, and that each form had two active sites, bound two molecules of pyridoxal phosphate, and had no esterified inorganic phosphate. Amino acid analyses of the two forms are shown in Table I, which shows that the two forms had essentially the same amino acid composition, with the possible exception that form I, the more electropositive of the two, contained one more lysine residue and one less proline residue than form II. It was possible that the difference in lysine residues might account for the difference in electrophoretic mobilities if the ε-amino group of this residue was in an easily ionized position.

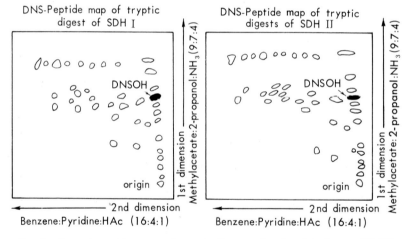

FIG. 2. Dansyl-peptide maps of tryptic digests of serine dehydratase (SDH) forms I and II
The darkened black spot refers to the Dansyl agent itself.

A fingerprint of the two forms of the enzyme is seen in Fig. 2, and it can be noted that they were almost identical, with the exception of one or two extra peptides seen in the map of form II. Based on the lysine and arginine content of the two enzymes, approximately 31 peptides would be expected from the tryptic hydrolysis of a 34,000 molecular weight subunit. Thus, the tryptic fingerprints, which yielded 37 and 39 peptides, respectively, were in general agreement with the chemical data and further substantiate the identical subunit nature of serine dehydratase. Obviously, the critical chemical differences between the two forms remained to be determined.

Regulation of the Two Forms of Serine Dehydratase

In Fig. 3 the elution pattern from DEAE-cellulose of the two forms of serine dehydratase obtained from the livers of animals fed a 90% protein diet (A) and animals fed a laboratory ration and given glucagon (B) is seen. It was thus ap-

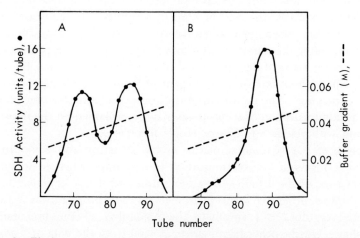

FIG. 3. Elution patterns of serine dehydratase (SDH) from DEAE-cellulose from livers of animals fed a 90% protein diet (A) for one week and animals fed a laboratory ration and given glucagon (B) at a dose of 250 µg per rat, 6 hr prior to sacrifice

TABLE II. Levels of the Two Isozymic Forms of Serine Dehydratase (SDH) in Liver and Hepatoma under Various Dietary and Hormonal Conditions

Treatment	Type of SDH (%)		Activity of serine dehydratase (units/g liver)
	I	II	
Intact rats [a]			
Chow	0	100	11.5
90% protein diet (1 week)	55	45	105.0
Diabetes (1 week)	2	98	78.4
Newborn (1 week)	0	100	7.8
0% protein diet (5 days)			0.9
0% protein+glucagon (10 hr)	3	97	8.2
0% protein+tryptophan (9 hr)	56	44	4.0
Adrenalectomized rats [b]			
12% protein diet (3 days)	0	100	8.4
12% protein+tryptophan (9 hr)	0	100	9.2
12% protein diet+cortisone (6 hr)	31	69	19.1

Rats were sacrificed at the end of the time period shown in the table. Four to six animals were used for each condition.

[a] Diabetes was produced in rats by the intravenous injection of alloxan (40 mg/kg). Rats on the protein-free diet were fed for 5 days and then starved for 10 hr. Glucagon (0.5 mg/rat, intraperitoneally) and tryptophan (100 mg/rat, intubation) were given at 0 and 6 hr.

[b] Rats were fed a 12% protein diet for 3 days and then starved for 10 hr. Tryptophan (40 mg/rat) and cortisone acetate (10 mg/rat) were given at 0 hr.

parent that the amount of each form synthesized differed under these two different environmental conditions. Extending these studies, the data of Table II became apparent. From the data of this table it can be seen that, in contrast to animals fed a 90% protein diet, animals on a chow diet showed virtually all of the enzymes in form II. This was also true of newborn animals and alloxan-diabetic animals. When tryptophan was administered to animals on a protein-free diet, both forms of the enzyme were synthesized in similar amounts. In contrast, in adrenalectomized animals when tryptophan was administered there was no synthesis of form I, while after the administration of cortisone, form I appeared in the liver extract. On the basis of data presented in this table, it is reasonable to conclude that the primary hormonal inducing agent of form II is glucagon, while that of form I is cortisone. The increase in form I occurring after feeding a high-protein diet or the administration of tryptophan may be explained on the basis of adrenal stimulation with release of corticosteroids *in vivo*.

Because of the extreme similarity of the structure and composition of the two enzymic forms, it became important to determine whether or not one form was a precursor of the other. In earlier studies (9) it was demonstrated that the rate of synthesis of the two forms in animals given tryptophan was virtually equal. This, however, does not eliminate the possibility of the precursor-product, relationship. On the other hand, since, as seen from the data of Table II, form II of the enzyme may be synthesized in the complete absence of form I, it is reasonable to conclude that the more electronegative form of serine dehydratase may be synthesized independently. In order to determine whether the converse was true, the experiment outlined in Table III was performed. Adrenalectomized animals were maintained on a Purina chow diet for 5 days. After a 10-hr fast, animals were given cortisone at 0 time, 40 μCi of L-leucine-^{3}H at 6 hr, and sacrificed 30 min later. Control animals received saline instead of the steroid. It might be noted that the administration of cortisone doubled the serine dehydratase activity and produced a significant amount of form I of the enzyme. Serine dehydratase was then precipitated by a specific antibody and

TABLE III. ^{3}H-Leucine Incorporation into Two Forms of Serine Dehydratase (SDH) in Adrenalectomized Rats Treated with Cortisone *in vivo*

	−Cortisone		+Cortisone	
	I	II	I	II
SDH activity (units/g liver)	21.2		47.0	
Percentage of each form	0	100	26	74
SDH units for immunoprecipitation		25.1	21.0	27.3
^{3}H-Leucine (dpm) in antigen-antibody precipitate		3,290	5,544	4,778
SDH specific radioactivity (dpm/unit SDH)		131	264	175

Adrenalectomized rats (15), weighing 180 to 200 g, were fed chow for 5 days. After 10 hr of starvation, rats were injected with 10 mg cortisone acetate at 0 hr and leucine-^{3}H (40 μCi, 15 nm) at 6 hr. Control rats (−cortisone) were injected with saline instead of cortisone. Rats were killed at 6.5 hr. Separation of the two forms of SDH and immunoprecipitation were carried out as described elsewhere (9, 10).

the specific activity of the antigen-antibody precipitate determined. As seen in the table, in the animal administered cortisone, the specific activity of form I was greater than that of form II. This is true for both the antigen-antibody precipitate and the enzyme itself, based on the units utilized for the antigen-antibody reaction. On this basis, one must conclude that form II could not be the precursor of form I, since its specific activity is certainly not greater than that of form I and the rate of synthesis of the two forms is equal (9), which is a prerequisite if such a precursor-product relationship exists. From these data, one may then conclude that the two forms of the enzyme are synthesized independently of each other.

Regulation of Multiple Forms of Tyrosine Aminotransferase

As indicated, studies by Kenney (12) and Oliver and associates (8, 16) had shown the existence of multiple forms of tyrosine aminotransferase in liver, and the latter authors had indicated the possibility of regulation of these forms by different environmental conditions. In our laboratory we were more successful in separating multiple forms of tyrosine aminotransferase by the use of hydroxylapatite chromatography than by electrophoresis on gels, as carried out by Holt and Oliver (8). The technique employed was as follows:

Livers of male albino Holtzman rats (130–150 g) were homogenized, using a Potter-Elvehjem homogenizer, in 3 volumes of 0.1 M potassium phosphate buffer (pH 7.0) containing 0.4 mM pyridoxal 5'-phosphate and 1 mM dithiothreitol. Other tissues were homogenized in the same media by means of a Polytron homogenizer. The homogenate was centrifuged at 105,000g for 60 min and the supernatant (S_3) obtained was used directly. Tyrosine aminotransferase activity was assayed by a modification of the method described by Diamondstone

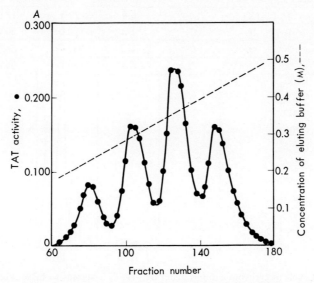

FIG. 4. Elution pattern of tyrosine aminotransferase (TAT) activity from hydroxylapatite column

(4). The reaction mixture contained 12 μmoles tyrosine, 60 μmoles α-ketoglutarate, 1.2 μmoles pyridoxal 5′-phosphate, 3 μmoles EDTA, and 300 μmoles potassium phosphate in a total volume of 3.0 ml, pH 7.6. When necessary, 12 μmoles of diethyldithiocarbamate was added to the above reaction mixture. The reaction was started by the addition of the enzyme and allowed to continue for 10 min at 37°, at which time it was stopped by rapid mixing with 0.3 ml of 10 N KOH. The reaction mixture, with the addition of 0.3 ml of 10 N KOH before adding the enzyme, was used as a blank. The absorbance was measured against a blank 30 min after the addition of KOH at 331 nm. One unit of enzyme was defined as the amount of enzyme that forms 1 μmole of p-hydroxyphenylpyruvate per hr at 37°. An extinction coefficient of 19,900 M^{-1} was used (4).

The supernatant of the homogenate (S_3) was applied to a hydroxylapatite column (7×20 cm) which had been equilibrated with 0.1 M potassium phosphate buffer (pH 6.9) containing 1 mM dithiothreitol from an initial concentration of 0.1 M to a final concentration of 0.5 M. The flow rate was maintained at about 1.0 ml/min. Usually from 100 to 1,000 units of enzyme was applied to the column, and the recovery of the activity was greater than 85% in most instances. Each peak was separately collected and its activity assayed. In Fig. 4 is seen the elution pattern of tyrosine aminotransferase activity from a high-speed supernatant (S_3) of rat liver. There were four distinct and easily separable peaks of activity

Tissue		Heart	Kidney	Liver		
Treatment		Adx.	Adx.	Adx. scheduled feeding of 12.5% protein diet for 2 weeks. Killed at midnight	Adx. scheduled feeding of 60% protein diet for 12 days. Killed at 14:00 pm	Adx. 60% protein scheduled feeding. Killed 12:00 midnight
Total activity of TAT (μmole/hr/g tissue)		15.0	14.3	32.0	621.4	135.0
Activity of TAT isozymes (μmole/hr/g tissue)	I	13.1	14.7	6.8	22.4	17.5
	II	0	0	6.4	158.0	30.4
	III	0	0	12.3	247.0	56.5
	IV	0	0	6.6	194.0	29.6

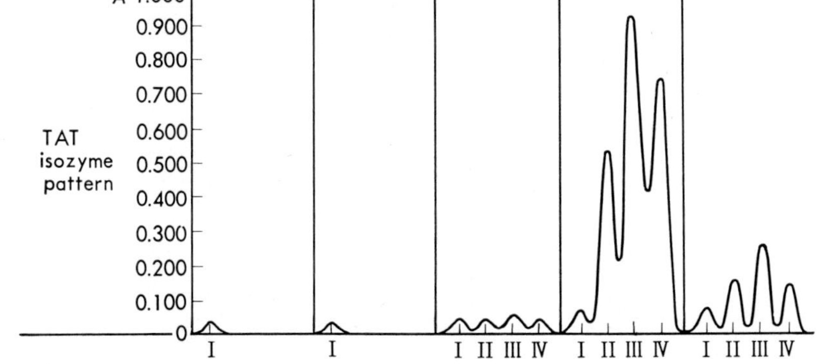

FIG. 5. Levels of tyrosine aminotransferase (TAT) isozymes in heart, kidney, and liver of adrenalectomized animals fed various protein diets

The methodology is essentially that of Fig. 4.

eluting from the hydroxylapatite column, arbitrarily designated I, II, III, and IV. Studies have demonstrated that the peaks may be isolated separately and, in fact, peaks II, III, and IV have been separately purified essentially to homogeneity.

In order to determine the effect of various environmental manipulations on the levels of the four forms, adrenalectomized animals were placed on various feeding regimens. Figure 5 shows the level of these forms in heart, kidney, and liver of adrenalectomized animals. The activity of tyrosine aminotransferase in heart and kidney was extremely low and only occurred as form I. In contrast, in the liver of adrenalectomized animals maintained on a 12.5% protein diet, all four forms were present but at extremely low levels. When animals were fed a 60% protein diet, the levels of II, III, and IV increased significantly ($P<0.01$). However, the change in form I was insignificant. The variations in the levels of forms II, III, and IV between noon and midnight on a reversed lighting schedule (13) was quite extreme. On the other hand, form I did not significantly change during this period. The overall variation in tyrosine aminotransferase in liver with the time of day has been reported by several authors (3, 6, 19, 20). The effect of the administration of several hormones and cyclic AMP is seen in Fig. 6. It can be noted that cortisone acetate administered to adrenalectomized animals caused a

Treatment		Adx. 12.5% protein diet for 2 weeks. Cortisone acetate (10mg/100BW) at 10:00am and killed at 15:00pm	Adx. 12.5% protein diet for 2 weeks. Glucagon (0.3mg/100BW) at 10:00am and killed at 13:00pm	Adx. 12.5% protein diet for 2 weeks. cAMP (5mg/100BW) at 12:00 and killed 11/2 hr later	Adx. 12.5% protein diet for 2 weeks. Epinephrine at 10:00 am and 12:00 killed at 14:00 pm
Total activity of TAT (μmole/hr/g tissue)		891.0	287.0	122.0	144.0
Activity of TAT isozymes (μmole/hr/g tissue)	I	19.7	14.2	14.8	11.4
	II	355.0	97.3	65.5	82.4
	III	378.0	127.0	37.9	43.6
	IV	137.0	47.9	4.8	7.2

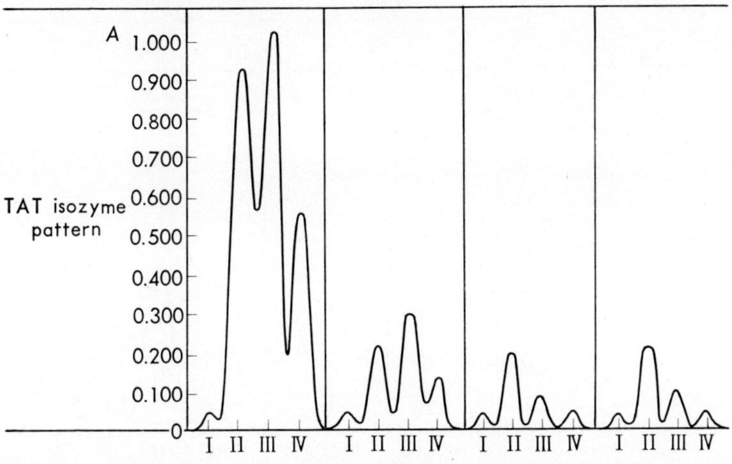

FIG. 6. Activity of tyrosine aminotransferase (TAT) isozymes in adrenalectomized animals given cortisone, glucagon, cyclic AMP, or epinephrine

marked increase in forms II, III and IV. Glucagon likewise stimulated an increase in these forms, but to a much lesser degree. In contrast, administration of cyclic AMP or epinephrine caused an increase mostly in form II with a slight concomitant increase in form III. Again, form I did not significantly change during these environmental alterations. While these experiments do not clearly indicate which hormone controls the level of which form, as was seen in the case of glucagon and cortisone with the two forms of serine dehydratase, there is a suggestion that the level of form II may be related to cyclic nucleotide levels in the liver. On the other hand, as seen from the data of Fig. 7, adrenalectomized animals made diabetic by the administration of alloxan showed an almost complete disappearance of form IV. Administration of insulin resulted, within 3 hr, in a marked increase (15-fold) in the level of form IV. These data strongly suggest that the synthesis of this form is regulated by insulin.

As was seen from the data of Table II, the neonatal rat contained only one form of serine dehydratase, that inducible by glucagon, in its liver. In Table IV is seen the sequence of development of the four forms of tyrosine aminotransferase

Treatment		Adx. 12.5% protein diet for 2 weeks. Alloxan 15mg/100 BW injected intraperitoneally and sacrificed 5 days later	Adx. 12.5% protein alloxan diabetic insulin(1 unit/100 BW) intraperitoneally killed and 3hr later
Total activity of TAT (μmole/hr/g tissue)		96.0	121.0
Activity of TAT isozymes (μmole/hr/g tissue)	I	12.5	13.0
	II	53.6	28.6
	III	24.9	30.3
	IV	3.0	45.6

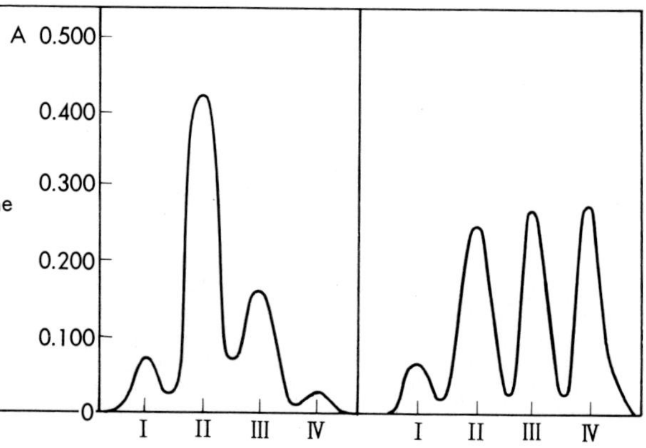

FIG. 7. Activity of tyrosine aminotransferase (TAT) isozymes in adrenalectomized diabetic animals compared with similar animals given insulin

TABLE IV. Development of the Four Forms of Tyrosine Aminotransferase (TAT) in Rat Liver

Age	TAT activity (μmole/hr/g/liver)	TAT activities of each form (μmole/hr/g/liver)			
		I	II	III	IV
−1 day	10.3±3.3	9.6±2.6	—	—	—
6 hr	49.7±8.0	7.9±3.1	36.5±6.6	4.2±1.1	—
3 days	44.5±5.1	12.3±4.5	27.8±5.4	4.9±2.6	—
14 days	26.8±2.4	8.4±2.9	15.0±2.6	4.4±1.7	—
21 days	46.2±4.8	10.6±3.2	13.2±4.8	17.5±2.3	2.7±1.8
28 days	41.4±4.6	10.0±3.6	7.6±4.5	16.9±3.1	7.4±2.0

Values represent mean ± standard error of 3 groups of littermates. TAT activity was assayed on the original preparation.

TABLE V. Tyrosine Aminotransferase (TAT) Isozymes in Morris Hepatomas

Morris hepatoma	TAT activity (μmole/hr/g tumor)	TAT activities of each form (μmole/hr/g tumor)			
		I	II	III	IV
9618A	31.6±6.4	8.9±1.1	16.9±2.5	4.4±0.7	—
9121	84.5±12.0	14.0±3.4	54.0±8.4	9.2±2.4	—
7800	245.0±16.8	10.5±3.6	192.1±7.7	42.9±3.5	—
5123C	492.6±22.3	8.3±5.0	406.4±25.8	86.2±10.6	—

Values are mean ± standard error for three animals.

from fetal life to beyond weaning. It can be noted that the non-inducible form I occurred during fetal life, but on the day of birth forms II and III appeared, with a preponderance of the former. This pattern was maintained until the 21st day, when the animal was weaned, and form IV then appeared. Finally, the level of the various isozymes of tyrosine aminotransferase in four different Morris hepatomas is seen in Table V. It could be noted that in none of these hepatomas was the presence of form IV demonstrated.

DISCUSSION

It is apparent from the data presented in this paper that serine dehydratase and tyrosine aminotransferase exist in liver in multiple forms, each form being regulated by a different environmental condition, presumably a different hormone. The major question to be answered is the difference in structure between these two forms. Serine dehydratase, forms I and II, appears to be distinguished by two amino acid differences, although even these may be open to some question because of the level of significance based on so many residues. On the other hand, in view of the fact of the two peptide differences, it may be reasonable to suggest that these may account for the differences seen. We have no knowledge yet of the structural differences among the four forms of tyrosine aminotransferase, although preliminary studies indicate that form I is significantly different from the other three forms, both immunochemically and kinetically. It should be mentioned that the four forms of the enzyme are quite different from the mitochondrial form of tyrosine aminotransferase.

Initially, it was felt that the differences between the two forms of serine dehydratase might be due to phosphate esterified to the protein of form II, which could account for the difference seen, as well as for its regulation by glucagon *via* a possible protein kinase activated by cyclic nucleotides. This was not the case (*10*), however, but one still must consider that the differences may be due to modifications occurring in the protein after transcription. Such may also be the case with tyrosine aminotransferase, and the possibility is quite distinct that modifying groups are inscribed into the molecule during or just after translation. Particularly with respect to the regulation of serine dehydratase by glucagon, if cyclic nucleotides are involved, it is not inconceivable that other enzymes modifying primary protein structure, such as acetylases, amidases, methylases, etc., may be activated by cyclic nucleotides in the same fashion that cyclic AMP activates protein kinases (*10*). Preliminary investigations in this laboratory have suggested the existence of a cytoplasmic protein acetylase activated by cyclic nucleotides.

A New Classification of Multiple Forms of Enzymes

Although the picture is not clear as to the differences in these multiple forms, one may theorize that the primary structure in the case of each enzyme is identical, and that modifications occur post-transcriptionally. This is somewhat analogous to the dependent and independent forms of glycogen synthetase (*7*). If one considers this, then one must also consider a broadening of the term " isozyme " as it was originally defined and utilized by Markert and Moller (*13*). Therefore, we have proposed the modification seen in Table VI.

The nomenclature seen in Table VI is patterned after that defining immunologic genotypes in immunobiology. Comparable terms are syngeneic, isogeneic, allogeneic, and xenogeneic, which refer respectively to tissues within the same organism, within identical twins, within organisms of the same species, and from organisms of different species. The classification system seen in Table VI, while not completely analogous to the immunobiologic system, is similar in many ways.

TABLE VI. Classification of Multiple Forms of Proteins Having the Same Catalytic Activity

Synzymes	Proteins having the same basic amino acid sequence coded by the same gene but differing in conformation or structural modification occurring post-transcriptionally, *e.g.*, -PO_4, -acetyl, -methyl, *etc.*
Isozymes	Proteins having dissimilar amino acid compositions, coded by different genes within the same species.
Allozymes	Proteins resulting from the expression of a number of alleles of the same gene within a species.
Xenozymes	Proteins having the same catalytic activity in two or more different species.

The synzymes occur within a single organism and are coded for by one gene, their structural differences being due to changes in conformation or modifications occurring after transcription, such as phosphate esters, *n*-acetyl or methyl groups being added to the protein. An example of synzymes would be the dependent and independent form of glycogen synthetase (7). Isozymes are defined largely as most people consider them at present. These are proteins having dissimilar amino acid compositions, *i.e.*, being coded for by different genes, occuring within the same species, and catalyzing the same reaction. The isozymes of hexokinase, lactate dehydrogenase, pyruvate kinase, *etc.*, are such examples. Allozymes would be the term used for the allelic forms of an enzyme within one species. The allelic forms of glucose-6-phosphate dehydrogenase seen in the human (2) are one of the best examples of allozymes. Finally, the term xenozyme may be reserved for those proteins occurring in different species but having the same catalytic function, *e.g.*, threonine deaminase in *E. coli* and rat liver.

While this nomenclature suffers from possible misinterpretations, it does offer the advantage of indicating more specifically the nature of the form of the protein having catalytic activities identical to other proteins. We would thus propose that the advantages of this system outweigh its disadvantages and that it can be useful in more explicit discussions of the multiple forms of enzyme proteins seen in biology.

REFERENCES

1. Auricchio, F., Valeriote, F., Tomkins, G., and Riley, W. D. Studies on the structure of tyrosine aminotransferase. *Biochim. Biophys. Acta*, **221**, 307–313 (1970).
2. Beutler, E., Mathai, C. K., and Smith, J. E. Biochemical variants of glucose-6-phosphate dehydrogenase giving rise to congenital non-spherocytic hemolytic disease. *Blood*, **31**, 131–150 (1968).
3. Black, I. B. and Exelrod, J. Regulation of the daily rhythm in tyrosine transaminase activity by environmental factors. *Proc. Natl. Acad. Sci. U.S.*, **61**, 1287–1291 (1968).
4. Diamondstone, T. I. Assay of tyrosine transaminase activity by conversion of *p*-hydroxyphenylpyruvate to *p*-hydroxybenzaldehyde. *Anal. Biochem.*, **16**, 395–401 (1966).
5. Fritz, P. J., Vesell, E. S., White, E. L., and Pruitt, K. M. The roles of synthesis and degradation in determining tissue concentrations of lactate dehydrogenase-5. *Proc. Natl. Acad. Sci. U.S.*, **62**, 558–565 (1969).
6. Fuller, R. W. and Snoddy, H. D. Feeding schedule alteration of daily rhythm in tyrosine alpha-ketoglutarate transaminase of rat liver. *Science*, **159**, 738 (1968).
7. Hers, H. G., Dewulf, H., and Stalmans, W. The control of glycogen metabolism in the liver. *FEBS Letters*, **12**, 73–82 (1970).
8. Holt, P. G. and Oliver, I. T. Multiple forms of tyrosine aminotransferase in rat liver and their hormonal induction in the neonate. *FEBS Letters*, **5**, 89–91 (1969).
9. Inoue, H. and Pitot, H. C. Regulation of the synthesis of serine dehydratase isozymes. *Adv. Enzyme Regulation*, **8**, 289–296 (1970).
10. Inoue, H., Kasper, C. B., and Pitot, H. C. Studies on the induction and repression of enzymes in rat liver. VI. Some properties and the metabolic regula-

tion of two isozymic forms of serine dehydratase. *J. Biol. Chem.*, **246**, 2626–2632 (1971).
11. Katzen, H. M. and Schimke, R. T. Multiple forms of hexokinase in the rat: Tissue distribution, age dependency, and properties. *Proc. Natl. Acad. Sci. U.S.*, **54**, 1218–1225 (1965).
12. Kenney, F. T. Induction of tyrosine-α-ketoglutarate transaminase in rat liver. II. Enzyme purification and preparation of antitransaminase. *J. Biol. Chem.*, **237**, 1605–1609 (1962).
13. Markert, C. L. and Moller, F. Multiple forms of enzymes: Tissue, ontogenetic, and species specific patterns. *Proc. Natl. Acad. Sci. U.S.*, **45**, 753–763 (1959).
14. Nakagawa, H. and Kimura, H. The properties of crystaline serine dehydratase of rat liver. *J. Biochem. (Tokyo)*, **66**, 669–683 (1969).
15. Niemeyer, H., Perez, N., and Codoceo, R. Liver glucokinase induction in acute and chronic insulin insufficiency in rats. *J. Biol. Chem.*, **242**, 860–864 (1967).
16. Sadleir, J. W., Holt, P. G., and Oliver, I. T. Fractionation of rat liver tyrosine aminotransferase during the course of purification. Further evidence for multiple forms of the enzyme. *FEBS Letters*, **6**, 46–48 (1970).
17. Salas, M., Vinuela, E., and Sols, A. Insulin-dependent synthesis of liver glucokinase in the rat. *J. Biol. Chem.*, **238**, 3535–3538 (1963).
18. Sharma, C., Manjeshwar, R., and Weinhouse, S. Effects of diet and insulin on glucose-adenosine triphosphate phosphotransferases of rat liver. *J. Biol. Chem.*, **238**, 3840–3845 (1963).
19. Watanabe, M., Potter, V. R., and Pitot, H. C. Systematic oscillations in tyrosine transaminase and other metabolic functions in liver of normal and adrenalectomized rats on controlled feeding schedules. *J. Nutr.*, **95**, 207–227 (1968).
20. Wurtman, R. J. and Axelrod, J. Daily rhythmic changes in tyrosine transaminase activity of the rat liver. *Proc. Natl. Acad. Sci. U.S.*, **57**, 1594–1598 (1967).

MAMMALIAN RIBONUCLEOTIDE REDUCTASE AND CELL PROLIFERATION

Howard L. ELFORD

*Department of Medicine and the Department of Physiology and Pharmacology, Duke University Medical Center**

Evidence is presented to support the contention that reduction of ribonucleotides to deoxynucleotides is a crucial and rate-limiting step in the pathway to DNA synthesis and cell replication. One main piece of evidence is the demonstration that there is a close correlation between ribonucleotide reductase activity and tumor growth rates utilizing a series of rat hepatomas of varying growth rates. Enzyme activity is also shown to increase appreciably during liver regeneration but does not reach the levels seen in the malignant state as exhibited by the fast-growing hepatomas. Ribonucleotide reductase activity was also examined during the physiological development of several organs. Results are reported for spleen development that show sharp changes in activity during the first week of neonatal development. The results suggest that these high levels of ribonucleotide reductase activity may be associated with an intense period of cellular replication related to a process of differentiation or response of the newborn to the extramaternal environment.

These dramatic changes in ribonucleotide reductase activity are shown to be due to *de novo* enzyme synthesis and not to enzyme activation. Both a protein synthesis inhibitor, cycloheximide, and a DNA-directed RNA synthesis inhibitor, Actinomycin-D, prevent increases of enzyme activity in developing spleen, but only the protein synthesis inhibitor causes a decrease in enzymatic activity in Novikoff tumors.

Ribonucleotide reductase is shown to be associated with a smooth membrane component of the cytoplasm of the proliferating mammalian cell.

The term " malignozyme " is used to describe ribonucleotide reductase.

Deoxynucleotides, in contrast to ribonucleotides, are normally found at low levels in mammalian cells (*3*). The biosynthesis of deoxynucleotides indicates that a cell is committed to replication. Deoxynucleotides are formed by direct reduction of ribonucleotides. This reductive event occupies a regulatory command or switching position which, when activated, channels ribonucleotides away from normal cellular processes to serve as precursors for DNA synthesis.

---

* Durham, North Carolina 27706, U.S.A.

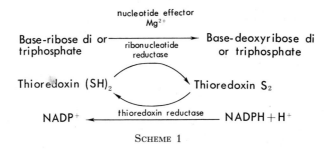

Scheme 1

Ribonucleotide reductase systems of *Escherichia coli* and *Lactobacillus leichmannii* have been well characterized, but less is known about mammalian ribonucleotide reductase (6, 13). The general scheme for ribonucleotide reduction is shown in Scheme 1.

Thus far, in all of the systems studied, one enzyme complex has been responsible for the reduction of the four possible ribonucleotide substrates. A reduced protein, "thioredoxin," serves as the physiological hydrogen donor. It is a low-molecular weight, disulfhydryl protein which is part of a complex that includes a second protein thioredoxin reductase that catalyzes the reduction of the oxidized form of thioredoxin utilizing NADPH. The activity of the ribonucleotide reductase is modified by an elegant and complex regulatory system involving ATP and deoxyribonucleotides which serve as both positive and negative allosteric effectors.

It is the aim of this report to present experimental evidence obtained in my laboratory that substantiates the hypothesis that ribonucleotide reduction is intimately associated with mammalian cell proliferation rate and malignant state. The large increments in enzyme activity observed in replicating cells are a consequence of increased rates of *de novo* enzyme synthesis rather than of enzyme activation.

Ribonucleotide Reductase and Tumor Growth Rate

A correlation between tumor growth rate and ribonucleotide reductase activity would provide significant evidence that this reaction plays an important role in controlling the rate of DNA synthesis and cell proliferation.

Therefore the level of ribonucleotide reductase activity was measured in extracts from a series of rat hepatomas of different growth rates including several hepatomas of the Morris series, and in extracts of livers from normal and tumor-bearing animals. The results are shown in Fig. 1. An excellent correlation was found between tumor growth rates, as measured by transplantation time, and specific activity of ribonucleotide reductase. We observed differences of over 200-fold in enzyme-specific activities between the very fast growing hepatomas, Novikoff and 3683A, and the slow-growing tumors (4). This represents one of the largest enhancements of any enzyme activity measured in these tumors. No detectable activity was found in either crude extracts (25,000g for 30 min) or ultracentrifuge extracts (104,000 g for 60 min) of normal rat liver or liver from

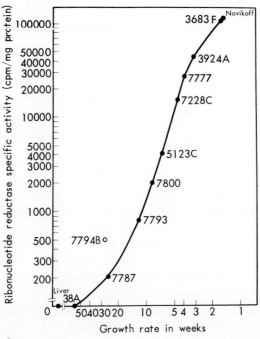

FIG. 1. Relationship between tumor growth rate and ribonucleotide reductase specific activity

Activity was measured by monitoring the reduction of CDP to dCDP. Enzyme assay was patterned after Reichard *et al.* (*12*) as descirbed by Elford *et al.* (*4*). Tumor growth rate was defined and measured by Morris and Wagner (*8*). Data are from Fig. 1, Elford *et al.* (*4*), with permission from the Journal of Biological Chemistry.

tumor-bearing animals. The assay was such that 1/500 of the activity of the fast-growing tumor extracts or 1/1,000 of a fetal liver extract activity could have been detected. In contrast, the differences in measurable thymidylate synthetase and thymidine kinase activity between the fast- and slow-growing tumor extracts were much less, only about 20-fold in the case of thymidylate synthesis and 50-fold in thymidine kinase.

The reducing complex, thioredoxin-thioredoxin reductase, is not the limiting factor that accounts for these large variations of ribonucleotide reductase activity among the various hepatomas. It also is not responsible for the absence of detectable activity in adult rat liver. The evidence for this is 2-fold. First, the assays were run in the presence of dithiothreitol which could substitute for the natural hydrogen donor complex. Second, when utilizing a partially purified ribonucleotide reductase from Novikoff extracts, it was possible to demonstrate that the thioredoxin-thioredoxin reductase complex is present at comparable levels in both tumor extracts and normal adult liver (Table I). This is because the partially purified ribonucleotide reductase is completely devoid of ribonucleotide reductase activity unless an auxiliary reducing system, such as a chemical-reducing agent, *i.e.*, dithiothreitol or the physiological NADPH-thioredoxin-thioredoxin reductase system, is supplemented to it in the assay. Both thioredoxin systems

TABLE I. Evidence for Presence of Thioredoxin-Thioredoxin Reductase Complex in Tumor and Rat Liver

Reducing system	Deoxyribonucleotide formed (nmoles)
Dithiothreitol	19.4
NADPH	0
NADPH+Novikoff tumor $(NH_4)_2SO_4$, 40–70%	6.8
Novikoff tumor $(NH_4)_2SO_4$, 40–70%	0
NADPH +rat liver $(NH_4)_2SO_4$, 40–70%	7.7
Rat liver $(NH_4)_2SO_4$, 40–70%	0

The ribonucleotide reductase used for this experiment was a 0–40% $(NH_4)_2SO_4$ fraction from Novikoff tumor extracts. The assay, description, and data are from Table III, Elford et al. (4) (reprinted with permission from Journal of Biological Chemistry).

from Novikoff and rat liver can be purified in a similar manner and appear to have similar properties.

Treatments to remove nucleotides, such as dialysis, charcoal adsorption, and Dowex and molecular gel chromatography, had no appreciable effect ($\leq 20\%$) on any of the tumor or liver extracts. I interpret this to mean that the allosteric control of this reaction by the various nucleotides is not responsible for the large differences in activities seen between the hepatomas or for the absence of activity in the liver. It is possible that the nucleotides are too tightly bound for the methods utilized to have removed them.

Ribonucleotide Reductase and Development

An examination of ribonucleotide reductase activity during the neonatal development of several rat organs has been undertaken in my laboratory. In this report I would like to comment on the results found for two of these organs, liver and spleen. Particular attention was given to the functional development of the organ.

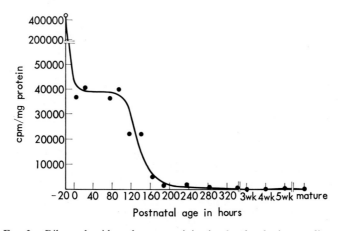

FIG. 2. Ribonucleotide reductase activity in the developing rat liver

One of the purposes of this study was to obtain information about this crucial reaction in DNA synthesis during development in order, possibly, to relate changes in ribonucleotide reductase levels with stages of differentiation of the organ and to the possible implication that a block in differentiation may lead to malignant transformation. Potter, in his " oncogeny as a blocked ontogeny " thesis, has recently reformulated the concept that malignancy may arise from a block in differentiation and/or from the reactivation of latent genes (*11*). Abundant evidence has been presented at this conference that in many hepatomas there is an expression of isozymes resembling that of the fetal state.

If one could equate ribonucleotide reductase activity with DNA synthesis and cell replication rates, one could see whether during a particular key stage in differentiation an intense period of cell replication occurs to achieve the differentiated state of the next plateau of maturation.

The profile of ribonucleotide reductase activity in the developing rat liver is shown in Fig. 2. There is a considerable variation in enzyme activity within the relatively short time span of 3 weeks. The fetal liver extract exhibits a very high level of enzyme activity which is several times higher than that of the fast-growing undifferentiated hepatomas. There is a very sharp decline in activity at the time of birth and then progressive decreases in activity until at one week after birth there is only a very small amount of activity remaining—1/20 of the newborn liver activity and only 1/200 of that of the fetal liver extract. By 3 to 4 weeks after birth there is no activity, although 1/1000 of the amount of fetal liver activity could have been detected.

In contrast, ribonucleotide reductase activity in the developing spleen has a much different chronological profile as is shown in Fig. 3. Enzyme activity is low at birth and then dramatically increases, reaching a peak approximately 6–7 days after birth. This peak in activity is about 5-fold greater than in other rat tissues tested, including Novikoff hepatomas, and occurs at a time when the onset of white cell production is initiated in the rat spleen; therefore, a higher mitotic activity is occurring. This burst in lymphocyte production may be a response

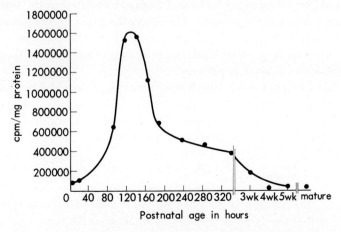

FIG. 3. Ribonucleotide reductase activity in the developing rat spleen

to the exposure of the newborn to antigens of the extramaternal environment. Activity then declines, but there is detectable activity in the adult spleen which is about 1/20 that of a fast-growing hepatoma.

Regenerating Liver

An examination of the change in ribonucleotide reductase in regenerating liver was undertaken. Two other papers on this subject (5, 7) appeared during the progress of this work. However, I have included some of our data in order that a comparison in activities can be drawn between controlled rapid cell replication as exemplified by regenerating liver, and malignant cell proliferation, illustrated by the poorly differentiated, fast-growing hepatoma. In contrast to the finding of Larsson that maximum activity occurs at 50 hr after hepatectomy (7), we found that maximum activity was reached between 30 and 40 hr (Fig. 4). The paper

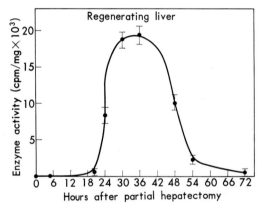

FIG. 4. Ribonucleotide reductase activity in regenerating liver

of King and Van Lancker (5) had values only for 18, 24, and 36 hr. The reason for the difference in the time of maximum activity observed in the Larsson paper and in the time we found is not obvious but may be related to the age of the animals and diurnal or dietary effects. However, the main observation is that this maximum activity in regenerating liver is much lower than the specific activity seen in the more rapidly growing tumors, and it falls in the range of the moderately growing, well-differentiated hepatomas.

It has been pointed out a number of times that normal liver may undergo cell proliferation without the concomitant severe imbalance in metabolic pathways observed in neoplastic transformation. The normal liver still has the regulatory mechanism which halts cell replication at a prescribed time.

De novo Enzyme Synthesis

Although this reaction is modified by ATP and deoxyribonucleotides, it is my contention that the large differences in ribonucleotide reductase activity between tumors of different growth rates and in the neonatal development of several

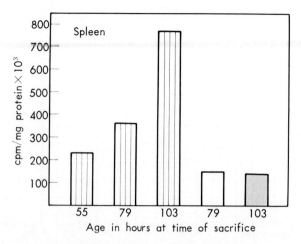

Fig. 5. Effect of cycloheximide on ribonucleotide reductase activity in the developing rat spleen

Cycloheximide (2.5 μg) was injected intraperitoneally.
▥ control; ☐ cycloheximide injected at 55 hr;
▨ cycloheximide injected at 55 and 79 hr.

rat organs is a consequence of varying rates of *de novo* enzyme synthesis and degradation rather than the result of enzyme activation mediated by allosteric agents. To test this hypothesis cycloheximide, a protein synthesis inhibitor, was injected into a neonatal rat at a time of greatly increasing ribonucleotide reductase activity in the spleen. As can be seen in Fig. 5, with the concentration of cycloheximide used, there is not only a prevention of the increased enzyme activity expected 24 hr later, but the level of activity diminishes below the level found at the time of cycloheximide treatment. This finding supports the contention that *de novo* enzyme synthesis is required for increased enzymatic activity and that there is a rapid turnover of the enzyme. When Actinomycin-D, a DNA-

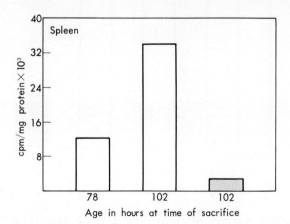

Fig. 6. Effect of Actinomycin-D on ribonucleotide reductase activity in the developing rat spleen

Actinomycin-D (50 μg) was injected intraperitoneally.
☐ control; ▨ Actinomycin D treated at 78 hr.

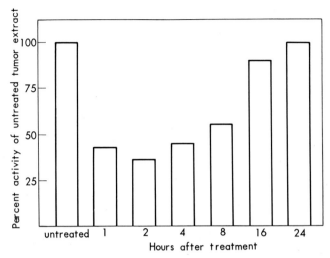

Fig. 7. Effect of cycloheximide on ribonucleotide reductase activity in Novikoff tumor(s)

Cycloheximide (1.5 mg/kg body wt.) was injected intraperitoneally.

directed RNA synthesis inhibitor, was tested, similar results to those found with cycloheximide were observed (Fig. 6). These results also suggest a short half-life of ribonucleotide reductase messenger in the developing spleen. The use of cycloheximide and Actinomycin-D yielded similar results in the developing thymus and heart.

To obtain information about the rate of *de novo* enzyme synthesis and enzyme turnover in a neoplastic tissue, fast-growing hepatomas for example, a series of

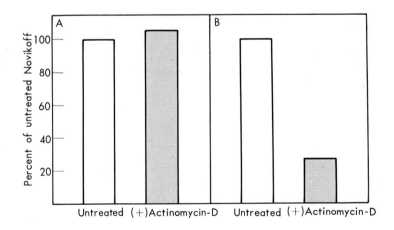

Fig. 8. Effect of Actinomycin-D on ribonucleotide reductase activity in Novikoff tumor(s)

Actinomycin-D was injected intraperitoneally with 2 dosages of 200 μg each 3 hr apart. The tumor-bearing animal was killed 7 hr after the first injection of Actinomycin-D. Uridine-^{3}H was injected 1 hr prior to sacrifice.

A: Ribonucleotide reductase activity; B: Uridine-^{3}H incorporation into RNA.

experiments was performed utilizing cycloheximide and Actinomycin-D. The results of a single intraperitoneal injection of cycloheximide into a Novikoff tumor-bearing animal at a time of rapid tumor growth are shown in Fig. 7. Within 1 hr there was a sharp drop in enzyme activity, reaching a minimum after 2 hr. The activity remained 40% less than that of the uninjected tumor until 8 hr after the injection, when an increase in enzyme activity was noted. There was a continued increase in activity until, by 24 hr after the cycloheximide injection, activity returned to the uninjected control level. Actinomycin-D has no effect on ribonucleotide reductase levels in Novikoff tumor extracts (Fig. 8).

Evidence of a Natural Inhibitor in Liver

Although the preceding portion of this presentation has been directed towards providing evidence to show that the high levels of ribonucleotide reductase enzymatic activity seen in rapid cell proliferation are achieved through *de novo* enzyme synthesis, some initial experiments indicate that there may be a natural factor in liver which negatively affects ribonucleotide reductase activity. This factor is distinct from nucleotides or other low-molecular weight substances and is heat labile. The inclusion of a supernatant fraction from ultracentrifugation of adult rat liver caused a significant inhibition of activity (75%) in a partially purified Novikoff tumor ribonucleotide reductase preparation (Table II). Apparently,

TABLE II. Evidence for a Natural Inhibitor of Ribonucleotide Reductase in Liver

Enzyme	Addition	Activity (cpm)
Purified Novikoff reductase	—	9,200
"	Rat liver extract	2,000
"	Rat liver extract charcoal treated	4,400
"	Rat liver extract Sephadex treated	4,000
"	Regenerating liver extract: 5 hr	2,000
"	Regenerating liver extract: 24 hr	7,800
"	Fetal (day 16) liver	12,000

The ribonucleotide reductase used for this experiment was a partially purified preparation (0–40% $(NH_4)_2SO_4$) of Novikoff tumor.

less of this inhibitory substance is present in 24-hr regenerating liver because inhibition is only 20%. The 5-hr regenerating liver extract is as inhibitory as the adult liver. On the other hand, fetal liver preparation is not inhibitory. These results are only preliminary and are being extended to include neonatal liver and to ascertain that material in liver is not destroying the substrate or product of the reaction.

Subcellular Localization of Ribonucleotide Reductase

A close examination of the subcellular localization of ribonucleotide reductase in animal cells was undertaken. This was prompted by the reports of Baril

TABLE III. Intracellular Localization of Enzyme Activity

Fraction	Total units	Specific activity (units/mg)
600g supernatant (S-1)	2,055	2.21
Nuclear extract (P-1)	0	0
Postmitochondrial supernatant (S-2)	3,270	3.65
Mitochondrial extract (P-2)	0	0
Postmicrosomal supernatant (S-3)	3,190	5.44
Microsomal extract (P-3)	0	0
Centrifugation of S-3		
Supernatant (S-4)	1,950	5.21
Pellet (P-4)	1,070	10.53

The subcellular fractions were obtained by the procedure described by Baril et al. (1, 2). A unit of activity is the formation of 10,000-cpm deoxynucleotide in 1 hr.

et al. (1, 2) that there is DNA polymerase associated with smooth membranes from rat liver and hepatoma cytoplasm. Using differential centrifugation and sucrose density gradient separation techniques, nuclei, microsomes, mitochondria, ribosomes, smooth membrane of the cytoplasm, and a soluble fraction obtained after centrifugation for 15 hr of the postmicrosomal supernatant were isolated from Novikoff hepatoma and 24-hr old rat regenerating liver cells. The survey of the subcellular distribution of ribonucleotide reductase activity from Novikoff cells revealed (Table III) that activity is found only in the soluble fraction (S-4) and the 78,000 g for 15-hr pellet (P-4). However, the specific activity of the pellet was twice that of the soluble fraction which indicates that the activity found in the cell sap probably originated in the pellet material. When this pellet was further purified by discontinuous sucrose density gradient contrifugation which yields free ribosomes, membrane fractions, and a soluble fraction, over 95% of the activity was found to be associated with the membrane fractions, and the rest was associated with the ribosomes. The small amount of activity found to be associated with the free ribosomes was attributed to a small amount of contamination by membrane material adhering to the ribosomes. Subcellular localization fractionation using 24-hr regenerating liver cells yield identical results but with much lower specific activities in the membrane and cell sap. We also found thymidine kinase activity in the membrane fraction.

DISCUSSION

The results presented in this report support the contention that ribonucleotide reduction is a rate-limiting step in DNA synthesis and cell division and is closely correlated with mammalian cell proliferation rates.

The high degree of correlation of ribonucleotide reductase with tumor growth rates in a series of hepatomas is an additional piece of evidence in support of the " molecular correlation concept " expressed by Weber which predicts that certain key enzymes and metabolic pathways are correlated with tumor growth rates (15,16). He further proposes that the resulting metabolic imbalance reveals a distinct pattern at the molecular level which can be useful in the diagnosis and chemotherapy of the cancer cell.

The data on the enzyme profile in the developing liver fits the postulated enzyme pattern that Potter (*11*) envisions for crucial enzymes involved in DNA synthesis and cell division in the neoplastic cell in which the expression of the enzyme pattern would resemble some early stage in development and become "locked in" at this immature state and so be prevented it from differentiating into a mature cell that no longer divides. The question of whether this change occurs during differentiation or involves the reactivation of repressed genes is problematical.

I would like to invoke the term "malignozyme" to describe ribonucleotide reductase. By this term I mean those enzymes which fit the following 3 criteria:

1) The level of enzyme activity in the differentiated, nondividing cell is insufficient to support DNA synthesis for rapid cell proliferation in uncontrolled growth.
2) Activity of the enzyme increases greatly in the neoplastic state in order to catalyze the formation of necessary precursors for DNA synthesis and cell division.
3) A high level of enzyme is mandatory to maintain the malignant state.

Therefore certain enzymes in rate-limiting steps in DNA synthesis and cell replication must be expressed anew. This increase in activity could result from an increase in enzyme synthesis, from decreased catabolism, or from enzyme activation.

I envision the allosteric control of the reaction performed by the nucleotide effectors, ATP and deoxynucleotide triphosphates, to play an important role in the fine tuning of the reaction, that is, maintaining a balanced supply of deoxyribonucleotides and preventing overproduction of any single deoxynucleotide. However, to meet the challenge of supplying adequate quantities of deoxynucleotides for DNA synthesis in rapid cell proliferation, additional quantities of enzyme must be synthesized. Enzyme synthesis, rather than enzyme activation, plays the dominant role. This is supported by the experimental evidence discussed earlier that showed that a protein synthesis inhibitor prevents expected increases in enzyme activity, seen during the neonatal development of spleen and thymus in the rat, and also causes a marked decrease in activity in a neoplastic tissue that has a high capacity for ribonucleotide reduction. Actinomycin-D, a DNA-directed RNA synthesis inhibitor, has the same effect on ribonucleotide reductase activity in the neonatal development situation but has no effect on the tumor synthesis of enzyme. This might imply that neoplastic transformation confers increased stability to the RNA messenger for ribonucleotide reductase. Pitot has advanced the theory of altered template stability as an explanation for the defective mechanisms regulating the control of genetic expression in neoplastic tissue (*10*). A selective growth advantage results in those cells whose templates for enzymes important to DNA synthesis are more stable than the normal cell's. In addition it has been shown that ribonucleotide reductase activity detected only in the S-phase in synchronized cultures of mouse L-cells decreases rapidly with a half-life of 2 hr when cycloheximide is added to the cultures (*14*).

Also, cycloheximide and Actinomycin-D prevent an increase in ribonucleotide reductase activity as Chinese hamster fibroblast cells enter the S-phase (9). Furthermore, King and Van Lancker (5) showed that X-irradiation and Actinomycin-D inhibit increases in ribonucleotide reductase in regenerating liver.

It seems that the enzyme pattern in the developing rat spleen is associated with functional development of the organ. The peak in enzyme activity occurs at the time the spleen becomes competent and begins to increase the white cell production. It would be interesting to see if there is an increase in ribonucleotide reductase activity in the spleen after the administration of an antigen.

The data reported here, that ribonucleotide reductase as well as thymidine kinase is associated with a smooth membrane organelle in proliferating cells, in conjunction with the finding of Baril et al., that there is a DNA polymerase associated with this smooth membrane fraction from hepatomas and rat liver cytoplasm (1, 2) suggest that there may exist a complex of several DNA synthetic enzymes in a functional unit.

Acknowledgements

I express my gratitude to the following: The National Science Foundation for travel funds to attend this seminar, Dr. E. Baril for critical review of this report and for assistance in the subcellular fractionation, and Mrs. R. Elford for aid in the preparation of this paper. The cooperation of Dr. H.P. Morris in providing the Morris hepatomas is greatly appreciated. Portions of this work were supported by Grant P-407 from the American Cancer Society and by Grant CA 10441 and CA 11978 from the United States Public Health Service.

REFERENCES

1. Baril, E. F., Jenkins, M. D., Brown, O. E., and Laszlo, J. DNA polymerase activities associated with smooth membranes and ribosomes from rat liver and hepatoma cytoplasm. *Science*, **169**, 87–89 (1970).
2. Baril, E. F., Brown, O. E., Jenkins, M. D., and Laszlo, J. Deoxyribonucleic acid polymerase with rat liver ribosomes and smooth membranes. Purification and properties of the enzymes. *Biochemistry*, **10**, 1981–1992 (1971).
3. Behki, P. M. and Schneider, W. C. Intracellular distribution of deoxyribosedic compounds in normal and regenerating liver and in Novikoff hepatoma. *Biochim. Biophys. Acta*, **61**, 663–667 (1962).
4. Elford, H. L., Freese, M., Passanami, E., and Morris, H. P. Ribonucleotide reductase and cell proliferation. I. Variations of ribonucleotide reductase activity with tumor growth rate in a series of rat hepatomas. *J. Biol. Chem.*, **245**, 5228–5233 (1970).
5. King, C. D. and Van Lancker, J. L. Molecular mechanisms of liver regeneration. VII. Conversion of cytidine to deoxycytidine in rat regenerating liver. *Arch. Biochem. Biophys.*, **129**, 603–608 (1969).
6. Larsson, A. and Reichard, P. Enzymatic reduction of ribonucleotides. *Progr. Nucleic Acid Res.*, **7**, 303–347 (1967).
7. Larsson, A. Ribonucleotide reductase from regenerating rat liver. *Eur. J. Biochem.*, **11**, 113–121 (1969).

8. Morris, H. P. and Wagner, B. P. Induction and transplantation of rat hepatoma with different growth rate (including minimal deviation hepatomas). *Methods Cancer Res.*, **4**, 125–152 (1968).
9. Murphree, S., Stubblefield, E., and Moore, E. C. Synchronized mammalian cell cultures. III. Variation of ribonucleotide reductase activity during the replication cycle of Chinese hamster fibroblasts. *Exp. Cell Res.*, **58**, 118–124 (1969).
10. Pitot, H. C. Altered template stability: the molecular mask of malignance? *Perspect. Biol. Med.*, **8**, 50–70 (1964).
11. Potter, V. R. Recent trends in cancer biochemistry. *Can. Cancer Conf.*, **8**, 9 (1969).
12. Reichard, P., Baldesten, A., and Rutberg, L. Formation of deoxycytidine phosphates from cytidine phosphates in extracts from *Escherichia coli*. *J. Biol. Chem.*, **236**, 1150–1157 (1961).
13. Reichard, P. The biosynthesis of deoxyribonucleotides. *Eur. J. Biochem.*, **3**, 259–266 (1968).
14. Turner, M. K., Abrams, R., and Lieberman, I. Levels of ribonucleotide reductase activity during the division cycle of the L cell. *J. Biol. Chem.*, **243**, 3725 (1968).
15. Weber, G. The molecular correlation concept: studies on the metabolic pattern of hepatomas. *GANN Monograph*, **1**, 151–178 (1966).
16. Weber, G. Carbohydrate metabolism in cancer cells and the molecular correlation concept. *Naturwissenschaften*, **55**, 418–429 (1968).

MULTIMOLECULAR FORMS OF PYRUVATE KINASE AND PHOSPHOFRUCTOKINASE IN NORMAL AND CANCER TISSUES

Takehiko TANAKA, Kiichi IMAMURA, Tosen ANN, and
Koji TANIUCHI

*Department of Nutrition and Physiological Chemistry, Osaka University Medical School**

It has been reported from this laboratory that at least two types of pyruvate kinase are present in rat liver, type L (liver) and type M (muscle). Liver is the only organ which contains both types. These two types are distinctly different in immunochemical and kinetic nature. Type M can be electrophoretically separated into two types, M_1 and M_2, which are not distinguishable immunochemically. It is still not conclusive whether these types are interconvertible or are different in their primary structure. Type M_1 is found in the muscle, heart, and brain; its nature has already been well documented by Boyer (1). Type M_2 is widely distributed in glycolytic tissues such as liver, lung, intestine, spleen, stomach, adipose tissues, kidney, testis, ovary, and cancer tissues. The kinetic characteristics of type M_2 are intermediate between those of type L and type M_1 with respect to fructose 1,6-diphosphate activation, ATP inhibition and p-chloromercuribenzoic acid inhibition. The amount of type M_2 is markedly increased in newborn child tissues, even in muscle, heart, and brain. Type M_2 could be regarded as a prototype of pyruvate kinase in mammalian tissues.

Phosphofructokinase is classified into four types based on behavior in TEAE-cellulose column chromatography. These four types are designated as types I, II, III, and IV in order of their elution. Type I is distributed in the muscle, heart, and brain; type II in kidney, testis, spleen, and cancer tissues, and in small quantities in the brain and the thymus; type III in spleen, stomach, kidney, testis, thymus, and cancer tissues; and type IV, which could be called type L, in liver, erythrocyte, and all cancer tissues tested so far. These four types of phosphofructokinase differ in their kinetic and immunochemical properties.

Sugimura *et al.* reported in detail the distribution patterns of four types of hexokinase in various strains of tumor cells (11). From results on pyruvate kinase and phosphofructokinase by the present authors and from results on hexokinase by Sugimura *et al.*, it was concluded that hexokinase II, phosphofructokinase IV, and pyruvate kinase M_2 are major components in all tumor strains tested thus far.

---

* Joancho 33, Kita-ku, Osaka 530, Japan (田中武彦，今村喜一，安 斗宜，谷内孝次).

The combined pattern consisting of the three types of enzymes could be regarded as tumor specific since it was not found in normal tissues except in regenerating liver and fetal liver.

Multimolecular forms of all three key enzymes in glycolysis have been identified in mammalian tissues. In 1963 it was reported simultaneously and independently by Sols and his associates (20) and by Walker (21) that at least two types of hexokinase had been identified in rat liver. They were designated as hexokinase and glucokinase. Katzen and Schimke (5) further studied hexokinase in rat tissues and reported that at least four types of hexokinase were electrophoretically and kinetically separable. In 1964, Tanaka et al. (13) reported that at least two types of pyruvate kinase (Types M and L) had been identified in rat tissues. In 1970, Tanaka et al. (14, 15) reported that at least four types of mammalian tissue phosphofructokinase were separable by TEAE-cellulose chromatography. They were named type I, II, III, and IV.

In 1963, Schapira et al. (9) reported evidence suggesting that the isozyme pattern of aldolase in human hepatic cancer tissues differs significantly from the normal tissue pattern; aldolase A becomes predominant in the cancer tissue. Later, Sugimura and his associates studied in more detail the aldolase isozyme pattern deviation in experimental cancer tissues. Sugimura et al. (11) and Weinhouse et al. (23) reported that hexokinase isozyme patterns deviated in experimental hepatomas. It was reported by Tanaka et al. (16) in 1965 that only pyruvate kinase type M was detected in Yoshida ascites hepatomas (AH-130), whereas type L was predominant in normal liver tissue (16). These results indicate that deviations in isozyme pattern of more than one key enzyme in glycolysis have occurred simultaneously in one strain of cancer cells.

In the present paper, enzyme-chemical nature of multimolecular forms of pyruvate kinase and phosphofructokinase is outlined. In addition, whether or not tumor-specific patterns of multimolecular forms of three key enzymes exist is discussed.

Enzyme assay: Two methods were used to assay pyruvate kinase. The 2,4-dinitrophenylhydrazone method was routinely employed and the latic dehydrogenase coupling method was used for the kinetic studies. The details of these two methods have been previously published (17). Phosphofructokinase was assayed using the method of Mansour and Ahlfors (7).

Animals: Walker carcinosarcoma 256 was transplanted to Sprague-Dawley albino rats and Yoshida ascites hepatoma to Wistar albino rats. Patterns of the multimolecular forms of pyruvate kinase and of phosphofructokinase were the same in the two strains.

Polyacrylamide gel electrophoresis: The method for polyacrylamide gel electrophoresis of pyruvate kinase has been described by Imamura and Tanaka (3).

Other experimental conditions are described in the legend to each figure.

Multimolecular Forms of Pyruvate Kinase

1. Dietary and hormonal responses of enzyme activity in liver

It was reported independently and almost simultaneously by Tanaka et al., Krebs et al., and Weber et al. that the level of liver pyruvate kinase changes in response to various dietary and hormonal conditions. As shown in Table I, liver

TABLE I. Pyruvate Kinase Levels of Liver and Muscle of Rats in the Gluconeogenetic State (16)

Treatment	No. of animals	Pyruvate kinase activity (μmoles/mg/min)	
		Liver	Muscle
None (normal)	4	0.633±0.10	25.7±1.4
Fasted for 24 hr	5	0.530±0.02	26.5±1.6
Fasted for 48 hr	6	0.360±0.04	29.6±0.7
Diabetic	6	0.130±0.01	27.9±2.5
Insulin-treated diabetic	5	0.635±0.14	
Fed on a 90% casein diet	4	0.340±0.05	28.2±1.7
Fed on a 10% casein diet	4	0.965±0.06	24.5±2.2

pyruvate kinase increased in animals fed high-carbohydrate diets and decreased in animals that were fasted, were diabetic, and were fed high-protein diets. In contrast, the enzyme level in muscle did not show any response to dietary or hormonal conditions. This suggested that the enzymes in these two organs were different from each other. Zone electrophoresis is the most reliable way to test this possibility.

2. Zone electrophoresis of pyruvate kinase from various organs and in cancer cells

Tanaka et al. demonstrated that at least two types of pyruvate kinase can be identified by starch zone electrophoresis and by immunochemical reactivities (16, 17). However, it has not been finally concluded whether or not type M pyruvate kinase in liver and in muscle are the same. Both pyruvate kinases showed quite similar mobilities in starch block electrophoresis and cross immunochemical reactivities to the antibody for muscle pyruvate kinase. Susor and Rutter showed that type M pyruvate kinases in liver and in muscle showed slightly different mobilities from each other when electrophoresis was carried out in the presence of fructose 1,6-diphosphate in the buffer (12). Figures 1 and 2 show electrophoretic zymograms of pyruvate kinase in various rat tissues using 3.75% polyacrylamide gel as the supporting medium. The enzyme activity was detected by the slightly modified method of Susor and Rutter (12). In these zymograms at least three major spots, types M_1, M_2, and L, are detectable. Type M_1 is not as widely distributed as expected and is found only in the muscle, brain, and heart, and perhaps in the intestine in a small amount. Type M_2 is widely distributed in various tissues, but muscle and erythrocyte are the only tissues tested so far which do not contain the type M_2 enzyme. Type L is found in liver and erythrocyte, and in the kidney in small amounts. Distribution of the three types of pyruvate kinase

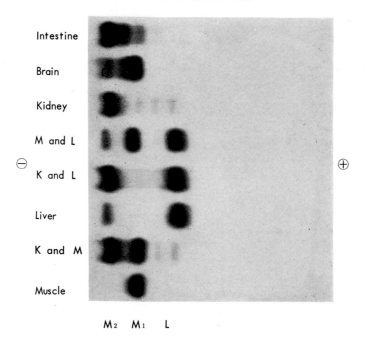

FIG. 1. Electrophoretic zymograms of rat tissue pyruvate kinase (I)
M, L, and K show muscle, liver, and kidney, respectively. The experimental procedures are described in Ref. (3).

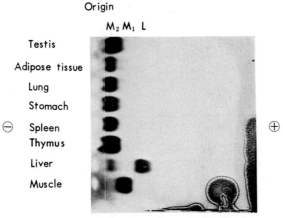

FIG. 2. Electrophoretic zymograms of rat tissue pyruvate kinase (II)
The experimental procedures are described in Ref. (3).

in various tissues is summarized in Table II. Normal rat tissues can be separated into three groups based on the distribution patterns of pyruvate kinase. The first group contains type M_1 as the main component, the second group contains type M_2, and the third group contains type L. It should be pointed out here that the distribution patterns of multimolecular forms of phosphofructokinase showed very similar groupings of tissues, as will be mentioned later.

As shown by Suda *et al.* in this volume, type M_2 is the predominant enzyme

TABLE II. Tissue Distributions of the Three Types of Pyruvate Kinase

Tissue	Type M_1	Type M_2	Type L
Muscle	++		
Heart	++	+	
Brain	++	+	
Spleen		++	
Stomach		++	
Lung		++	
Adipose tissue		++	
Testis		++	
Ovary		++	
Intestine	+	++	
Kidney		++	+
Liver		+	++
Yoshida ascites hepatoma AH-130		++	
Walker carcinosarcoma 256	+	++	

+ small amount; ++ large amount; blank, not detectable.

in regenerating liver and in tissues of newborn children (10). Rapidly growing tumor cells, such as Walker carcinosarcoma 256 and Yoshida ascites hepatoma AH-130, contain only the type M_2 enzyme, as shown in Fig. 7. In addition to these cancer strains all cancer strains tested so far, including the Yoshida sarcoma, several strains of Yoshida ascites hepatoma and the Ehrlich ascites tumor contain the type M_2 enzyme.

3. Kinetic properties of three types of pyruvate kinase

As reported from this laboratory, type M_1 and type L pyruvate kinase are different in their kinetic properties (17, 18). Type L enzyme is an allosteric enzyme whereas the muscle enzyme (M_1) is not. Generally speaking, type M_2 shows properties intermediate between those of type M_1 and type L. Various enzyme-chemical differences among these types of pyruvate kinase are summarized in Table III.

It was reported by Hess et al. that yeast pyruvate kinase is activated by fructose 1,6-diphosphate and is inhibited by ATP (2). The feedforward activation and feedback inhibition are competitive with each other, that is K_i for ATP increased from 0.16×10^{-3} M to 2.2×10^{-3} M in the presence of fructose 1,6-diphosphate (18), as seen in Fig. 3. It seems likely that the type L pyruvate kinase reaction is regulated by the balance between concentrations of both effects.

Type M_2 showed different kinetic properties from those of type L. Type M_2 is activated by fructose 1,6-diphosphate and is inhibited by ATP, although the extent of the activation and inhibition was smaller than that in type L. However, ATP inhibition was not reversed by fructose 1,6-diphosphate in the case of type M_2 (4).

Wood reported that inhibition of muscle pyruvate kinase by ATP was reversed with the increase of magnesium ion concentration (24). The same experiments

TABLE III. Summary of Differences among the Three Types of Pyruvate Kinase (4)

	Type M_1	Type M_2	Type L
Increase of the enzyme observed in liver		Regenerating and newborn child liver, and tumors	Insulin and high-carbohydrate diet
Distribution	Muscle, heart, brain, intestine	Kidney, spleen, lung, testis intestine, tumors	Liver, kidney, erythrocyte
Molecular weight	250,000		208,000
S Value	9.5	9.3	10.1
Antibody for muscle enzyme	Neutralized	Neutralized	Not neutralized
Antibody for type L enzyme	Not neutralized	Not netralized	Neutralized
K_m for PEP	$0.75 \cdot 10^{-4}$ M	$0.40 \cdot 10^{-3}$ M	$0.83 \cdot 10^{-3}$ M
Hill constant for PEP	1.0	1.4–1.5	1.8
FDP activation	—	+	++
ATP inhibition	+	++	++, Cooperative
Reversal of ATP inhibition by Mg^{2+}	++	+	—
PCMB inhibition	+	++	++
Inhibition by alanine and phenylalanine	—	++	++, Cooperative

PEP, phosphoenolpyruvate; FDP, fructose 1,6-diphosphate; PCMB, p-chloromercuribenzoic acid

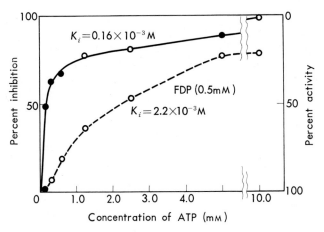

FIG. 3. Effect of fructose 1,6-diphosphate on ATP inhibition of type L pyruvate kinase

The experimental procedures are described in Ref. (18). ADP: 0.2 mM; PEP: 1.0 mM.

were carried out with three types of pyruvate kinase. As seen in Fig. 4, inhibition of type L by ATP was not reversed and inhibition of type M_1 was reversed with the increase of magnesium ion concentration. From this result it was concluded

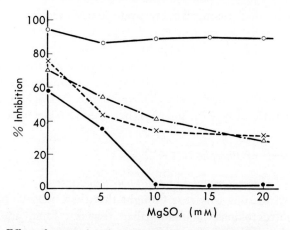

FIG. 4. Effect of magnesium ion concentration on ATP inhibition of the three types of pyruvate kinase

The enzyme was assayed in the presence of 1.0 mM ATP. Other assay conditions are the same as in Fig. 3. Type M_2 was purified from an extract of Yoshida ascites hepatoma AH-130 cells. ○ type L (liver); × type M_2 (AH-130); △ AH-130 (bearer); ● type M (muscle).

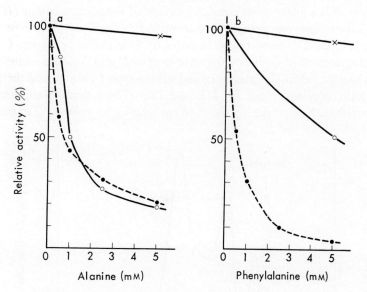

FIG. 5. Inhibitory effects of alanine and phenylalanine on the three types of pyruvate kinase

The experimental conditions are the same as in Fig. 4. a: Alanine; b: Phenylalanine. × type M (muscle); ● tyep M_2 (AH-130); ○ type L (liver).

that type L was allosterically inhibited by ATP and type M_1 was not. Type M_2 showed properties intermediate between those of type L and M_1, that is, inhibition of type M_2 by ATP was partially reversed with the increase of magnesium ion concentration.

Weber reported that mammalian pyruvate kinase was inhibited by alanine and phenylalanine (*22*). As seen in Fig. 5, type M_1 was not inhibited by these inhibitors, type M_2 and L were inhibited by alanine to almost the same extent, and type M_2 was much more strongly inhibited by phenylalanine than was type L.

Type M_2 pyruvate kinase has been purified from Yoshida ascites hepatoma AH-130 cells to an ultracentrifugically and electrophoretically single peak (*4*). Comparison of the structures of type M_1 and M_2 is being undertaken in our laboratory.

Multimolecular Forms of Phosphofructokinase

1. TEAE-cellulose chromatography of phosphofructokinase

There are two reasons why it is thought that there may be multimolecular forms of phosphofructokinase in mammalian tissues. First, among three key enzymes in glycolysis, phosphofructokinase is the only enzyme in which multimolecular forms have not been identified. Second, as reported by Tarui *et al.* (*19*), the enzyme level decreases only partially in red blood cells in muscle deficiency. The fact that muscle phosphofructokinase antiserum neutralizes the enzyme activity in various tissue extracts to quite variable extents has stimulated the present authors to study phosphofructokinase.

TEAE-cellulose column chromatography of phosphofructokinase was carried out by elution with a linear concentration gradient of ammonium sulfate (*15*). At least four peaks of enzyme activity were identified in rat tissues and were named types I, II, III, and IV, respectively, in order of their elution. Figure 6 shows the chromatograms of phosphofructokinase in muscle and Yoshida ascites hepatoma AH-130 cells. Muscle extracts contained the type I enzyme and the ascites hepatoma cells contained types II, III, and IV. These peaks were not interconvertible with each other during the chromatography procedure. For example,

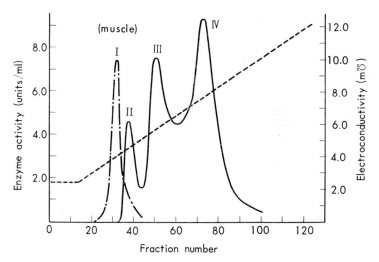

Fig. 6. TEAE-cellulose chromatography of phosphofructokinase of muscle and Yoshida ascites hepatoma AH-130 cells

The experimental procedures are described in Ref. (*15*).

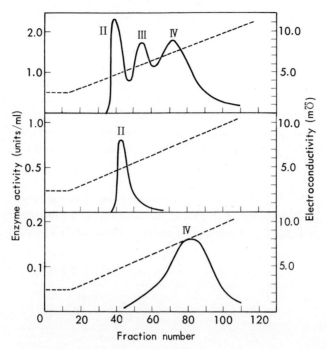

FIG. 7. TEAE-cellulose rechromatography of phosphofructokinase of Yoshida ascites hepatoma AH-130 cells

The procedure for chromatography was the same as in Fig. 6. The peak of type II or type IV was collected, the enzyme was precipitated at 60% saturation of ammonium sulfate and the precipitate was reapplied to the chromatography after dialysis, as described in Ref. (15).

as shown in Fig. 7, type II and type III were eluted at their proper concentrations of the buffer by repeating the chromatography.

2. *Immunochemical characteristics of the four types of phosphofructokinase*

Type I phosphofructokinase was purified from rat muscle extract; type II and IV were purified from Yoshida ascites hepatoma AH-130 cells by repeating the TEAE-cellulose chromatography until a single peak appeared in the enzyme activity. The antiserum for muscle-type enzyme (type I) was very sensitive in the precipitation of the homologous antigen. The same antiserum was used for

TABLE IV. Summary of Immunoreactivity of the Four Types of Phosphofructokinase (6)

Enzyme	Antibody		
	I	II	IV
I	+	−	−
II	−	+	−
III	−		−
IV	±	±	+

+ neutralized; − not neutralized; ± partially neutralized.

enzymes of the heart and brain, but more than 10 times as much was required to precipitate type IV enzyme. Types II and III showed almost no cross-reaction with type I antiserum. The type II antiserum was very sensitive in the precipitation of its corresponding antigen but reacted with type IV to a lesser extent. The antiserum, however, did not react with type I. The antibody for type IV was highly specific for type IV and did not react with other types of phosphofructokinase. The immunoreactivities of the four type of phosphofructokinase in rat tissues are summarized in Table IV; these results show that the four types not only differed in their behavior on TEAE-cellulose column chromatography, but also differed in their immunochemical reactivities.

3. *Distribution of the four types of phosphofructokinase in rat tissue*

The tissue distribution patterns of the four types of phosphofructokinase are summarized in Table V. Each type of enzyme was identified by the immunochem-

TABLE V. Tissue Distribution of the Four Types of Phosphofructokinase

Tissue	Type			
	I	II	III	IV
Muscle	++			
Heart	++			
Brain	++	+		
Spleen		+	++	
Stomach			++	
Testis		++	+	
Thymus		+	++	
White blood cell		+	++	
Kidney		++	+	+
Liver				++
Red blood cell				++

+ small amount; ++ large amount; blank, not detectable.

ical reactivity and electroconductivity at which it was eluted with TEAE-cellulose chromatography. Rat tissues can be separated into three groups. The first group, the muscle, heart, and brain, contains type I as the main phosphofructokinase component; the second group (most tissues belong to this group) contains type II and/or type III as the major components; and the third group, the liver and perhaps the kidney, contains type IV. This grouping of rat tissues based on distribution patterns of phosphofructokinases is completely equivalent to the grouping based on distribution patterns of pyruvate kinase, pointed out in the above section on pyruvate kinase.

4. *Distribution of the four types of phosphofructokinase in various cancer tissues*

Distribution patterns of the phosphofructokinase type in various cancer tissues are summarized in Table VI. It should be noted here that all tumor tissues tested thus far contained significant amounts of type IV. Among the tumor tissues

TABLE VI. Distribution of the Four Types of Phosphofructokinase
in Various Tumor Cells

Tissue	Type			
	I	II	III	IV
AH-66F		+	+	++
AH-130		++	++	++
Yoshida sarcoma		++	++	++
Walker carcinosarcoma 256	++	++	++	++
LY-7		?	+	++
LY-5		?	+	++
Human gastric cancer		+	±	++
Metastatic lymph node		+	±	++

AH-66F and AH-130 are the strains of Yoshida ascites hepatoma.
+ small amount; ++ large amount; blank, not detectable; ?, questionable.

listed in Table VI, Walker carcinosarcoma 256, human gastric cancer, and metastatic lymph nodes with gastric cancer originated from tissues other than the liver. In connection with this, Ehrlich ascites tumor cells, which also did not originate from the liver, contained type IV. As described above, in normal tissue type IV is distributed only in the liver, red blood cells, and kidney; therefore, this type could be called the liver type. Distribution patterns of the four phosphofructokinases in regenerating liver did not differ from those of normal liver. In the liver of newborn children, type IV was found as a major peak. These results strongly contrast with the fact that type M_2 pyruvate kinase increases and becomes a predominant component in cancer tissues and regenerating liver.

5. *Differences in kinetic properties among the four types of phosphofructokinase*

Phosphofructokinase is a strong allosteric regulatory enzyme whose numerous effectors have been reported. Kinetic properties of the four phosphofructokinases should be studied to clarify the specific control mechanisms of glycolysis in various tissues. Type I was purified from muscle extract and types II and IV from Yoshida ascites hepatoma AH-130 cells.

Figure 8 plots enzyme activity against ATP concentration in the presence and absence of 3′,5′-cyclic AMP at pH 7.1. Type IV was inhibited most by ATP and was the most significantly deinhibited in the presence of 3′,5′-cyclic AMP. Type II was not significantly inhibited by ATP at a concentration lower than 2 mM. Type I showed intermediate sensitivity. At a higher concentration, AMP and ADP showed inhibitory effects similar to those of ATP.

Figure 9 plots percentage activation against nucleotide concentration. Type IV was more sensitive to activation (or deinhibition) by 3′,5′-cyclic AMP than by AMP or by ADP.

Mansour reported that a change of pH markedly affected the allosteric properties of muscle phosphofructokinase (7). According to Mansour *et al.*, the enzyme was not inhibited by ATP when the pH was higher than 7.1. The pH dependency of ATP inhibition and its reversal by 3′,5′-cyclic AMP are shown

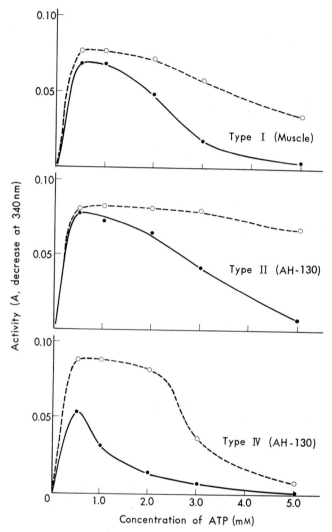

FIG. 8. Plots of phosphofructokinase activity against ATP concentration

The enzyme was assayed at pH 7.1 (50 mM HCl-imidazole buffer). Type I was purified from muscle. Type II and type IV were purified from Yoshida ascites hepatoma AH-130 cells by repeating the TEAE-cellulose chromatography, as shown in Fig. 7. —— enzyme activity in the absence of 3′,5′-cyclic AMP; - - - enzyme activity in the presence of the cyclic AMP; ● F6P: 0.046 mM; ○ F6P: 0.046 mM + cyclic AMP: 0.2 mM.

in Fig. 10. It should be pointed out here that type IV was inhibited by ATP even at a higher pH than 7.2 and that this inhibition was reversed by mono- and dinucleotides. It seems likely, therefore, that the activity of type IV enzyme is regulated by the balance of ATP and the di- and mononucleotides at physiological range of pH. Type II was the least allosteric enzyme of the three and was not significantly activated by 3′, 5′-cyclic AMP at a pH range of 6.9 to 8.0 and at an ATP concentration of 1.0 mM.

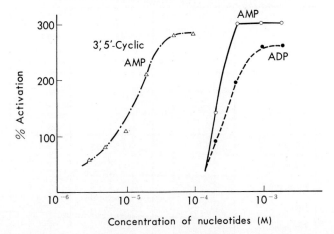

FIG. 9. Activation of purified phosphofructokinase type IV by adenine nucleotides

Percentage activation was plotted against a logarithmic scale of adenine nucleotide concentration.

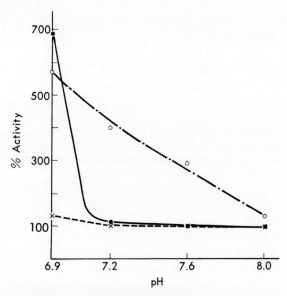

FIG. 10. pH dependence of activation of phosphofructokinase by cyclic AMP

The enzyme was assayed at 1 mM ATP, 0.046 mM fructose 1,6-diphosphate, and 0.2 mM 3′,5′-cyclic AMP. HCl-imidazole buffer was used at pH 6.8 and 7.2, and Tris-HCl buffer was used at pH 7.6 and 8.0. The concentration of both buffers used was 50 mM. ● type I (muscle); × type II (AH-130); ○ type IV (AH-130).

DISCUSSION

Identification and enzyme-chemical characterization of the multimolecular forms of phosphofructokinase and pyruvate kinase in rat tissues have been outlined in this paper. Distribution patterns of the multimolecular forms of three

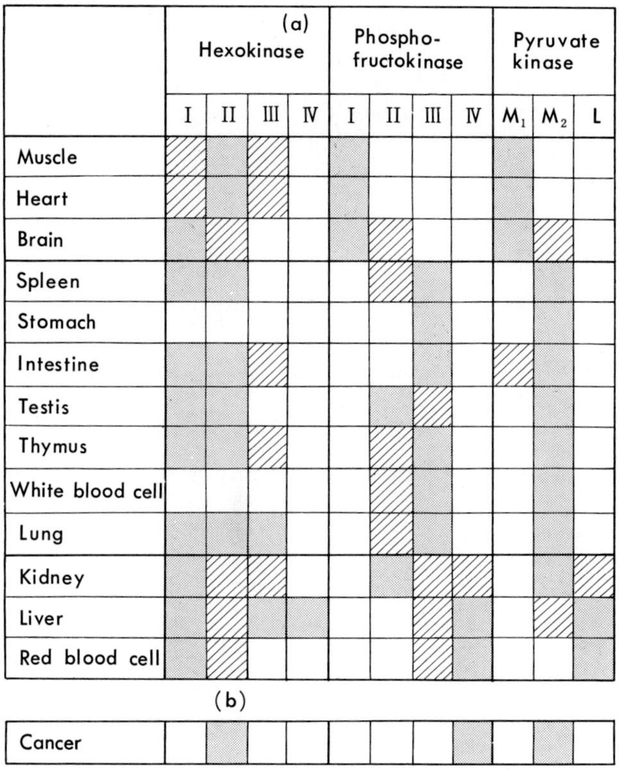

	(a) Hexokinase				Phospho-fructokinase				Pyruvate kinase		
	I	II	III	IV	I	II	III	IV	M_1	M_2	L
Muscle	▨		▨		▨				▨		
Heart	▨	▨			▨				▨		
Brain		▨			▨				▨		
Spleen							▨			▨	
Stomach										▨	
Intestine		▨								▨	
Testis							▨			▨	
Thymus		▨					▨			▨	
White blood cell							▨			▨	
Lung	▨						▨			▨	
Kidney	▨	▨					▨	▨		▨	▨
Liver	▨	▨			▨			▨			▨
Red blood cell	▨	▨			▨		▨			▨	
(b)											
Cancer		▨						▨		▨	

FIG. 11. Tissue distributions of multimolecular forms of the three glycolytic key enzymes

Distribution patterns of hexokinase (a) were modified to the simpler style of Katzen (6). Data regarding hexokinase in cancer tissues (b) are quoted from the paper by Sato *et al.* (8). Distribution of hexokinase in the stomach and in white blood cells was not described in Katzen's paper.

key enzymes in glycolysis are shown in Fig. 11. The distribution of hexokinase in Fig. 11 is from the paper by Katzen (6); his results have been somewhat modified to the simpler style of the figure. As shown in Fig. 11, rat tissues contain many various combinations of multimolecular forms of key enzymes. The authors classified rat tissues into three groups based on distribution patterns of pyruvate kinase and phosphofructokinase.

Sato *et al.* (8) reported that the distribution of the four types of hexokinase varies in different cancer cells, as shown in Fig. 12. However, it is noteworthy that their results show that hexokinase II was distributed, *i.e.*, it increased or appeared, in all of the cancer tissues they examined. As mentioned above, phosphofructokinase IV and pyruvate kinase M_2 were also found in all of the cancer tissues that we examined. This particular combination of three key enzymes was not found in normal tissues (Fig. 11) except regenerating liver and fetal liver. It should also be noted here that the same combination was found even in the cancer tissues which did not originate in the liver.

Two questions can be raised about this fact; first, what is the physiological

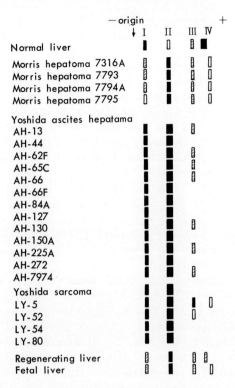

FIG. 12. Distributions of the four types of hexokinase in various hepatomas. This figure is cited from the paper by Sato et al. (8). ■ high; ▌ intermediate or low; □ questionable or absent.

significance for cancer cells of the combination of hexokinase II, phosphofructokinase IV, and pyruvate kinase M_2; second, how do the key enzyme patterns in glycolysis deviate to those patterns seen in young cells? It has been well established since the discovery by Warburg that activity in glycolysis increases in growing cells. In order to answer the above questions mechanisms of regulation of cell growth in relation to the regulation of glycolysis should be investigated.

REFERENCES

1. Boyer, P. D. Pyruvate kinase. *In* "The Enzymes," (2nd Ed.), ed. by P. D. Boyer, H. Lardy, and K. Myrbäck, Vol. VI, Academic Press Inc., New York, p. 95 (1962).
2. Hess, B., Haeckel, R., and Brand, K. FDP activation of yeast pyruvate kinase. *Biochem. Biophys. Res. Commun.*, **24**, 824–831 (1966).
3. Imamura, K. and Tanaka, T. Multimolecular forms of pyruvate kinase from rat and other mammalian tissues. I. Electrophoretic studies. *J. Biochem. (Tokyo)*, **71**, 1043–1051 (1972).
4. Imamura, K., Taniuchi, K., and Tanaka, T. Multimolecular forms of pyruvate kinase. II. Purification of M_2-type pyruvate kinase and enzymatic properties of three types of pyruvate kinases, L, M_1, and M_2 (in preparation).

5. Katzen, M. H. and Schimke, R. T. Multiple forms of hexokinase in the tissue distribution, age-dependency, and properties. *Proc. Natl. Acad. Sci. U.S.*, **54**, 1218–1225 (1965).
6. Katzen, H. M. The multiple forms of mammalian hexokinase and their significance to the action of insulin. *Adv. Enzyme Reguation*, **5**, 335–356 (1967).
7. Mansur, T. E. Studies on heart phosphofructokinase: purification, inhibition, and activation. *J. Biol. Chem.*, **238**, 2285–2292 (1963).
8. Sato, S., Matsushima, T., and Sugimura, T. Hexokinase isozyme patterns of experimental hepatomas of rats. *Cancer Res.*, **29**, 1437–1443 (1969).
9. Schapira, F., Dreyfus, G. C., and Schapira, G. Anomaly of aldolase in liver cancer. *Nature*, **200**, 995–997 (1963).
10. Suda, M., Tanaka, T., Yanagi, S., Hayashi, S., Imamura, K., and Taniuchi, K. Differentiation of enzymes in the liver of tumor-bearing animals. *GANN Monograph*, **13**, 79–93 (1972).
11. Sugimura, T., Matsushima, T., Kawachi, T., Hirata, Y., and Kawabe, S. Molecular species of aldolase and hexokinase in experimental hepatoma. *GANN Monograph*, **1**, 143–149 (1966).
12. Susor, W. A. and Rutter, W. J. Some distinctive properties of pyruvate kinase purified from rat liver. *Biochem. Biophys. Res. Commun.*, **30**, 14–20 (1968).
13. Tanaka, T., Harano, T., and Mori, R. Regulation of pyruvate kinase (in Japanese). *Seikagaku*, **36**, 532–533 (1964).
14. Tanaka, T., Sakaue, Y., and Ann, T. Multimolecular forms of phosphofructokinase in rat tissues (in Japanese). *Seikagaku*, **40**, 527 (1968).
15. Tanaka, T., Ann, T., and Sakaue, Y. Studies on multimolecular forms of phosphofructokinase in rat tissues. *J. Biochem. (Tokyo)*, **69**, 609–612 (1971).
16. Tanaka, T., Harano, Y., Morimura, H., and Mori, R. Evidence for the presence of two types of pyruvate kinase in rat liver. *Biochem. Biophys. Res. Commun.*, **21**, 55–60 (1965).
17. Tanaka, T., Harano, Y., Sue, F., and Morimura, H. Crystallization, characterization and metabolic regulation of two types of pyruvate kinase isolated from rat tissues. *J. Biochem. (Tokyo)*, **62**, 71–91 (1967).
18. Tanaka, T., Sue, F., and Morimura, H. Feed-forward activation and feed-back inhibition of pyruvate kinase type L of rat liver. *Biochem. Biophys. Res. Commun.*, **29**, 444–449 (1967).
19. Tarui, S., Okuno, G., Ikura, Y., Tanaka, T., Suda, M., and Nishikawa, M. Phosphofructokinase deficiency in skeletal muscle. A new type of glycogenolysis. *Biochem. Biophys. Res. Commun.*, **19**, 517–523 (1965).
20. Vinuêla, E., Salas, M., and Sols, A. Glucokinase and hexokinase in liver in regulation to glycogen synthesis. *J. Biol. Chem.*, **238**, PC 1175–1177 (1963).
21. Walker, D. G. On the presence of two soluble glucose-phosphorylating enzymes in adult liver and the development of one of these after birth. *Biochim. Biophys. Acta*, **77**, 209–226 (1963).
22. Weber, G. Regulation of pyruvate kinase. *Adv. Enzyme Regulation*, **7**, 15–40 (1969).
23. Weinhouse, S. Glycolysis, respiration, and enzyme deletions in slow-growing hepatic tumors. *GANN Monograph*, **1**, 99–115 (1966).
24. Wood, T. The inhibition of pyruvate kinase by ATP. *Biochem. Biophys. Res. Commun.*, **31**, 779–785 (1968).

BIOCHEMICAL STUDIES ON THE PRENEOPLASTIC STATE[*1]

Hideya ENDO, Masao EGUCHI, Susumu YANAGI, Takehiko TORISU, Yukio IKEHARA, and Tomoya KAMIYA

Cancer Research Institute, Faculty of Medicine, Kyushu University[*2]

The early stage of hepatocarcinogenesis was investigated using pyruvate kinase and aldolase isozymes as markers. The muscle-type isozyme activity of these enzymes in the liver of rats fed a diet containing 0.06% 3'-methyl-4-dimethylamino-azobenzene for 60 days and then a basal diet for 60 days was found to be elevated when compared to normal liver activity, although no appreciable differences were observed in total activity, mitotic index, histological architecture, and soluble protein and DNA content. A similar enzymic alteration was induced by another hepatocarcinogen, N, N'-2, 7-fluorenylenebisacetamide, whereas a noncarcinogen, 2-methyl-4-dimethylamino-azobenzene, failed to induce this change. These enzymic changes were already observed after administrating azo dye for 15 days. The increased level of muscle-type isozyme activity of these enzymes was maintained without decrease during a long observation period after discontinuation of azo dye administration. Partial hepatectomy of the azo dye-treated liver revealed that the muscle-type isozyme activities of both enzymes returned, through a transient increase, to their preoperative elevated levels after completion of regeneration, but never to any intermediate levels between normal and azo dye-fed liver. The elevated level of the muscle-type isozyme activity scarcely responded to dietary and hormonal controls, while the total activity varied greatly. Based on these findings, properties of the preneoplastic state was discussed.

A true understanding of the preneoplastic state is considered to offer an essential clue not only for analyzing carcinogenic mechanisms but also for contributing to the practical problem of early diagnosis of cancer. Hence, abundant research has been carried out to explore some crucial properties of the process of chemical carcinogenesis. Biochemical studies (*34*), mainly dealing with changes in various enzyme patterns during the process of hepatocarcinogenesis seem to be limited because the analysis can only express average changes in

---

[*1] This work was supported by a grant from the Society for the Promotion of Cancer Research and by a Grant-in-Aid for Scientific Research from the Ministry of Education of Japan. A part of this work was published in *Cancer Res.*, 30, 743–752 (1970) and in *GANN*, 62, 283–291 (1971).

[*2] Katakasu 1276, Fukuoka 812, Japan (遠藤英也, 江口正雄, 柳　進, 鳥巣岳彦, 池原征夫, 神谷知弥).

heterologous deviation appearing in the liver which has a heterogenous cell population. Biological studies, on the other hand, have provided evidence (7, 10, 20, 28, 32) indicating a considerable frequency of cancer incidence in animals when the administration of a submanifestational dose of a certain chemical carcinogen is followed, at a sufficient time interval, by an application of a submanifestational dose of the same or different carcinogen affecting the same target organ. Similar phenomena were also observed with the combination of two different carcinogens which affect the different target organs (2, 18, 21). Moreover, evidence (3, 5, 22) has been presented which indicates that certain compounds, when applied after the initial administration of a submanifestational dose of a carcinogen, assist the production of cancer even when such compounds are noncarcinogenic. Centering around this evidence, several interpretations (4, 9, 19, 36) have been proposed which at least seem to coincide with the consideration that the first exposure to the carcinogenic stimuli causes a somewhat irreversible change in the cell and that this change is maintained in the cell during a subsequent long period in which a carcinogen-free diet is administered. Based on this consideration, we attempted to see whether such an irreversible change involved in the transformation of normal cells to the malignant state does indeed exist, and, if so, how it might be expressed biochemically. Thus, experiments were carried out during the early stage of hepatocarcinogenesis.

Pyruvate Kinase and Aldolase Isozymes as Markers

Suitable markers seem to be essential for biochemical analysis; for the present study, we chose pyruvate kinase and aldolase isozymes. Tanaka et al. (33) found that pyruvate kinase in the liver is composed of two isozymes, type M and L. Quite recently this same group (13) showed that type M is different from muscle pyruvate kinase, which in our study is designated as muscle-type pyruvate kinase, since the activity of type M was shown to be totally inhibited by antimuscle pyruvate kinase antibody. Muscle-type pyruvate kinase was shown to differ from type L (liver-type pyruvate kinase) in electrophoretic pattern, immunological and kinetic properties, molecular weight, and responses to various physiological conditions. Meanwhile, aldolase in the liver has been found by Rutter et al. (26) to be composed of two isozymes, muscle-type and liver-type, which differ from each other in substrate specificity and immunological properties. Especially noteworthy are the findings of Tanaka et al. (33) on pyruvate kinase, and those of Rutter et al. (26), Schapira et al. (27), Sugimura et al. (30), Matsushima et al. (14), and Adelman et al. (1) on aldolase, which showed that the activity of the liver-type isozyme of pyruvate kinase or aldolase is much higher than that of the muscle-type in normal liver, and that the converse is true in poorly differentiated malignant liver tumors. These findings led us to the assumption that the process of hepatocarcinogenesis might eventually be analyzed quantitatively by using the pattern of pyruvate kinase or aldolase isozymes.

The abbreviation used are: 3′-Me-DAB, 3′-methyl-4-dimethylamino-azobenzene; 2-Me-DAB, 2-methyl-4-dimethylamino-azobenzene; 2,7-FAA, N,N′-2,7-fluorenylenebisacetamide.

Examination of the Change in Histological Architecture, Mitotic Rate, and Content of DNA and Soluble Protein in Rat Liver Treated with Azo Dye

In analyzing the biochemical changes occurring during chemical carcinogenesis, it seemed especially desirable to eliminate the problem of the nonspecific toxic effect of the carcinogen applied to the target organ. Otherwise, the possibility cannot be excluded that the results obtained might be a manifestation of the toxic effect of the carcinogen and unrelated to the transformation of the cells. The toxic effect of drugs in general, however, would disappear gradually after cessation of administration of the carcinogen. With this in mind, biochemical analyses of the target organ have been conducted in animals at a sufficient time interval after stopping administration of the carcinogen, so that only irreversible changes would be observed. It was therefore nesessary to know the time interval required for the reversal of the toxic effects of 3'-Me-DAB which was used in our study mainly as a hepatocarcinogen. For this purpose, histological and cytological examinations were performed on the livers at various time intervals after discontinuing the azo dye administration.

In comparison with normal liver (Photo 1), severe damage was observed in the specimen prepared from the livers of male albino rats (Wistar strain, 150–200 g) fed a diet containing 0.06% 3'-Me-DAB for 60 days (Photo 2). This damage was still observed in the specimen prepared 14 days after the cessation of azo dye feeding (Photo 3), but considerable recovery was seen in the specimen at 30 days after the discontinuation of azo dye treatment (Photo 4). The specimen prepared at 60 days after the end of azo dye treatment (Photo 5) was practically indistinguishable from that of normal liver.

TABLE I. Changes in Histological Architecture and Mitotic Rate, and in Content of DNA and Soluble Protein in Liver of Rats Fed a 3'-Me-DAB Diet for 60 Days and Then a Basal Diet for Various Periods

Days fed 3'-Me-DAB diet[a]	Days fed Basal diet	Distribution of cell types[b] (%)					Mitotic rate[c]	Content (mg wet liver) of DNA[d]	Content (mg wet liver) of Protein[d]
		I	II	III	IV	V			
Basal diet only (control)		64.6	27.9	2.7	2.0	2.8	0–3	2.32±0.16 (6)[e]	95.5±14.5 (35)[e]
60	0	26.5	32.0	25.0	10.5	6.0	12–20		
60	14	44.0	36.0	11.0	7.0	2.0	5–12		
60	30	54.0	31.0	8.0	5.0	2.0	2–8		
60	60–90	62.0	29.8	2.6	3.5	2.1	0–6	2.35±0.11 (6)[e]	92.6±11.0 (35)[e]
60	150	58.8	30.5	3.7	3.1	3.9	0–3		

[a] 3'-Me-DAB diet contained 0.06% 3'-Me-DAB in basal diet.
[b] Cell types composing liver tissue: I, parenchymal cells; II, littoral cells; III, cells of bile ducts; IV, cells of connective tissue; V, cells of blood vessel walls.
[c] Mitotic rate was expressed as the No. of mitotic figures/4,000 parenchymal cells.
[d] Data were expressed as the mean value±SD.
[e] No. of rats assayed.

The percentage distribution of the five types of cells in liver tissues was measured according to the classification of Daoust et al. (8). As shown in Table I, a remarkable decrease of parenchymal cells, as well as a concomitant increase in bile duct and connective tissue cells, was observed immediately after administration of azo dye for 60 days. These changes also seemed to reverse during the recovery period after discontinuing the azo dye feeding, and the percentage distribution of the cells in the liver measured at 60 to 90 days after the termination of azo dye feeding was found to be practically the same as that in normal liver. As shown in Table I, the mitotic rate which is normally 0 to 3/4000 parenchymal cells in liver was found to be elevated to 12 to 20/4000 cells immediately after the cessation of 60 days of azo dye administration. This change, however, seemed to reverse during the interval after the discontinuation of azo dye feeding. Thus, the mitotic rate in the liver was found to be practically at the same level as that in normal liver when measured at 60 to 90 days after the end of azo dye feeding. The content of DNA and soluble protein per wet weight of liver was measured in normal liver and in liver from rats 60 days subsequent to the end of azo dye administration. As shown in Table I, no appreciable difference could be seen between normal rat liver and the livers of rats fed with azo dye for 60 days and then with normal diet for 60 days. These results indicate that the nonspecific toxic effect of 3′-Me-DAB can be eliminated by the feeding of a normal diet for 60 days after 2 months of azo dye administration.

It must also be mentioned that, during the entire course of the present study, neither tumor formation nor any appreciable macroscopic changes could be seen in the livers of all animals fed 3′-Me-DAB for 60 days, presumably due to the high content of riboflavin in the diet, while administration of azo dye for 4 months produced liver tumors by the 200th day of observation in about 80% of the animals.

Total and Muscle-type Isozyme Activity of Pyruvate Kinase and Aldolase in the Liver and Malignant Liver Tumors of Rats Fed Azo Dye

The liver tissues of rats fed 3′-Me-DAB for 7, 15, 30, and 60 days were assayed for total and muscle-type isozyme activity of pyruvate kinase and aldolase on the 60th day after stopping the azo dye administration. The activity was compared with that found in normal liver and in azo dye-induced malignant liver tumors. The total and muscle-type isozyme activity of both enzymes in normal liver was fairly constant during a period of 180 days on the basal diet. With respect to the total activity of pyruvate kinase, no appreciable difference could be seen in all azo dye-treated liver tissues except for azo dye-induced malignant liver tumors as shown in Table II. The total activity of aldolase was virtually unchanged even in the azo dye-induced malignant tumors as shown in Table III. On the contrary, the muscle-type isozyme activity of both enzymes was found to be greatly elevated in liver tumors as compared with normal liver. The increase in muscle-type isozyme activity was already seen in the livers of rats fed azo dye for only 15 days. A similar increase was observed in the group fed azo dye for 30 days. The livers of rats fed azo dye for 60 days showed about a 2-fold increase

TABLE II. Changes in Total and Muscle-type Isozyme Activity of Liver Pyruvate Kinase in Rats Fed a 3'-Me-DAB Diet and Later Fed a Basal Diet

Days fed		No. of rats	Pyruvate kinase activity (units/mg protein at 37°)[b]	
3'-Me-DAB diet	Basal diet		Total	Muscle-type
Basal diet only (control)		17	0.809±0.091	0.087 ±0.023
7	60	5	0.735±0.016	0.084 ±0.019
15	60	9	0.872±0.075	0.119[a]±0.034
30	60	9	0.879±0.083	0.128[a]±0.031
60	60	9	0.829±0.065	0.165[a]±0.054
60	120	5	0.803±0.021	0.156[a]±0.016
60	150	5	0.842±0.043	0.163[a]±0.044
Malignant liver tumor		8	1.554±0.091	1.448[a]±0.088

Data are expressed as the mean value±SD.
[a] Differences from the control value were statistically significant ($P < 0.001$).
[b] Units mean μmoles pyruvate kinase formed/min.

TABLE III. Changes in Total and Muscle-type Isozyme Activity of Liver Aldolase in Rats Fed a 3'-Me-DAB Diet and Later Fed a Basal Diet

Days fed		No. of rats	Aldolase activity (units/mg protein at 30°)[c]	
3'-Me-DAB diet[a]	Basal diet		Total	Muscle-type
Basal diet only (control)		25	0.094±0.014	0.0115 ±0.0027
15	60	5	0.088±0.013	0.0172[b]±0.0011
30	60	8	0.104±0.014	0.0169[b]±0.0038
60	60–90	14	0.092±0.015	0.0203[b]±0.0034
60	150	4	0.128±0.009	0.0242[b]±0.0036
60	210	3	0.112±0.004	0.0212[b]±0.0030
60	300	4	0.108±0.012	0.0204[b]±0.0012
Malignant liver tumor		9	0.090±0.029	0.0567[b]±0.0312

Data are expressed as the mean value ±SD.
[a] 3'-Me-DAB diet contained 0.06% 3'-Me-DAB in basal diet.
[b] Differences from the control value were statistically significant ($P < 0.001$).
[c] Units mean μmoles fructose 1,6-di-phosphate cleaved/min.

in the muscle-type isozyme activity of both enzymes as compared with that of normal liver (Tables II and III). This increase was found to be statistically significant ($P=0.001$) when compared with control values.

Total and Muscle-type Isozyme Activity of the Liver Measured at Various Time Intervals after Discontinuation of Azo Dye Administration for 60 Days

An attempt was then made to see how the elevated muscle-type isozyme activity of both enzymes in the liver, observed on the 60th day after the cessation of 60 days of azo dye feeding, was influenced by further administration of the basal diet. As shown in Tables II and III, the elevated levels of muscle-type pyruvate kinase and aldolase activity were found to be unchanged in the liver for as long as 150 days for the former isozyme and 300 days for the latter after 60 days of administration of 3'-Me-DAB. The total activity of both enzymes varied only slightly.

Effect of the Noncarcinogenic Azo Dye, 2-Me-DAB, and Another Hepatocarcinogen, 2,7-FAA, on Total and Muscle-type Isozyme Activity of Pyruvate Kinase and Aldolase in the Liver of Rats and Mice

It appeared necessary to determine whether the observed increase in muscle-type isozyme activity in the liver after short-term administration of carcinogenic azo dye was specific for hepatocarcinogenesis. For this purpose, 2-Me-DAB and 2,7-FAA were chosen; 2-Me-DAB is known to bind with protein (16) despite its inability to produce cancer (15). Thus, the effect of 2-Me-DAB was examined with the same experimental procedure as was used with 3'-Me-DAB. Although 2,7-FAA induces malignant tumors in various organs other than liver in low frequency, it is known to be a potent hepatocarinogen.

The results on aldolase are shown in Table IV. Compared with normal liver, no appreciable increase in muscle-type aldolase activity was seen in the livers when 60 days of 2-Me-DAB administration were followed by administration of the basal diet for another 60 days. On the other hand, an increase in muscle-type aldolase activity in the liver was observed in the 2,7-FAA-fed animals. The degree of this increase was found to be of the same order as that observed with 3'-Me-DAB-fed animals.

Experiments on pyruvate kinase were carried out using not only another inbred strain of rat, but also mice, in order to see a possible generality of the

TABLE IV. Total and Muscle-type Isozyme Activity of Liver Aldolase in Rats Fed a Diet Containing 2-Me-DAB or 2,7-FAA, and Fed a Basal Diet Later

Days fed		No. of rats	Aldolase activity (units/mg protein at 30°)[d]	
			Total	Muscle-type
Basal diet only (control)		9	0.095 ± 0.0076	0.0125 ± 0.0032
3'-Me-DAB diet 60	Basal diet 60	10	$0.091 + 0.0092$	$0.0174^b \pm 0.0027$
2-Me-DAB[a] diet 60	Basal diet 60	13	0.093 ± 0.0134	0.0119 ± 0.0027
2,7-FAA[c] diet 60	Basal diet 60	10	$0.097 + 0.0102$	$0.0174^b \pm 0.0045$

Data are expressed as the mean value $\pm SD$.

[a] 2-Me-DAB diet contained 0.06% 2-Me-DAB in basal diet.

[b] Differences in the muscle-type aldolase activity between the 3'-Me-DAB group and its control, and the 2,7-FAA group and its control were statistically significant ($P < 0.01$ and $P < 0.02$, respectively).

[c] 2,7-FAA diet contained 0.025% 2,7-FAA in basal diet.

[d] Units mean μmoles fructose 1,6-di-phosphate cleaved/min.

TABLE V. Total and Muscle-type Isozyme Activity of Liver Pyruvate Kinase in SW-F_1[a] Rats Fed a Diet Containing 2-Me-DAB or 3'-Me-DAB

Days fed			No. of rats	Pyruvate kinase activities (units/mg protein at 37°)[b]		
2-Me-DAB	3'-Me-DAB	Basal diet		Total	Muscle-type	Muscle-type (%)
		30–60	9	0.578±0.119	0.039±0.025	6.4± 2.9
30			4	0.589±0.010	0.043±0.016	7.4± 2.0
60			5	0.469±0.099	0.039±0.058	6.6± 9.1
60		115	8	0.572±0.088	0.040±0.031	7.7± 6.3
	30		5	0.573±0.070	0.332±0.097	57.9±14.5
	60		5	0.542±0.046	0.071±0.044	14.2± 8.8
	60	115	6	0.621±0.069	0.096±0.017	15.4± 1.6

Data are expressed as the mean value±SD.
[a] F_1 strain between Sprague-Dawley and Wistar rats from Fuji Animal Farm.
[b] Units mean μmoles pyruvate formed/min.

TABLE VI. Total and Muscle-type Isozyme Activity of Liver Pyruvate Kinase in Mice Fed a Diet Containing 2,7-FAA

Days fed		No. of rats	Pyruvate kinase activity (units/mg protein at 37°)[a]			Soluble protein (mg/g liver)
2,7-FAA diet	Basal diet		Total	Muscle-type	Muscle-type (%)	
0 (control)	30–60	8	0.429±0.074	0.106±0.044	21.8± 8.6	76.2±6.7
30	0	8	0.449±0.056	0.203±0.051[b]	45.5±10.5[b]	73.7±6.6
60	0	9	0.415±0.060	0.126±0.044	29.9± 9.3	77.7±9.3
0	90–120	9	0.383±0.058	0.089±0.037	23.6±10.1	77.9±8.2
30	60	8	0.387±0.053	0.102±0.031	26.8± 7.1	69.5±4.0
60	60	7	0.477±0.076	0.176±0.064[b]	38.3± 7.3[c]	77.0±6.0

Data are expressed as the mean value ±SD.
[a] Units mean μmoles pyruvate formed/min.
Differences from the control value were statistically significant.
[b] $P < 0.01$.   [c] $0.01 < P < 0.02$.

phenomenon. As shown in Table V, 2-Me-DAB feeding showed no increase in muscle-type pyruvate kinase activity even during its administration, whereas 3'-Me-DAB did greatly enhance the muscle-type isozyme activity immediately after 1 month's administration. Two months of 3'-Me-DAB administration showed a lower level of muscle-type isozyme activity than that of 1 month of feeding, and the level was significantly elevated again when measured at a 115-day interval after cessation of 2 months of 3'-Me-DAB administration. This means that the biphasic change in the enzymic pattern during hepatocarcinogenesis and the increase of muscle-type isozyme activity observed in the livers of a certain inbred strain of rat at a sufficient time interval after 2 months of 3'-Me-DAB administration are also observable in another inbred strain in which total liver pyruvate kinase is considerably lower. Table VI shows similar but more distinct biphasic changes in the muscle-type pyruvate kinase activity appearing in mouse liver during administration of 2,7-FAA. Differences between this value and the control value were also statistically significant. Thus, it seems likely that 2,7-FAA has the same effect as 3'-Me-DAB on the enhancement of muscle-type isozyme activity. All these results on pyruvate kinase were quite similar to those on the aldolase isozymes described above.

Change in Total and Muscle-type Isozyme Activity of Pyruvate Kinase or Aldolase after Partial Hepatectomy in the Liver of Rats Fed Azo Dye

The next problem to be considered concerns the irreversible fixation of the increased level of muscle-type isozyme activity of the respective enzymes. Most easily proven is the possibility that the event may be a reflection of the injury still remaining in the liver. To verify this possibility, partial hepatectomy was performed on the rats fed the azo dye diet for 60 days and then fed the normal diet for 60 days, and the total and muscle-type isozyme activity of both enzymes was measured in various regeneration stages after operation. The total activity of pyruvate kinase and aldolase, which was shown to be virtually equal to that in normal liver, was almost unchanged during the 20 days. On the other hand, the muscle-type pyruvate kinase activity in normal and azo dye-fed rats was elevated transiently on the 2nd day but returned to the preoperative level by the 5th to the 10th day and remained unchanged until the 20th day after operation (Fig. 1). It is noteworthy that after the transient increase, muscle-type isozyme activity appearing in the azo dye-treated liver returned to its preoperative level after completion of the regeneration but never to that in normal liver or to any intermediate level between normal and azo dye-fed liver. As shown in Fig. 2, similar findings were also obtained by using aldolase isozyme as a marker. If the elevation of muscle-type isozyme activity were due to the injured portion still remaining in the azo dye-treated liver, it would be minimized by the value of the newly formed cells covering a 60% deficit of the liver due to partial hepatectomy and would

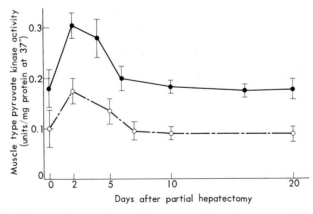

Fig. 1. Changes in muscle-type isozyme activity of pyruvate kinase in the livers of normal and azo dye-fed rats after partial hepatectomy
● 3'-Me-DAB; ○ control.

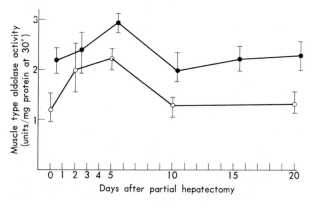

FIG. 2. Changes in muscle-type isozyme activity of aldolase in the livers of normal and azo dye-fed rats after partial hepatectomy
● 3′-Me-DAB; ○ control.

never return to the preoperative value since the newly formed cells would not have the injured portion. However, this was not the case. It seems likely, therefore, that the irreversible fixation of the increased level of muscle-type isozyme activity is not due to the injured portion remaining in the azo dye-treated liver. Possibly, the phenomenon observed may be a reflection of the derepressed gene expression induced by azo dye. It is now assumed that each liver cell is equally responsible for the enhancement of muscle-type isozyme activity induced by azo dye feeding, and each newly formed cell is considered to possess the same increased muscle-type isozyme activity as that of its parental cell. If so, the results of partial hepatectomy also imply the inheritance of the event induced by azo dye.

Effect of Dietary and Hormonal Controls on the Change of Enzyme Pattern in Rat Liver Induced by Azo Dye

Next, effects of dietary and hormonal conditioning on this event were studied to explore the biological meaning of the enhancement of muscle-type isozyme activity. As shown in Fig. 3, the total activity of pyruvate kinase responded sensitively to the dietary conditioning; that is, the administration of a high-carbohydrate diet for 4 days resulted in about a 3-fold elevation of total activity. No difference was observed in this response between normal and azo dye-fed rats. It must be pointed out, however, that the muscle-type activity was unaffected in this case. When the normal and azo dye-fed rats were fasted for 48 hr, the total activity was markedly reduced but the muscle-type isozyme activity was found to be unaffected (Table VII). The reduced activity obtained by treating the normal and azo dye-fed rats with alloxan injection, however, was restimulated by the insulin treatment; whereas the muscle-type pyruvate kinase activity was never affected by alloxan with or without insulin. The total activity in normal or azo dye-treated rats that received 25 mg/day hydrocortisone acetate for 5 days was shown to be slightly decreased, but the muscle-type isozyme activity was quite unchanged in this case too (Table VII). These findings clearly indicate and confirm the previous findings of Tanaka *et al.* (*33*) and Suda *et al.* (*29*) that the

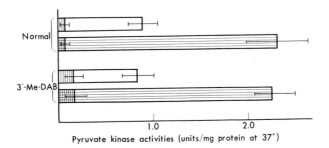

FIG. 3. Effect of a high-carbohydrate diet on total and muscle-type isozyme activity of pyruvate kinase in the livers of normal and azo dye-fed rats

The high-carbohydrate diet contained 7.0 g dextrin, 15 g sucrose, 10 g casein, 5 g of a mixture of salts, vitamins, and choline chloride, in 100 g, and was administered for 4 days.

☐ basal diet
☐ high-carbohydrate diet } total ;
▥ basal diet
▦ high-carbohydrate diet } muscle type.

TABLE VII. Effect of Dietary and Hormonal Control on Pyruvate Kinase Activity in the Liver of Rats Fed a Diet Containing 3′-Me-DAB and Fed a Basal Diet Later

Treatment	No. of rats	Pyruvate kinase activity[b] (units/mg protein at 37°)	
		Total	Muscle-type
None (control)			
Normal rats	6	0.742±0.082	0.094[a]±0.019
Azo dye-treated rats	4	0.773±0.073	0.149[a]±0.021
Fasted for 48 hr			
Normal rats	4	0.378±0.065	0.089±0.023
Azo dye-treated rats	4	0.389±0.034	0.141±0.017
Alloxan diabetic			
Normal rats	3	0.473±0.058	0.091±0.038
Azo dye-treated rats	3	0.426±0.099	0.152±0.027
Insulin-treated diabetic			
Normal rats	3	0.698±0.078	0.099±0.019
Azo dye-treated rats	4	0.713±0.048	0.141±0.063
Steroid-treated			
Normal rats	4	0.705±0.043	0.096±0.017
Azo dye-treated rats	4	0.689±0.071	0.149±0.041

Data are expressed as the mean value ±SD.
[a] Differences from the control value were statistically significant ($P < 0.001$).
[b] Units mean μmoles pyruvate formed/min.

muscle-type isozyme of pyruvate kinase is far more insensitive to dietary or hormonal control compared with the liver-type isozyme.

Regarding aldolase, administration of a high-carbohydrate diet elicited no appreciable response in the muscle-type isozyme activity or total activity either

TABLE VIII. Effect of a High-Carbohydrate Diet and a High-fructose Diet on Aldolase Activity in the Liver of Rats Fed Diet Containing 3′-Me-DAB and Fed a Basal Diet Later

Treatment	No. of rats	Aldolase activity (units/mg protein at 30°)[a]	
		Total	Muscle-type
Basal diet (control)			
Normal rats	4	0.100[b]±0.015	0.0120±0.0035
Azo dye-treated rats	3	0.109[e]±0.007	0.0205±0.0025
High-carbohydrate diet			
Normal rats	4	0.127[c]±0.011	0.0113±0.0018
Azo dye-treated rats	4	0.136[f]±0.011	0.0186±0.0044
High-fructose diet			
Normal rats	4	0.192[d]±0.023	0.0152±0.0013
Azo dye-treated rats	4	0.188[g]±0.019	0.0250±0.0014

Data are expressed as the mean value ±SD.

[a] Units mean μmoles fructose 1,6-di-phosphate cleaved/min.

Differences from the respective controls were statistically significant.

(c) > (b), (f) > (e)   ($P < 0.05$)
(d) > (b), (g) > (e)   ($P < 0.01$)

in normal liver or in the liver of rats fed azo dye for 60 days and then fed a basal diet for 60 more days. A high-fructose diet administered for 4 days, on the other hand, resulted in about a 2-fold elevation of total activity in both liver tissues, while the muscle-type isozyme activity was slightly changed (Table VIII).

All of these results indicate that the change in muscle-type isozyme activity of both enzymes appearing in the early stage of hepatocarcinogenesis is a change characterized by the increase of enzyme insensitivity to dietary and hormonal controls; in other words, it is a change towards deviation from a certain regulatory system of normal cells.

SUMMARY AND CRITIQUE

In summarizing the experimental results mentioned above, the following five points should be noted:

1. In the early stage of azo dye hepatocarcinogenesis, when no appreciable morphological changes could be seen, a biochemical change was observed.

2. This biochemical change seems to be a specific event related to hepatocarcinogenesis, since the phenomenon could also be induced by the use of the other hepatocarcinogen but not by a noncarcinogen.

3. This biochemical change was maintained irreversible during a long observation period after cessation of azo dye administration.

4. This biochemical event is suggested to be the derepressed gene expression induced by azo dye which might eventually be transmitted by the parental cells to their progeny during cell division.

5. This early stage event is considered to be a change toward deviation from a certain regulatory system of normal cells.

The first question raised by the present study is concerned with markers, especially aldolase, used for the performance of biochemical analysis. We have assayed total and muscle-type isozyme activity of aldolase according to the methods of Racker (*25*) and Blostein and Rutter (*6*), respectively. The muscletype aldolase was estimated by neutralizing it with antimuscle aldolase antibody and by subtracting the remaining activity from the total activity. We did not adopt an ordinary differential assay method for measurement of the fructose 1,6-di-phosphate/fructose 1-phosphate ratio because the ratio does not change linearly but hyperbolically, responding to percentage of increase in muscle-type aldolase activity, and varies slightly even with a considerable change of the muscle-type aldolase content, especially at a low level (*11*). This property was not considered adequate for the purpose of the present study in which only slight increases in muscle-type aldolase activity were observed, and, on the other hand, changes in the muscle-type aldolase activity measured by the antisera method were expressed linearly. This is why the antisera method was used in this experiment. In the liver, however, three kinds of hybrids exist (*23*) in addition to the muscle-type and liver-type isozymes. Matsushima *et al.* (*14*) reported that these hybrids are greatly involved in the difference in isozyme pattern between normal liver and malignant liver tumor. It seemed to be very important in the present study, therefore, to determine whether the inhibition by antimuscle aldolase antibody of the activity of each hybrid is parallel to the content of the muscle-type subunits. Using rat kidney extract, Penhoet *et al.* (*23*) showed that antimuscle aldolase serum inhibits the activity of the muscle-type aldolase completely, and, in decreasing fashion, the three hybrids, but antimuscle aldolase sera had no effect on the liver-type isozyme. The same results were obtained in our laboratory with hybrids reconstituted artificially and fractionated by isoelectric focusing electrophoresis (unpublished data). The activity of A_3C hybrid isolated from rat brain was demonstrated by Sugimura *et al.* (*31*) to be inhibited 89% by antimuscle aldolase antibody. These results imply that the antimuscle aldolase antibody blocks the activity of the muscle-type subunit of each hybrid selectively without interfering with the liver-type subunit, and thus the muscle-type aldolase activity measured by the use of antisera is the sum of the activities of pure muscle-type isozyme and muscle-type subunits in each hybrid. Quite recently, however, Penhoet and Rutter (*24*) reported that the activity of A-C and B-C heterotetramer of rabbit aldolase showed essentially complete inhibition by the antimuscle aldolase antibody under certain conditions. Detailed immunochemical analysis is therefore needed to solve this discrepancy since the accuracy of the present hypothesis on the muscle-type aldolase activity seems to depend on this point. Regarding pyruvate kinase, the assay of total and muscle-type isozyme activity was carried out according to the method of Tanaka *et al.* (*33*), and the obtained results also seem to be relevant, since possible existence of the hybrids was suggested by isoelectrofocusing between type M and L isozymes (*35*).

The second concern is what kind of cells in the liver are responsible for the biochemical changes observed in the present study. There are two possibilities. One is that the observed event may be the sum of changes occurring in each cell of liver tissue. The other is that the event may be due to a change in the population of cells responsible for the muscle-type isozyme synthesis.

According to our preliminary experiment (*12*) using fluorescent antimuscle pyruvate kinase antibody, the specimen prepared from normal liver was found to be slightly but diffusely stained except for the cell nuclei. Diffuse but fairly intense staining was also observed in the liver of rats fed azo dye for 60 days and then fed basal diet for 60 days, whereas the specimen prepared from the liver treated with azo dye for 90 days showed considerable matrix staining. Tumor tissues were intensely stained to the same extent as muscle. It seems likely, therefore, that the observed increase in muscle-type pyruvate kinase activity may not be due to the change in the population of cells as far as the liver treated with azo dye for 60 days is concerned.

Finally, the relationship between the biochemical changes observed in this study and hepatocarcinogenesis must be explained. The answer to this question, however, must await further study.

Acknowledgments

We are grateful to Dr. H. Otsuka, Department of Pathology, Faculty of Medicine, Tokushima University, and the late Dr. N. Takizawa, Department of Pathology, Faculty of Medicine, Chiba University, for their help in studying the histological specimens.

REFERENCES

1. Adelman, R. C., Morris, H. P., and Weinhouse, S. Fructokinase, triokinase, and aldolases in liver tumors of the rat. *Cancer Res.*, **27**, 2408–2413 (1967).
2. Baba, T., Misu, Y., and Takayama, S. Induction of cancer of the glandular stomach in a rat: a new form of experiment. *GANN*, **53**, 381–387 (1962).
3. Berenblum, I. Carcinogenesis and tumor pathogenesis. *Adv. Cancer Res.*, **2**, 129–175 (1954).
4. Berenblum, I. Some new implications of the two-stage mechanism in the study of skin carcinogenesis. *Ciba Found. Symp. Carcinogenesis*, 55–69 (1959).
5. Berenblum, I. and Haran, N. The significance of the sequence of initiating and promoting actions in the process of skin carcinogenesis in the mouse. *Brit. J. Cancer*, **9**, 268–271 (1955).
6. Blostein, R. and Rutter, W. J. Comparative studies of liver and aldolase. II. Immunochemical and chromatographic differentiation. *J. Biol. Chem.*, **238**, 3280–3285 (1963).
7. Clayton, C. C. and Baumann, C. A. Diet and azo dye tumors: effect of diet during a period when the dye is not fed. *Cancer Res.*, **9**, 575–582 (1949).
8. Dauost, R. and Cantero, A. The numerical properties of cell types in rat liver during carcinogenesis by 4-dimethylaminoazobenzene (DAB). *Cancer Res.*, **19**, 757–762 (1959).
9. Druckrey, H. Pharmacological approach to carcinogenesis. *Ciba Found. Symp. Carcinogenesis*, 110–130 (1959).
10. Druckery, H. and Küpfmüller, K. Quantitative Analyse der Krebsentstehung. *Z. Naturforsch.*, **3B**, 254–266 (1948).

11. Endo, H., Eguchi, M., and Yanagi, S. Irreversible fixation of increased level of muscle-type aldolase activity appearing in rat liver in the early stage of hepatocarcinogenesis. *Cancer Res.*, **30**, 743–752 (1970).
12. Ikehara, Y. and Endo, H. Changes of enzyme activities during carcinogenic process; an approach by the fluorescent antibody technique. *Proc. Jap. Cancer Assoc. 26th Gen. Meet.*, 184 (1967).
13. Imamura, K., Taniuchi, K., and Tanaka, T. Enzymatic properties of type M_2 pyruvate kinase of Yoshida ascites hepatoma AH-130 of rat. *Proc. Jap. Cancer Assoc. 29th Gen. Meet.*, 11 (1970).
14. Matsushima, T., Kawabe, S., Sibuya, M., and Sugimura, T. Aldolase isozymes in rat tumor cells. *Biochem. Biophys. Res. Commun.*, **30**, 565–570 (1968).
15. Miller, J. A. and Baumann, C. A. The carcinogenicity of certain azo dye related to p-dimethylaminoazobenzene. *Cancer Res.*, **5**, 227–234 (1945).
16. Miller, E. C., Miller, J. A., Sapp, R. W., and Weber, G. M. Studies on the protein-bound aminoazo dyes formed *in vivo* from 4-dimethylaminoazobenzene and its C-monomethyl derivatives. *Cancer Res.*, **9**, 336–343 (1949).
17. Morris, H. P., Wagner, B. P., Ray, F. E., Snell, K. C., and Stewart, H. L. Comparative study of cancer and other lesions of rats fed N, N′-2,7-fluorenylenebisacetamide or N-2-fluorenylacetamide. *Natl. Cancer Inst. Monograph*, **5**, 1–53 (1961).
18. Muta, Y. Kombinationsversuch der karzinogenen Wirkungen verschiedener Substanzen. *GANN*, **37**, 298–300 (1943).
19. Nakahara, W. Critique of carcinogenic mechanism. *Progr. Exp. Tumor Res.*, **2**, 158–202 (1961).
20. Nakahara, W. and Fukuoka, F. Summation of carcinogenic effects of chemically unrelated carcinogen, 4-nitroquinoline N-oxide and 20-methylcholanthrene. *GANN*, **51**, 125–137 (1960).
21. Odashima, S. Combined effect of carcinogens with different actions. I. Development of liver cancer in the rat by the feeding of 4-dimethylaminostilbene following initial feeding of 4-dimethylaminoazobenzene. *GANN*, **53**, 247–257 (1962).
22. Odashima, S. and Fujimaki, C. Development of liver cancers in the rat by the feeding of 4-aminoazobenzene, 20-methylcholanthrene, β-naphthylamine, Food Red No. 5 (CI 12141) and Food Red No. 101 (CI 16150) following initial feeding of DAB. *Proc. Jap. Cancer Assoc. 22nd Gen. Meet.*, 205 (1963).
23. Penhoet, E., Rajkumar, T., and Rutter, W. J. Multiple forms of fructose diphosphate aldolase in mammalian tissues. *Proc. Natl. Acad. Sci. U.S.*, **56**, 1275–1282 (1966).
24. Penhoet, E. E. and Rutter, W. J. Catalytic and immunochemical properties of homomeric and heteromeric combinations of aldolase subunits. *J. Biol. Chem.*, **246**, 318–323 (1971).
25. Racker, E. Spectrophotometric measurement of hexokinase and phosphohexokinase activity. *J. Biol. Chem.*, **167**, 843–854 (1947).
26. Rutter, W. J., Blostein, R. E., Woodfin, B. M., and Weber, C. S. Enzyme variants and metabolic diversification. *Adv. Enzyme Regulation*, **1**, 39–56 (1963).
27. Schapira, F., Dreyfus, J. C., and Schapira, G. Anomaly of aldolase in primary liver cancer. *Nature*, **200**, 995–995 (1963).
28. Schmähl, D., Thomas, C., and König, K. Experimentelle Untersuchungen zur "Syncarcinogenese." I. Versuche zur Krebserzeugung an Ratten bei gleichzeitiger Applikation von Diäthylnitrosamin und 4-Dimethylamino-Azobenzol. *Z. Krebsforsch.*, **65**, 342–350 (1963).

29. Suda, M., Tanaka, T., Sue, F., Harano, Y., and Morimura, H. Dedifferentiation of sugar metabolism in the liver of tumor-bearing rat. *GANN Monograph*, **1**, 127–141 (1966).
30. Sugimura, T., Matsushima, T., Kawachi, T., Hirata, Y., and Kawabe, S. Molecular species of aldolases and hexokinases in experimental hepatomas. *GANN Monograph*, **1**, 143–149 (1966).
31. Sugimura, T., Sato, S., and Kawabe, S. The presence of aldolase C in rat hepatoma. *Biochem. Biophys. Res. Commun.*, **39**, 626–630 (1970).
32. Takayama, S. and Imaizumi, T. Sequential effects of chemically different carcinogens, dimethylnitrosoamine and 4-dimethylaminoazobenzene, on hepatocarcinogenesis in rats. *Int. J. Cancer*, **4**, 373–383 (1969).
33. Tanaka, T., Harano, Y., Sue, E., and Morimura, H. Crystallization, characterization and metabolic regulation of two types of pyruvate kinase isolated from rat tissues. *J. Biochem. (Tokyo)*, **62**, 71–91 (1967).
34. Weber, G. Behavior of liver enzymes in hepatocarcinogenesis. *Adv. Cancer Res.*, **6**, 403–494 (1961).
35. Yanagi, S., Kamiya, T., Ikehara, Y., and Endo, H. Isozymes in the liver from mice given hepatocarcinogen and from tumor-bearing mice. *GANN*, **62**, 283–291 (1971).
36. Yoshida, T. Recent results of cancer research in my institute. *Nouva Serie*, **XXV**, parte II, 1–28 (1967).

EXPLANATION OF PHOTOS

PHOTO 1. Section of normal rat liver (Wistar strain). Hematoxylin-Eosin stain (H-E), ×50 (A); ×125 (B).

PHOTO 2. Section of rat liver prepared immediately after cessation 60-day administration of a diet containing 0.06% 3'-Me-DAB. The severe damage caused by the azo dye is seen in the specimen. H-E, ×50 (A); ×125 (B).

PHOTO 3. Section of rat liver prepared at 14-day intervals after cessation of 60-day administration of a 3'-Me-DAB diet. The damage by the azo dye is still retained in the liver. H-E, ×50 (A); ×125 (B).

PHOTO 4. Section of rat liver prepared at 30-day intervals after cessation of 60-day administration of a 3'-Me-DAB diet. Considerable recovery from the damage can be seen in the specimen. H-E, ×125.

PHOTO 5. Section of rat liver prepared at 60-day intervals after cessation of 60-day administration of a 3'-Me-DAB diet. This specimen is practically indistinguishable from that of normal liver (Photo 1). H-E, ×50 (A); ×125 (B).

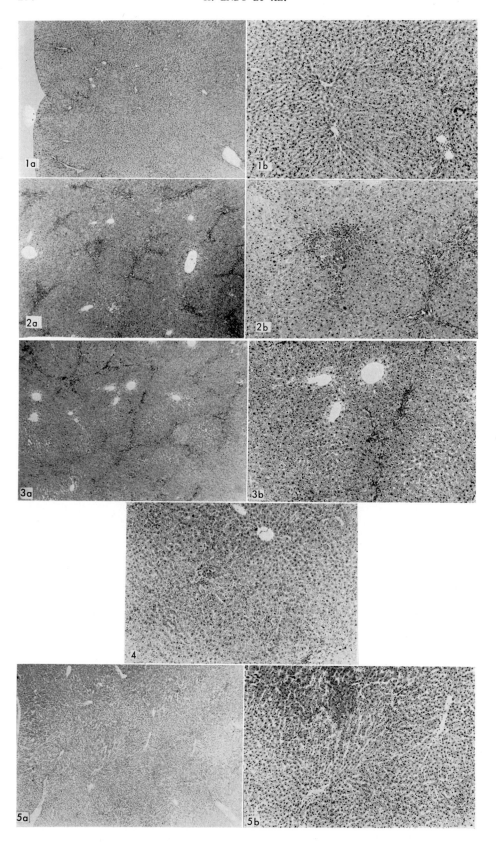

ENZYMOLOGY, ULTRASTRUCTURE, AND ENERGETICS OF MITOCHONDRIA FROM THREE MORRIS HEPATOMAS OF WIDELY DIFFERENT GROWTH RATE[*1]

Peter L. PEDERSEN

Department of Physiological Chemistry, Johns Hopkins University School of Medicine[*2]

Mitochondria were isolated from three Morris hepatomas (9618A, 7800, and 3924A) of widely different growth rates and studied with respect to their enzyme content, ultrastructure, and capacity for energy production. Mitochondria from the well-differentiated hepatomas, 9618A and 7800, respire in the presence of both β-hydroxybutyrate and succinate, exhibit high acceptor control ratios, and appear morphologically and ultrastructurally intact. Mitochondria isolated from the more rapidly growing hepatoma, 3924A, respire only in the presence of succinate, elicit low acceptor control ratios, and have diluted rather than condensed matrices. 2,4-Dinitrophenol enhances the rate of respiration of mitochondria from all three hepatomas.

The activities of five enzymes, monoamine oxidase, adenylate kinase, cytochrome oxidase, malate dehydrogenase, and ATPase, were measured. Monoamine oxidase activity is reduced in mitochondria from hepatomas 9618A and 7800, and absent in mitochondria from hepatoma 3924A. Adenylate kinase activity is near normal in mitochondria from hepatoma 9618A but reduced more than 40% in mitochondria from hepatomas 7800 and 3924A. Cytochrome oxidase activity and malate dehydrogenase activity are nearly normal in all of the hepatoma mitochondria examined. ATPase activity is markedly reduced in mitochondria from hepatomas 9618A and 7800.

Cytochrome oxidase activity was also measured in intact cells from hepatoma 7800 and 3924A. The activity of this terminal oxidase in hepatoma 7800 is somewhat higher than in control tissue, whereas in hepatoma 3924A it is about 60% lower than the control values.

These studies clearly show that mitochondria from the most rapidly growing hepatoma, 3924A, bear the least resemblance to host liver mitochondria, whereas mitochondria from hepatomas 9618A and 7800 are very similar to, but by no means identical with, control

---

[*1] This work was aided by Grants CA 10951, GM 12125, and CA 10729 from the National Institutes of Health, U.S.A., and by institutional grants to the Johns Hopkins School of Medicine from the American Cancer Society. The author is a recipient of a Research Career Development Award (1-K4-CA-23,333) from the National Cancer Institute, NIH.

[*2] Baltimore, Maryland 21205, U.S.A.

mitochondria. In addition, these studies strongly suggest that the content of mitochondria in hepatoma 3924A, as assessed by the specific activity of cytochrome oxidase, is markedly reduced.

Although mitochondria from several rapidly growing tumors have been examined in detail for their ability to carry out electron transport and oxidative phosphorylation (5, 7, 9, 24, 36, 39), few studies have dealt with the energetics of mitochondria from tumors of widely different growth rates. Experiments summarized in this report were carried out with this purpose in mind and with the purpose of laying the groundwork for future studies on the general enzymological properties of tumor mitochondria. This latter goal seemed particularly important in view of a number of recent studies which, when taken together, show that mitochondria contain in addition to those enzymes involved directly in ATP formation more than one hundred other enzymes as well (31).

This report will be divided into four related sections. In the first, a brief description of the enzymatic composition of mitochondria and the compartmentation of enzymes within these organelles will be given. In the second, a number of properties of freshly isolated mitochondria from Morris hepatomas 9618A, 7800, and 3924A will be summarized. In the third, information about the relationship between the content of mitochondria and the content of mitochondrial enzymes within hepatomas will be provided. In the fourth and final section, an attempt will be made to summarize what generalizations can be made about the properties of hepatoma mitochondria at this time.

Sources of materials and details of experimental procedures have been described in previous publications (18, 19, 29).

Enzymic Composition of the Four Components of Normal Liver Mitochondria

As shown diagrammatically in Fig. 1 and in the electron micrographs (Photos 1, 3, and 5), rat liver mitochondria are composed of four distinct components, the outer membrane, the inner membrane, the space between the two membranes (intracristal or intermembrane space), and the matrix. In recent years, reliable methods have been developed in our laboratories (3, 26) and in Ernster's laboratory in Sweden (32) to subfractionate rat liver mitochondria into their component parts, thus making it possible to identify where, within mitochondria, various enzymes reside. By sequential use of the two nonionic detergents, digitonin and Lubrol WX, we have successfully fractionated rat liver mitochondria into their four distinct components (3, 26). As shown in Fig. 1, digitonin is employed to selectively remove the outer membrane and intracristal space. Separation of these two components is effected by centrifugation at $144,000g$ for 1 hr. Following removal of the outer compartment (outer membrane+intracristal space), Lubrol WX is employed to remove the matrix from the resultant mitoplast fraction (inner membrane+matrix). Another high-speed centrifugation step separates the inner membrane from the matrix.

The general enzymic composition of the four mitochondrial components, as determined by the digitonin-Lubrol method in our laboratories (3, 26, 27) and in

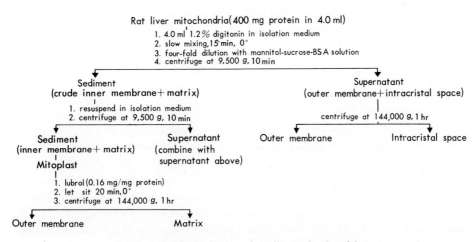

Fig. 1. Subfractionation of rat liver mitochondria

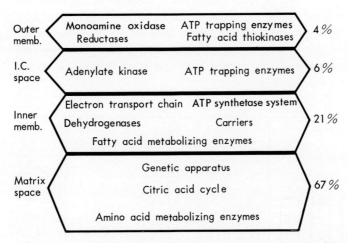

Fig. 2. Compartmentation of enzymes of rat liver mitochondria

others (*31*), and by the method of Sottocasa et al. (*31, 33*), is summarized in Fig. 2. The matrix constitutes as much as 67% of the total mitochondrial protein and contains most enzymes of the citric acid cycle and several enzymes involved in amino acid metabolism. The inner membrane, which by the digitonin-Lubrol procedure is found to constitute about 21% of the total mitochondrial protein, contains, in addition to the enzymes of oxidative phosphorylation and electron transport, some of the fatty acid metabolizing enzymes and a large part of the genetic apparatus. The outer membrane and intracristal space, which together constitute only about 10% of the total mitochondrial protein, contain ATP trapping enzymes, such as nucleoside diphosphokinase which are capable of converting ATP to other high-energy compounds required in extramitochondrial metabolism, and fatty acid thiokinases which are necessary to activate fatty acids prior to their oxidation in mitochondria. The outer membrane and intracristal space are marked also by enzymes whose relationship to mitochondrial metabolism is not yet clear. These include enzymes such as monoamine oxidase, kynurenine hydroxylase, NADH cytochrome c reductase, and adenylate kinase.

Properties of Mitochondria from Morris Hepatomas 9618A, 7800, and 3924A

1. Activity of marker enzymes

In view of the large number of enzymes localized within mitochondria and the importance of these enzymes to the overall metabolism of the liver cell, it became of interest to determine whether the specific activities of some of these enzymes remained constant or varied as a function of tumor growth rate. For our initial studies, at least one important enzyme from each of the four mitochondrial components was chosen for assay. Thus, cytochrome oxidase, the terminal enzyme of electron transport, and ATPase, the terminal enzyme of oxidative phosphorylation (20, 21, 34), were chosen as representatives of the inner membrane (1, 3, 26). This latter enzyme requires both 2,4-dinitrophenol and Mg^{2+} for maximal activity. Malate dehydrogenase, a member of the citric acid cycle, was chosen as a matrix representative (3, 14, 26), and adenylate kinase and monoamine oxidase as representatives of the outer membrane and intracristal space, respectively (25, 26). Each of these enzymes was assayed many times in freshly isolated mitochondria from control liver and from hepatomas 9618A, 7800, and 3924A.

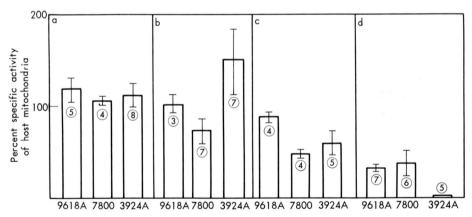

FIG. 3. Some enzymic activities of mitochondria isolated from host rat livers and from hepatomas 9618A, 7800, and 3924A

Numbers in circles represent the number of different mitochondrial preparations examined. For experimental details see Ref. (19).

    a: Cytochrome oxidase;   b: malate dehydrogenase;
    c: adenylate kinase;   d: monoamine oxidase.

As can be seen in Fig. 3, neither cytochrome oxidase nor malate dehydrogenase is markedly reduced in any of the tumor mitochondria examined. Adenylate kinase is near normal in mitochondria isolated from the slowly growing hepatoma, 9618A, but is reduced to less than 70% of host value in mitochondria isolated from the more rapidly growing hepatomas, 7800 and 3924A. Monoamine oxidase is low in all tumor mitochondria studied. Results of ATPase assays are not included here but will be considered below with information related to the energy capacity of mitochondria from these hepatomas.

2. Ultrastructure

It could be argued justifiably that the low content of adenylate kinase and monoamine oxidase in those tumors examined may not reflect a property of tumor mitochondria but rather a failure to isolate a homogeneous population of mitochondria containing an intact outer membrane. For this reason, freshly isolated mitochondria from control liver and from hepatomas 9618A, 7800, and 3924A were examined under the electron microscope. As shown in Photos 2, 4 and 6, mitochondria from the three hepatomas examined are similar to control liver mitochondria (Photos 1, 3, and 5) in that they all have intact outer and inner membranes. This finding renders unlikely the possibility that membrane damage suffered by the hepatoma mitochondria during isolation results in low levels of adenylate kinase and monoamine oxidase.

Interestingly, it was observed that only mitochondria from the most rapidly growing hepatoma studied, 3924A, differ significantly in their general ultrastructure from control liver mitochondria. Thus, almost every mitochondrion from hepatoma 3924A (Photo 6), in contrast to control liver mitochondria (Photo 5), has a matrix which looks diluted.

3. Capacity for energy production

A) Respiration: The respiratory activities of hepatoma mitochondria relative to those of control liver mitochondria are presented in Fig. 4. In order

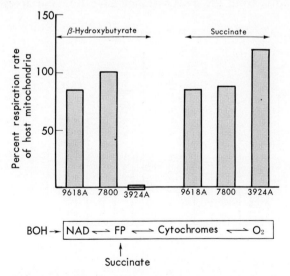

FIG. 4. Relative specific rates of respiration of mitochondria from host livers and from hepatomas 9618A, 7800, and 3924A

Respiration rates were measured, as described in Ref. (*19*), in the presence of substrate (10 mM sodium β-hydroxybutyrate or 5 mM succinate), ADP (142 μM), and phosphate (2.5 mM).

to obtain maximal rates in all cases, respiration was measured in the presence of ADP and phosphate. In the presence of β-hydroxybutyrate, a substrate that enters the electron transport chain near the first phosphorylation site, mitochondria from hepatomas 9618A and 7800 had rates of respiration similar to those of control liver mitochondria. However, mitochondria from the rapidly growing hepatoma, 3924A, fail to respire at all in the presence of this substrate. When succinate, a citric acid cycle intermediate which enters the electron transport chain near the second phosphorylation site, is used as a substrate, mitochondria from all hepatomas exhibit rates of respiration similar to control liver mitochondria. The uncoupling agent, 2,4-dinitrophenol, activates respiration in all cases (19). Inhibitors of respiration in normal liver mitochondria such as Antimycin A and cyanide also inhibit respiration in mitochondria from all three hepatomas (19). It appears, therefore, that the respiratory chain is unaltered from flavoprotein to oxygen in all hepatoma mitochondria examined, and defective prior to this point only in mitochondria from hepatoma 3924A.

B) *Phosphorylation and acceptor control:* As shown in Fig. 5, mitochondria from all tumors examined have the capacity to make ATP. In the presence of β-hydroxybutyrate as substrate, mitochondria from hepatomas 9618A and 7800 phosphorylate as well, or nearly as well, as control liver mitochondria (Fig. 5A). As shown in Fig. 5B, mitochondria from these slowly growing tumors also have acceptor control ratios very near control values, a finding which indicates that they phosphorylate ADP with a high efficiency.

Since mitochondria from hepatoma 3924A fail to respire in the presence of β-hydroxybutyrate, succinate was employed as substrate. In the presence of succinate, mitochondria should elicit a P/O ratio near 2.0. As shown in Fig. 5, however, mitochondria from the 3924A tumor have P/O and acceptor: control

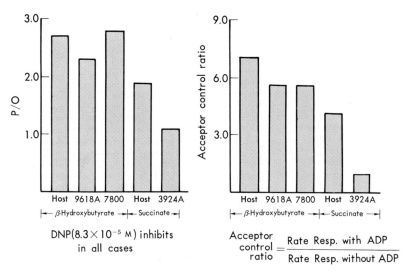

FIG. 5. Phosphorylation capacity and efficiency of mitochondria from host livers and from hepatomas 9618A, 7800, and 3924A
Experimental details are given in Ref. (19).

ratios significantly lower than control values. Thus, once again, mitochondria from the more rapidly growing tumor, 3924A, differ from mitochondria from the more slowly growing tumors, 9618A and 7800, in this case in their efficiency to couple substrate oxidation to phosphorylation.

C) *ATPase activity:* The reactions leading to ATP formation in normal liver mitochondria are thought to be reversible since 2,4-dinitrophenol and other uncoupling agents activate ATPase activity *(15, 37)*. As shown in Fig. 6, normal liver mitochondria have a low rate of ATP hydrolysis in the absence of 2,4-dinitrophenol, but in its presence they hydrolyze ATP at a rate between 300 and 400 nmoles P_i/min/mg protein. Disruption of the mitochondria with Lubrol WX followed by addition of Mg^{2+} has essentially the same effect as 2,4-dinitrophenol does. In sharp contrast to normal liver mitochondria, mitochondria from the two slowly growing hepatomas, 9618A and 7800, are not characterized by significant ATPase activity in the presence of 2,4-dinitrophenol and have only about 20 to 25% of the amount of control activity in the presence of Lubrol WX+Mg^{2+} (Fig. 6). This latter figure ranged from 20 to 37% in eight different mitochondrial preparations. Devlin and Pruss *(8)* observed a similarly low uncoupler-sensitive ATPase activity in the rapidly growing hepatoma 3924A. Taken together, these findings suggest that the reversibility of the energy coupling reactions in mitochondria from Morris hepatomas 9618A, 7800, and 3924A is markedly reduced.

Electrophoretic patterns of mitochondrial membrane proteins from control liver and hepatoma 7800, obtained by the sodium lauryl sulfate-mercaptoethanol procedure of Weber and Osborn *(35)*, clearly show the presence of three protein bands ($MW_A=62,000$, $MW_B=57,000$, and $MW_C=37,500$) *(2, 18)* with mobilities corresponding to those of the three subunit bands of the homogeneous ATPase

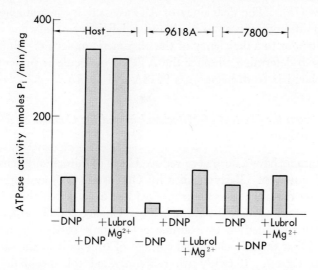

FIG. 6. ATPase activity of hepatomas 9618A and 7800 in the absence or presence of either 2,4-dinitrophenol (8.3×10^{-5} M) or Lubrol (0.160 mg/mg protein) +Mg^{2+} (4.0 mM)

ATP was present at a concentration of 10 mM. Details of the assay are given in Ref. *(18)*.

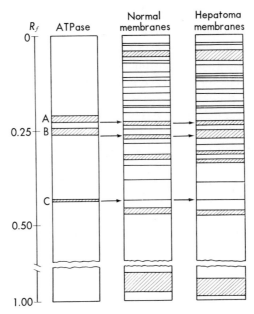

Fig. 7. Gel electrophoresis patterns of the purified rat liver ATPase and of membrane preparations of host liver and hepatoma 7800 mitochondria

ATPase (20 μg) and membranes (200 μg) were subjected to electrophoresis under denaturing conditions described by Weber and Osborn (*35*). See Ref. (*18*) for experimental details.

from rat liver mitochondria (Fig. 7). Moreover, the relative staining intensities of the A+B subunit bands are as great in the hepatoma membranes as in the control membranes (*18*). The deficiency of ATPase activity in mitochondria from hepatoma 7800 does not appear to be due to a deficiency of any one of the three ATPase subunits nor to a deficiency of the oligomeric molecule. Similar studies are in progress to determine whether the ATPase molecule is present at normal levels in mitochondria from hepatomas 9618A and 3924A.

Relationship between the Content of Mitochondria and the Content of Mitochondrial Enzymes in Hepatomas

Whether tumors have a normal or reduced level of mitochondria has remained a controversial question. Early studies by Greenstein (*12*) and others (*28, 30, 33*) suggested, on the basis of cytochrome oxidase assays, that the mitochondrial content of some rapidly growing tumors is low. Chance and Hess (*4, 6*), however, using a spectroscopic method to examine cytochrome oxidase in ascites tumor cells, concluded that the mitochondrial content of tumors is in the same range as that of normal tissues. Recently we reinvestigated this problem using both spectroscopic and enzymatic assays to examine cytochrome oxidase in control liver and hepatoma 3924A cells (*29*). The results of these studies, summarized in Fig. 8, clearly show that cytochrome oxidase is about 60% lower in the 3924A tumor than in the control liver. This reduction is best attributed to a low con-

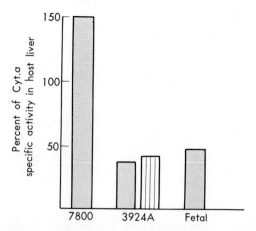

FIG. 8. Relative specific cytochrome oxidase activities in host livers and in hepatomas 7800 and 3924A

For experimental details see Ref. (*29*). ☐ enzymic assay; ☒ spectroscopic assay.

tent of mitochondria in the tumor since the specific activity of cytochrome oxidase in hepatoma 3924A mitochondria is as high as in control liver mitochondria (Fig. 3).

The specific activity of cytochrome oxidase in fetal liver, which has a growth rate similar to hepatoma 3924A, and in hepatoma 7800, which has a growth rate two to three times slower than that of the 3924A tumor, were also measured. As shown in Fig. 8, the specific activity of cytochrome oxidase in fetal liver is about as low as in hepatoma 3924A. In contrast, the specific activity of cytochrome oxidase in hepatoma 7800 is somewhat higher than in control liver.

A reasonable explanation for the above findings is that mitochondrial replication may not be able to keep up with cell division in hepatoma 3924A and in fetal liver but is able to do so in the more slowly growing hepatoma 7800. Such a hypothesis was suggested many years ago by Potter *et al.* (*22, 23*) for another tumor line. This hypothesis is now gaining support from more recent studies which show that the turnover rates of most mitochondrial components have half-lives of about 7 to 12 days (*11, 13*), a time considerably longer than that required for a 3924A hepatoma or fetal cell to divide (*10, 38*).

The relevance of these studies to the content of mitochondrial enzymes of tumors may not seem immediately obvious. However, it should be kept in mind that the specific activity of any mitochondrial enzyme in a tumor with a cell division rate of less than 8 to 12 days will be low. This is not necessarily as a result of repression of the synthesis of the active enzyme, but is more likely as a result of a low content of mitochondria within the tumor. Thus, it seems that the only unequivocal way to determine whether a given mitochondrial enzyme is repressed in a tumor is to assay for its activity in freshly isolated mitochondria which have been shown to be ultrastructurally and morphologically intact.

260   P. L. PEDERSEN

CONCLUSIONS

Conclusions that we feel can be drawn at this time about the properties of mitochondria of hepatomas 9618A, 7800, and 3924A are given below and are summarized diagrammatically in Fig. 9.

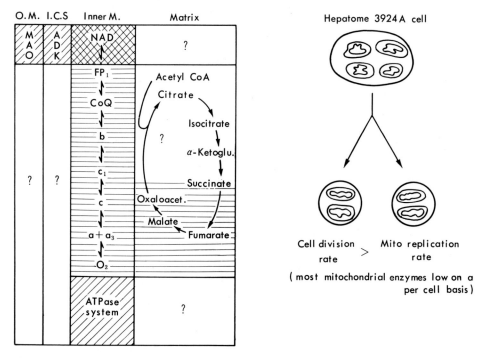

FIG. 9. Mitochondrial enzyme patterns of hepatomas 9618A, 7800, and 3924A relative to patterns of host liver mitochondria

A defect in the "NAD" region of the electron transport chain in mitochondria from hepatoma 3924A is indicated because these mitochondria fail to oxidize β-hydroxybutyrate. Whether this site is defective in its ability to oxidize other NAD-linked substrates has not been established. O.M., outer membranes; I.C.S., intracristal space.

☐ low or absent; ▨ defective in hepatoma 3924A; ☐ unaltered; ? not established.

The respiratory chain from flavoprotein to oxygen is unaltered in all hepatoma mitochondria examined and is defective prior to this point only in mitochondria from hepatoma 3924A. The defect in hepatoma 3924A (inability to oxidize β-hydroxybutyrate) may reside at the level of β-hydroxybutyrate dehydrogenase as Weinhouse et al. (17) have noted a marked deficiency of this enzyme in this rapidly growing tumor.

The Krebs cycle from succinate to oxaloacetate remains unaltered in all tumor lines examined. This conclusion is derived from the findings that all tumor mitochondria oxidize succinate at a high rate and have normal levels of malate dehydrogenase. Enzymes associated with the remaining part of the cycle were not examined.

All tumor mitochondria examined have the capacity to make ATP. The efficiency of ATP formation, as assessed by acceptor control ratios, is low only in mitochondria from hepatoma 3924A; this may be related to the abnormal ultrastructure of these mitochondria. Thus, the matrices of hepatoma 3924A mitochondria were diluted in appearance relative to control liver mitochondria, a finding that suggests that the inner membranes may be permeable to water. The presence of water at the site(s) of ATP formation may cause competition for high energy intermediates (16) and thus reduce the efficiency of oxidative phosphorylation.

The activity of some enzymes is reduced in all tumor mitochondria examined. These include monoamine oxidase, ATPase, and perhaps adenylate kinase the activity of which is only slightly reduced in mitochondria from hepatoma 9618A but markedly reduced in mitochondria from hepatomas 7800 and 3924A. Reduction of ATPase activity does not appear to result from a failure of the hepatomas to synthesize the ATPase molecule itself but rather from some hitherto unknown mechanism.

The mitochondrial content of the rapidly growing hepatoma 3924A appears to be markedly reduced to about the same level as mitochondria in fetal liver cells. This finding suggests that investigators should exercise caution in correlating enzyme patterns with growth rate when the enzyme in question is a mitochondrial enzyme and assays are carried out on cell homogenates. Thus, mitochondrial enzymes in tumors with a cell division rate which exceeds the rate of mitochondrial replication may be significantly reduced by a mechanism unrelated to genetic repression of the enzyme *per se*, but rather related to a low content of mitochondria within the tumor.

Finally, it seems reasonable to conclude that mitochondria from hepatoma 3924A have deviated considerably more from control liver mitochondria than have mitochondria from the more slowly growing hepatomas 9618A and 7800.

Acknowledgments

The work reported in this paper was carried out in part by T. L. Chan, Walter X. Balcavage, James Schreiber, William A. Catterall, and Terry Eska with the active collaboration of Dr. J. W. Greenawalt, Johns Hopkins School of Medicine, Baltimore, Maryland, and Dr. Harold P. Morris, Howard University School of Medicine, Washington, D. C.

REFERENCES

1. Brosner, P., Vogell, W., and Bucher, T. Morphologische und enzymatishe Muster bei der Entwicklung indirekter Flugmuskeln von *Locusta migratoria*. Biochem. Z., **338**, 854–910 (1963).
2. Catterall, W. A. and Pedersen, P. L. Adenosine triphosphatase from rat liver mitochondria. I. Purification, homogeneity and physical properties. *J. Biol. Chem.*, **246**, 4987–4994 (1971).
3. Chan, T. L., Greenawalt, J. W., and Pedersen, P. L. Biochemical and ultrastructural properties of a mitochondrial inner membrane fraction deficient in outer membrane and matrix activities. *J. Cell Biol.*, **45**, 291–305 (1970).

4. Chance, B. and Hess, B. Metabolic control mechanisms. I. Electron transfer in mammalian cells. *J. Biol. Chem.*, **234**, 2404–2412 (1959).
5. Chance B. and Hess B. Metabolic control mechanisms II. Crossover phenomena in mitochondria of ascites tumor cells. *J. Biol. Chem.*, **234**, 2413–2415 (1959).
6. Chance, B. and Hess, B. Spectroscopic evidence of metabolic control. *Science*, **129**, 700–708 (1959).
7. Devlin, T. M. Neoplastic tissue mitochondria. *Methods Enzymol.*, **10**, 110–114 (1967).
8. Devlin, T., M. and Pruss, M. P. Oxidative phosphorylation and ATPase activity in mitochondria from rat hepatomas. *Proc. Am. Assoc. Cancer Res.*, **3**, 315 (1962).
9. Emmelot, P., Bos, C. J., Brombacher, P. J., and Hampe, J. F. Studies on isolated tumor mitochondria: biochemical properties of mitochondria from hepatomas with special reference to a transplanted rat hepatoma of the solid type. *Brit. J. Cancer Res.*, **13**, 348–379 (1959).
10. Enesco, M. and Leblond, C. P. Increase in cell number as a factor in the growth of the organs and tissues of the young male rat. *J. Embryol. Exp. Morphol.*, **10**, 530–562 (1962).
11. Fletcher, M. S. and Sanadi, D. R. Turnover of rat liver mitochondria. *Biochim. Biophys. Acta*, **51**, 356–360 (1961).
12. Greenstein, J. P. Some biochemical characteristics of morphologically separable cancers. *Cancer Res.*, **16**, 641–653 (1956).
13. Gross, N. J., Getz, G. S., and Rabinowitz, M. Apparent turnover of mitochondrial deoxyribonucleic acid and mitochondrial phospholipids in the tissues of the rat. *J. Biol. Chem.*, **244**, 1552–1562 (1969).
14. Klingenberg, M. and Pfaff, E. "Regulation of Metabolic Processes in Mitochondria," ed. by J. M. Tager, S. Papa, E. Quagliariello, and E. C. Slater, Elsevier Publ. Co., New York, Vol. 7, pp. 180–201 (1966).
15. Lardy, H. A. and Wellman, H. The catalytic effect of 2,4-dinitrophenol on adenosine triphosphate hydrolysis by cell particles and soluble enzymes. *J. Biol. Chem.*, **201**, 357–370 (1953).
16. Lehninger, A. L. and Gregg, C. T. Dependence of respiration on phosphate and phosphate acceptor in submitochondrial systems. I. Digitonin fragments. *Biochim. Biophys. Acta*, **78**, 12–26 (19163).
17. Ohe, K., Morris, H. P., and Weinhouse, S. β-Hydroxybutyrate dehydrogenase activity in liver and liver tumors. *Cancer Res.*, **27**, 1360–1371 (1967).
18. Pedersen, P. L., Eska, T., Morris, H. P., and Catterall, W. A. Deficiency of uncoupler-stimulated adenosine triphosphatase activity in tightly coupled hepatoma mitochondria. *Proc. Natl. Acad. Sci. U.S.*, **68**, 1079–1082 (1971).
19. Pedersen, P. L., Greenawalt, J. W., Chan, T. L., and Morris, H. P. A comparison of some ultrastructural and biochemical properties of mitochondria from Morris hepatomas 9618A, 7800 and 3924A. *Cancer Res.*, **30**, 2620–2626 (1970).
20. Pedersen, P. L. and Schnaitman, C. A. The oligomycin-sensitive adenosine diphosphate-adenosine triphosphate exchange in an inner membrane-matrix fraction of rat liver mitochondria. *J. Biol. Chem.*, **244**, 5065–5073 (1969).
21. Penefsky, H. S., Pullman, M. E., Datta, A., and Racker, E. Partial resolution of the enzymes catalyzing oxidative phosphorylation. I. Purification and properties of soluble dinitrophenol-stimulated adenosine triphosphatase. *J. Biol. Chem.*, **235**, 3322–3329 (1960).
22. Potter, V. R. Biochemical uniformity and heterogeneity in cancer tissues (Further Discussion). *Cancer Res.*, **16**, 658–667 (1956).

23. Potter, V. R., Price, J. M., Miller, E. C., and Miller, J. A. Studies on the intracellular composition of livers from rats fed various aminoazo dyes. III. Effects on succinoxidase and oxalacetic acid oxidase. *Cancer Res.*, **10**, 28–35 (1950).
24. Sauer, L. A., Martin, A. P., and Stotz, E. Cytochemical fractionation of the lettre-Ehrlich ascites tumor. *Cancer Res.*, **20**, 251–256 (1960).
25. Schnaitman, C. A., Erwin, V. G., and Greenawalt, J. W. The submitochondrial localization of monoamine oxidase. *J. Cell Biol.*, **32**, 719–735 (1967).
26. Schnaitman, C. A. and Greenawalt, J. W. Enzymatic properties of the inner and outer membranes of rat liver mitochondria. *J. Cell Biol.*, **38**, 158–175 (1968).
27. Schnaitman, C. A. and Pedersen, P. L. Localization of oligomycin-sensitive ADP-ATP exchange activity in rat liver mitochondria. *Biochem. Biophys. Res. Commun.*, **30**, 428–433 (1968).
28. Schneider, W. C. and Potter, V. R. Biocatalyst in cancer tissue. III. Succinic dehydrogenase and cytochrome oxidase. *Cancer Res.*, **3**, 353–357 (1943).
29. Schreiber, J. R., Balcavage, W. X., Morris, H. P., and Pedersen, P. L. Enzymatic and spectral analysis of cytochrome oxidase in adult and fetal rat liver and Morris hepatoma 3924A. *Cancer Res.*, **30**, 2497–2501 (1970).
30. Shack, J. Cytochrome oxidase and d-amino acid oxidase in tumor tissue. *J. Natl. Cancer Inst.*, **3**, 389–396 (1943).
31. Smoly, J. M., Kuylenstierna, B., and Ernster, L. Topological and functional organization of the mitochondrion. *Proc. Natl. Acad. Sci. U.S.*, **66**, 125–131 (1970).
32. Sottocasa, G. L., Kuylenstierna, B., Ernster, L., and Bergstrand, A. An electron transport system associated with the outer membrane of rat liver mitochondria. *J. Cell Biol.*, **32**, 415–438 (1967).
33. Stotz, F. The estimation and distribution of cytochrome oxidase and cytochrome c in rat tissues. *J. Biol. Chem.*, **131**, 555–565 (1939).
34. Warshaw, J. B., Lam, K. W., Nagy, B., and Sanadi, D. R. Studies on oxidative phosphorylation. XV. Latent adenosine 5′-triphosphatase activity of factor A. *Arch. Biochem. Biophys.*, **123**, 385–396 (1968).
35. Weber, K. and Osborn, M. The reliability of molecular weight determinations by dodecyl sulfate-polyacrylamide gel electrophoresis. *J. Biol. Chem.*, **244**, 4406–4412 (1969).
36. Wenner, C. E. and Weinhouse, S. Metabolism of neoplastic tissue. III. Diphosphopyridine nucleotide requirements for oxidation by mitochondria of neoplastic and non-neoplastic tissues. *Cancer Res.*, **13**, 21–26 (1953).
37. Williamson, R. L. and Metcalf, R. L. Salicylanilides: a new group of active uncouplers of oxidative phosphorylation. *Science*, **158**, 1694–1695 (1967).
38. Winick, M. and Noble, A. Quantitative changes in DNA, RNA, and protein during prenatal and postnatal growth in the rat. *Dev. Biol.*, **12**, 451–466 (1965).
39. Wu, R. and Racker, E. Regulatory mechanisms in carbohydrate metabolism. III. Limiting factors in glycolysis of ascites tumor cells. *J. Biol. Chem.*, **234**, 1029–1035 (1959).

EXPLANATION OF PHOTOS

Preparations were fixed with glutaraldehyde and OsO_4, and stained with uranyl acetate and lead citrate, $\times 15,000$. The micron marker on Photo 6 also applies to the other photos.

Photo 1. Liver mitochondria of an animal bearing 9618A hepatomas
Photo 2. Mitochondria from hepatoma 9618A
Photo 3. Liver mitochondria of an animal bearing 7800 hepatomas
Photo 4. Mitochondria from hepatoma 7800
Photo 5. Liver mitochondria of an animal bearing 3924A hepatomas
Photo 6. Mitochondria from hepatoma 3924A

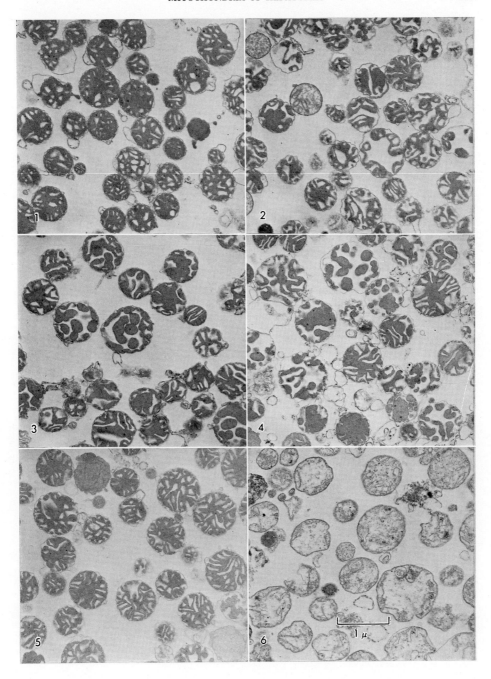

GLUCOSE 6-PHOSPHATE DEHYDROGENASE ISOZYMES IN CULTURED MORRIS HEPATOMA CELLS

Mochihiko OHASHI[*1] and Tetsuo ONO[*2]

Department of Biochemistry, Tokyo Metropolitan Institute of Gerontology[*1]
and Department of Biochemistry, Cancer Institute[*2]

From Morris hepatoma strain 7288C, a cultured cell line has been established and maintained *in vitro* for several years in our laboratory. Several clonal cell lines derived from a cultured hepatoma (7288tc) have also been well maintained. As in the original hepatoma, the glucose 6-phosphate dehydrogenase (G-6-PDH) activity of these cultured cells is fixed at a characteristically high level.

It is well known that G-6-PDH can be induced in rat liver by feeding the rat a high-glucose diet after a few days of fasting by the rat. Therefore, the isozyme pattern of G-6-PDH using different substrates and coenzymes in 7288tc cells was compared with that of induced rat liver by means of polyacrylamide gel electrophoresis. The isozymes of G-6-PDH were divided into two groups, H and G, according to their substrate specificity. The H-group takes both glucose 6-phosphate (G-6-P) and galactose 6-phosphate (H-6-P), and the G-group takes only G-6-P as substrate. When NADP was used as a coenzyme in the control liver of a rat, three distinct bands of H-group enzymes were demonstrated, while in the induced liver one of them was found to increase and in addition five bands of G-group enzymes were expressed. However, no isozyme of NAD-dependent G-6-PDH could be induced. In the case of cultured 7288tc cells, the isozyme pattern was almost identical with that from induced liver except it lacked certain bands and there were some differences in the patterns observed among the clonal lines. From these results, the schematic subunit constitutions for each band were tentatively proposed.

We have already reported that there were significantly different levels of glucose 6-phosphate dehydrogenase (G-6-PDH) activity in the tumor tissues of various hepatomas during a survey of enzyme patterns of these experimental hepatomas (*15*). As shown in Table I, the 7288C Morris hepatoma strain has an especially high level of G-6-PDH activity when compared to that of normal liver or other hepatomas, including minimal deviation hepatomas and Yoshida ascites hepatomas. Moreover, Tepperman and Tepperman reported that this G-6-PDH can be

---

[*1] 35-2 Sakaecho, Itabashi-ku, Tokyo 173, Japan (大橋望彦).
[*2] Kami-Ikebukuro 1-37-1, Toshima-ku, Tokyo 170, Japan (小野哲生).

TABLE I. Glucose 6-Phosphate Dehydrogenase Activities in the Tissues of Various Hepatomas

Hepatoma	G-6-PDH (μmoles/g tissue/hr)
5123 A	1,350
5123 B	381
5123 C	554
5123 D	252
5123tc	543
7316 A	676
7316 B	1,565
7793	446
7794 A	3,665
7794 B	803
7795	750
7800	842
7288 C	6,515
H-35TC	290
3924 A	740
3683	914
AH-7974	350
AH-130	250
Normal liver	
Buffalo	207
Donryu	301

induced in rat liver by feeding it a high-glucose diet after it has fasted for a few days. The induced level of the enzyme, however, returns to the original level after several days (22). This seems to indicate that the mechanism of metabolic regulation for the production of G-6-PDH in the liver may have been lost in 7288C heapatoma. The derepressed state of this enzyme in 7288C hepatoma seems to be fixed firmly to the cell by some other mechanism which may occur at the onset of carcinogenesis. Several cultured cell lines derived from this 7288C hepatoma have been established in our laboratory (11) and maintained *in vitro* for five years. These established cell lines of hepatoma were used to analyze the characteristic enzyme regulation patterns. A cultured hepatoma 7288C cell line has also been reported by Tomkins *et al.* (1–3, 4, 7, 13, 17, 18, 23, 24) who have studied the synthesis of tyrosine aminotransferase in cultured hepatoma cells stimulated by adrenal steroid hormones. Their HTC cells, however, differ from our 7288tc cells in some characteristics such as the number of chromosomes, inducibility of tyrosine aminotransferase (which does not increase the activity in our strain by the hormones), and growth rate. These differences may be due to the selection of a certain characteristic cell from the mixed population of original hepatoma tissue. In our cultured 7288tc cells, the G-6-PDH activity has been observed in as high a level as the original tumor tissue even though the cells have been cultured *in vitro* for several years. The results of the isozymic consideration of

G-6-PDH from various states of rat liver tissue and hepatoma cells will be described here.

Morphological Appearances of Cultured 7288tc Hepatoma Cells

At the present time, seven cloned cell lines have been isolated using the soft agar method; their colonies differ morphologically, as shown in Photo 1. In Photo 1 three typical shapes of colonies, diffused, unevenly surfaced, and evenly surfaced, are illustrated. Each colony removed from the agar plate was transferred into a small plastic dish of fresh culture fluid and maintained in a CO_2 incubator at 37°. Transfer was carried out by trypsinization once a week and diluted to a certain number of cells in order to maintain them well in a growing condition. The cloned cells were cultured in a suspended state but some of them adhered to the plastic dish. In Photo 2 the shapes of the growing cells of each clone are shown. It is clear that the cells of clone number 7 grew dispersely and maintained their original, characteristic, colony shape (diffused) in the agar plate while the other clones formed clumps. Clone numbers 14 and 16 also had the characteristic large clumps and firmly fused cells. The degrees of adhesiveness to the plastic dish of each cloned cell are summarized in Table II. The cells of clone number 7 were not found to be adhesive to the plastic dish.

TABLE II. Adhesiveness of 7288tc-2-1 Cloned Cells

Clone No.	No. adhesive cells / No. total cells	Ratio
1	24/274	8.8
4	9/216	4.2
7	0/124	0.0
8	47/348	13.5
9	16/270	5.9
14	37/195	19.0
16	57/233	24.5
Parent	17/251	6.8

Isozymic Pattern of G-6-PDH on Polyacrylamide Gel Electrophoresis

It seemed of interest to compare the G-6-PDH isozyme patterns of the cloned cell lines mentioned above with those of normal and induced rat liver tissues. In order to extract cellular particle-bound enzymes completely the packed cells were suspended in a 0.2% solution of deoxycholate, and freezing and thawing was carried out several times. The supernatant was collected after high-speed centrifugation and used for the isozyme pattern analysis. The use of polyacrylamide gel electrophoresis was convenient for the fine separation of the isozyme bands.

By the split gel electrophoresis in polyacrylamide gel, two different samples were manipulated in the same disc gel, and the color for enzyme reaction was developed under the same conditions. The resulting bands of enzymes in each sample can be exactly compared on the same disc gel. As another demonstration of different specific reaction, the cutting and staining of gel was also applied.

NADP and NAD Dependency of G-6-PDH Isozymes in Rat Tissues

In general, the enzyme activity of G-6-PDH depends on the presence of NADP as a coenzyme. However, Levy reported that the same enzyme is also dependent on NAD in about a tenth of that activity using NADP (*12*). In our experiment a similar result was confirmed for rat liver and kidney tissues, and the coenzyme dependency of each isozymic component separated in the gel electrophoresis was tested. The patterns are shown in Fig. 1. The NADP- and

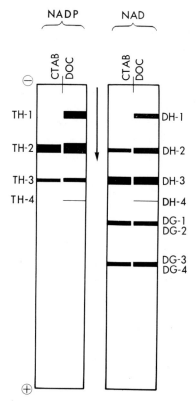

FIG. 1. Isozyme composition of G-6-PDH in the liver of a rat
The bands of enzymes extracted by two different detergents, CTAB (0.2% cetyltrimethylammonium bromide) and DOC (0.2% sodium deoxycholate), were compared by split gel electrophoresis. Each band was classified by a number. G-6-P was used as a substrate in this experiment.

NAD-dependent enzymes have almost common bands, but the position of the main band is different for each coenzyme. Two columns in Fig. 1 represent the diagrammatic patterns of the isozymes in split gels and compare them with patterns from different extraction methods. A number was given to classify each band. T stands for NADP-dependent enzymes and D for NAD-dependent ones. G and H indicate the different substrate specificities for glucose 6-phosphate and galactose 6-phosphate.

Substrate Specificity of G-6-PDH

Shaw and others (14, 19–21) reported that there are two forms of this enzyme, namely G and H, which were previously called A and B. They have different substrate specificities and it is thought that they are directed by genomes of different chromosomes, that is, autosomal and sexual. The G-group isozyme is specific for glucose 6-phosphate (G-6-P), and the H-group is reactive for both G-6-P and galactose 6-phosphate (H-6-P). In an electrophoresis, the H-group moves more slowly toward the anode than the G-group. The isozymes in the H-group region were stained by reaction mixtures containing either G-6-P or H-6-P as the substrate (see Fig. 2). The correlation of the position of the bands that are reactive to each substrate can be seen by vertically cutting the gel after electrophoresis and by staining each piece with different substrate reaction mixtures.

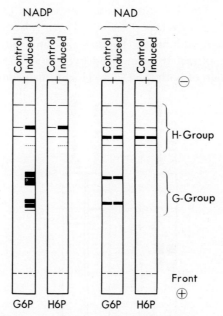

FIG. 2. G-6-PDH isozymes from normal rat liver tissue

The enzyme preparations were 0.2% DOC extract of 10% tissue homogenates which were centrifuged at 10,000 rpm for 20 min. The clear supernatant was diluted to the appropriate concentration with distilled water. The ordinary 7.5% polyacrylamide gel was prepared basically by Ornstein's method (16) and its split gel electrophoresis was carried out in the cold at a constant current of 2 mA/tube for 5–6 hr. After electrophoresis the site of each enzyme band in the gel was stained by the deposition of a reduced Nitro-Blue Tetrazolium dye formed by the enzymic reduction of NADP. The staining solution used was a mixture of Tris-HCl buffer of pH 8.6 containing $MgCl_2$, Nitro-Blue Tetrazolium, phenazinemethosulfate, a substrate (G-6-P or H-6-P), and a coenzyme (NADP or NAD) according to Tsao's directions (25), but with a slight modification. The substrate was D-glucose 6-phosphate (G-6-P) or D-galactose 6-phosphate (H-6-P). The H-group of the isozymes catalyzed both the G-6-P and the H-6-P oxidation. The G-group only catalized the G-6-P oxidation.

Isozymic Patterns of G-6-PDH in Normal Rat Liver and Its Induced State

In normal rat liver extract it was difficult to detect distinct NADP-dependent isozyme bands in the G-group region, although distinct bands of NAD-dependent enzymes could be found in both G and H regions as shown in Fig. 2. This pattern was also found in normal rat kidney extracts but with differences in band densities. The induction of G-6-PDH in liver was caused by feeding a high-glucose diet to a normal rat after it had fasted for 3 days (as mentioned earlier). The activity of G-6-PDH in the induced liver was 10- to 20-fold higher than that of normal liver. The NADP-dependent G-6-PDH isozyme pattern was remarkably changed by this induction. When G-6-P was used as a substrate, the G-group isozyme consisted of five components, and these components were divided into two slowly moving bands and three faster bands. The band moving the fastest was faint. With H-6-P as a substrate the H-group isozyme pattern was quite similar to that of G-6-P in the H-group region, but no band in the G-group region was detected. On the other hand, components of the NAD-dependent isozyme were not affected by this process. These relationships are represented in Fig. 3. The pattern with all the bands was found only in the induced liver. Therefore, using the extract from induced liver as a reference in split gel electrophoresis, the position of both the G and H isozyme groups can be clearly identified.

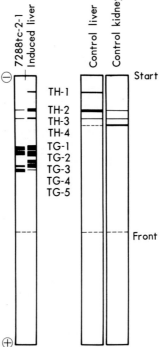

FIG. 3. NADP-dependent G-6-PDH isozymes in rat tissues
The two right-hand columns show the findings by single gel electrophoresis. The front line was marked by the mobility of BPB which was contained in the upper buffer system before electrophoresis. In this experiment, G-6-P was used as a substrate and the other conditions were similar to those described in the legend to Fig. 2.

TABLE III. Glucose 6-Phosphate Dehydrogenase Activities
of 7288tc-2-1 Cloned Cells

Clone No.	Passage	G-6-PDH/mg protein[a]
1	T-6	30.6
4	T-7	29.4
7	T-7	28.6
8	T-7	39.7
9	T-7	29.0
14	T-7	31.6
16	T-7	34.8
Parent	T-7 (T-172)[b]	22.4

The activity was determined on the 6th day after transfer.
[a] μmoles/mg protein/hr.   [b] The number of passage after initial cloning of 7288tc cell.

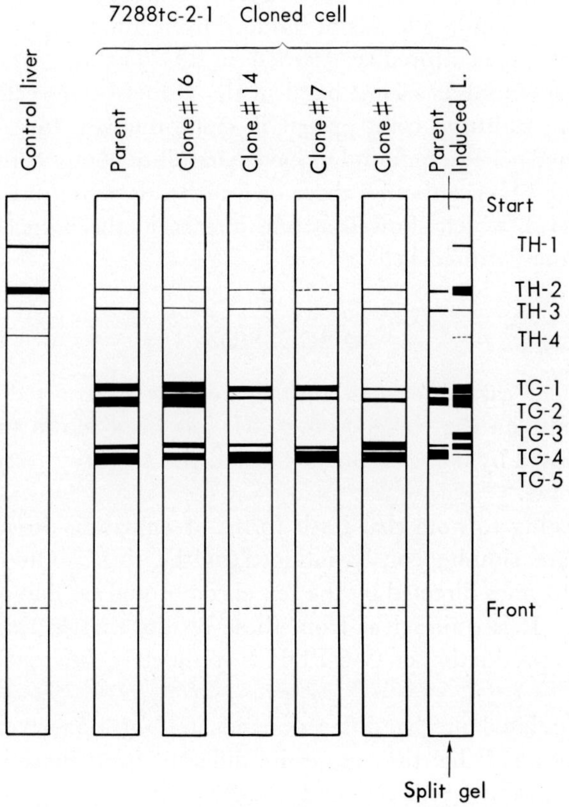

FIG. 4. NADP-dependent G-6-PDH isozymes in normal liver, in induced liver and in cultured hepatoma cells

The clonal cells of 7288tc hepatoma in culture were harvested 5 days after transfer and packed by low-speed centrifugation. The enzyme extraction from each cloned cell was carried out by several cycles of freezing and thawing in a small amount of 0.2% DOC. The supernatant, after centrifugation at 10,000 rpm for 20 min was used as a crude enzyme specimen and charged on top of the gel. Electrophoretic and staining conditions of the enzyme were the same as described in the legend to Fig. 2.

Isozyme Pattern of G-6-PDH in the Cultured 7288tc

When the isozyme pattern of G-6-PDH in the cultured hepatoma cells was analyzed by the split gel procedure there were clearly six bands in the G- and H-group regions, as shown in Fig. 3. The total activity of the enzyme in 7288tc cultured cells was comparable to that of induced liver. As shown in Table III, the activity of G-6-PDH in each clone of the hepatoma cells used in this experiment was at almost the same levels as that of the parent cells. Although the activity of the cultured normal liver cells could not be determined, tissue-cultured cells in general exhibited much lower levels of activity than the *in vivo* levels. Also, the primary cultured 7288C cell's activity decreased to an almost null level within a few days after transfer *in vitro* (*11*).

The isozyme patterns of this enzyme in each 7288tc clone are summarized in Fig. 4. The position of each band was common to all the cultured cells. In all of the cultured cells, the bands in the G-group isozymes were found to be as intense as those corresponding bands of induced liver, although the intensity varied in each. However, the cultured cells lacked the TG-3 band. The fastest moving, TG-5, which was found as a faint band in the induced liver cells, was the main component in the cultured cells, except in clone number 16. In the H-group region the cultured cells were found to have two minor components and these cells lacked TH-1 and TH-4, although three or four bands were found in the induced liver. Band TH-2, which showed strong intense in the induced liver, was not derepressed in the cultured cells.

DISCUSSION

As mentioned earlier, several investigators have reported the location of G-6-PDH genome on the chromosomes. It has been found that the G-group enzyme is expressed by a sex chromosome and the H-group enzyme by an autosomal chromosome.

It is interesting to note that both forms of enzymes, directed by different chromosomes, are simultaneously induced in the liver. On the other hand, only G-group enzymes directed by the sex chromosome are derepressed in 7288tc hepatoma cells. It is quite clear from these results that in this tumor cell the regulation of the production of G-6-PDH is completely different from that in the induced liver.

It was also pointed out that different bands in G-group enzymes are expressed in different clones and that they are quite different from those of induced liver, especially at the position of TG-3 and TG-5 bands.

Kirkman and co-worker (*8, 10*), Yoshida (*27*), and others (*5, 6, 26*) have already investigated the protein structures of G-6-PDH in highly purified preparations from human erythrocytes. Kirkman found that the isozymes of this enzyme in human erythrocytes are composed of three bands, monomer (A), dimer (AA) without bound NADP, and dimer with bound NADP (AAT). Recently his extensive experiments led to the finding that the last AAT band is replaced by AX, that is, a complex composed of a subunit (or monomer) and another protein

TABLE IV. Suggested Subunit Compositions of G-6-PDH Isozymes

No. of band	Subunit composition
TH-1	$B_1 B_1$
TH-2	$B_1 B_2$
TH-3	$B_2 B_2$
TH-4	B_1 and/or B_2
TG-1	AB_1
TG-2	AB_2
TG-3	AX or AAX
TG-4	AA
TG-5	A

(X) which is similar to such a subunit in size but has no active fragment (9). Yoshida proposed a schematic presentation of the relationship of various G-6-PDH configurations and activities based on the dissociation and association of six similarly sized subunits and NADP. According to his suggestion, this enzyme consists of hexomer or at least trimer by the similar subunits. Shaw (20) obtained some genetically variant forms of H-6-PDH from deer mouse and confirmed the presence of their hybrid molecular form in heterozygote. This seems to support the theory of a dimerizing structure in these isozyme components. Considering the results of our experiments and those of previous investigators, the isozymic components among the TH- and TG-groups, consisting of distinct subunits, could be explained by the dimer theory without any contradictions. Then, as shown in Table IV, the subunit constitutions of each band found in our experiments may be suggested by the following relationships.

In the isozymes among the TH-group, two distinct subunits (monomers), B_1 and B_2, are able to give three separable bands on the gel electrophoresis as dimer structure. In addition, although the monomer itself, *e.g.*, TH-4, including B_1 and/or B_2, has no activity, it is possible to detect it after electrophoresis because these subunits are highly concentrated into local positions and dimerized to the active form during incubation with NADP and substrate. As mentioned in Kirkman's proposal, only one kind of subunit (A) and another protein (X) exist in the TG-group. However, the properties of this specific protein (X) have not yet been clarified. The monomer in the TG-group, A, also seems to make an active dimer form (AA) after electrophoresis, as well as in the case of B subunits. Therefore, this fastest moving band of TG-5 is found only in faint form in the pattern of induced liver. Our knowledge of the subunit constitution of the H-6-PDH isozyme is poor and further information is needed.

The hybrid molecules between the TG- and TH-groups, described in Table IV as TG-1 and TG-2, have not yet been proposed by other researchers. However, we have suggested this hybrid hypothesis because these two bands exist only in the specimens which simultaneously contain both the TG- and TH-groups. Presently, the experiments to test this hypothesis are under way in our laboratory. It this suggestion is correct, it would be a little difficult to explain why they exhibit the same catalytic properties as that of the TG-group, despite the fact that they contain the TH-subunit.

From our experiments on the G-6-PDH isozyme, it has also been shown that the regulation of gene expression is considerably disturbed in the tumor cell. This alteration of regulation is not only revealed in the formation of the enzyme molecule itself, but also in the formation of another protein such as an X-molecule. The lack of a band corresponding to the AX molecular form in 7288tc hepatoma cells may be due to the repression of the X-molecules in the hepatoma cells.

Acknowledgments

The authors are indebted to Mr. A. Kibayashi of Iwate Medical College and to Mr. N. Watanabe, Mr. H. Kodama, and Miss S. Kuwahata of this laboratory for their cooperation and assistance.

This work was supported by grants for scientific research from the Ministry of Education and from the Princess Takamatsu Fund for Cancer Research.

REFERENCES

1. Auricchio, F., Martin, D., Jr., and Tomkins, G. M. Control of degradation and synthesis of induced tyrosine aminotransferase studied in hepatoma cells in culture. *Nature*, **224**, 806–808 (1969).
2. Ballard, P. L. and Tomkins, G. M. Hormone induced modification of the cell surface. *Nature*, **224**, 344–345 (1969).
3. Ballard, P. L. and Tomkins, G. M. Glucocorticoid-induced alteration of the surface membrane of cultured hepatoma cells. *J. Cell Biol.*, **47**, 222–234 (1970).
4. Baxter, J. D. and Tomkins, G. M. The relationship between glucocorticoid binding and tyrosine aminotransferase induction in hepatoma tissue culture cells. *Proc. Natl. Acad. Sci. U.S.*, **65**, 709–715 (1970).
5. Chung, A. E. and Langdon, R. G. Human erythrocyte glucose 6-phosphate dehydrogenase. I. Isolation and properties of the enzymes. *J. Biol. Chem.*, **238**, 2309–2316 (1963).
6. Chung, A. E. and Langdon, R. G. Human erythrocyte glucose 6-phosphate dehydrogenase. II. Enzyme-coenzyme interrelationship. *J. Biol. Chem.*, **238**, 2317–2324 (1963).
7. Granner, D. K., Hayashi, S., Thompson, E. B., and Tomkins, G. M. Stimulation of tyrosine aminotransferase synthesis by dexamethasone phosphate in cell culture. *J. Mol. Biol.*, **35**, 291–301 (1968).
8. Kirkman, H. N. Glucose 6-phosphate dehydrogenase from human erythrocytes. I. Further purification and characterization. *J. Biol. Chem.*, **237**, 2364–2370 (1962).
9. Kirkman, H. N. and Hanna, J. E. Isozymes of human red cell glucose-6-phosphate dehydrogenase. *Ann. N.Y. Acad. Sci.*, **151**, 133–148 (1968).
10. Kirkman, H. N. and Hendrickson, E. M. Glucose 6-phosphate dehydrogenase from human erythrocytes. II. Subactive states of the enzyme from normal persons. *J. Biol. Chem.*, **237**, 2371–2376 (1962).
11. Koyama, H., Yatabe, I., Ohashi, M., and Ono, T. The long-term cultivation of Morris hepatoma, 7288C—on its metabolic deviation (in Japanese). *Igaku no Ayumi (Progr. Med.)*, **65**, 666–673 (1968).
12. Levy, H. R. The pyridine nucleotide specificity of glucose 6-phosphate dehydrogenase. *Biochem. Biophys. Res. Commun.*, **6**, 49–53 (1961).

13. Martin, D., Tomkins, G. M., and Breslar, M. Control of specific gene expression examined in synchronized mammalian cells. *Proc. Natl. Acad. Sci. U.S.*, **63**, 842–849 (1969).
14. Ohno, S., Payne, H. W., Morrison, M., and Beutler, E. Hexose 6-phosphate dehydrogenase found in human liver. *Science*, **153**, 1015–1016 (1966).
15. Ono, T. Enzyme patterns and malignancy of experimental hepatomas. *GANN Monograph*, **1**, 189–205 (1966).
16. Ornstein, L. Disc electrophoresis. I. Background and theory. *Ann. N.Y. Acad. Sci.*, **121**, 321–349 (1964).
17. Peterkofsky, B. and Tomkins, G. M. Effect of inhibitors of nucleic acid synthesis on steroid-mediated induction of tyrosine aminotransferase in hepatoma cell cultures. *J. Mol. Biol.*, **30**, 49–61 (1967).
18. Samuels, H. H. and Tomkins, G. M. Relation of steroid structure to enzyme induction in hepatoma tissue culture cells. *J. Mol. Biol.*, **52**, 57–74 (1970).
19. Shaw, C. R. Glucose-6-phosphate dehydrogenase: homologous molecules in deer mouse and man. *Science*, **153**, 1013–1015 (1966).
20. Shaw, C. R. and Barto, E. Autosomally determined polymorphism of glucose 6-phosphate dehydrogenase in *Peromyscus*. *Science*, **148**, 1099–1100 (1965).
21. Shaw, C. R. and Koen, A. L. Glucose 6-phosphate dehydrogenase and hexose 6-phosphate dehydrogenase of mammalian tissues. *Ann. N.Y. Acad. Sci.*, **151**, 149–156 (1968).
22. Tepperman, H. M. and Tepperman, J. On the response of hepatic glucose 6-phosphate dehydrogenase activity to changes in diet composition and food intake pattern. *Adv. Enzyme Regulation*, **1**, 121–136 (1963).
23. Thompson, E. B., Tomkins, G. M., and Curran, J. F. Induction of tyrosin α-ketoglutarate transaminase by steroid hormones in a newly established tissue culture cell line. *Proc. Natl. Acad. Sci. U.S.*, **56**, 296–303 (1966).
24. Tomkins, G. M., Gelehrter, T. D., Granner, D. K., Martin, D., Jr., Samuels, H. H., and Thompson, E. B. Control of specific gene expression in higher organisms. *Science*, **166**, 1474–1480 (1969).
25. Tsao, M. U. Heterogeneity of tissue dehydrogenases. *Arch. Biochem. Biophys.*, **90**, 234–238 (1960).
26. Tsutsui, E. A. and Marks, P. A. A study of the mechanism by which triphosphopyridine nucleotide affects human erythrocyte glucose-6-phosphate dehydrogenase. *Biochem. Biophys. Res. Commun.*, **8**, 338–341 (1962).
27. Yoshida, A. Glucose 6-phosphate dehydrogenase of human erythrocytes. I. Purification and characterization of normal (B^+) enzyme. *J. Biol. Chem.*, **241**, 4966–4976 (1966).

EXPLANATION OF PHOTOS

PHOTO 1. Typical shape of colonies of 7288tc cells in agar plate. Unstained, ×40.
    a: Diffused; b: uneven surfaced; c: even surfaced.
PHOTO 2. Cloned cell lines of 7288tc cells in a fluid culture. Unstained, ×40. Note an adhesive cell on the plate and clump formation of some clones. The picture was taken at 5 days after transfer.
    a, clone #1; b, clone #4; c, clone #7; d, clone #16; e, clone #8; f, clone #9; g, clone #14.

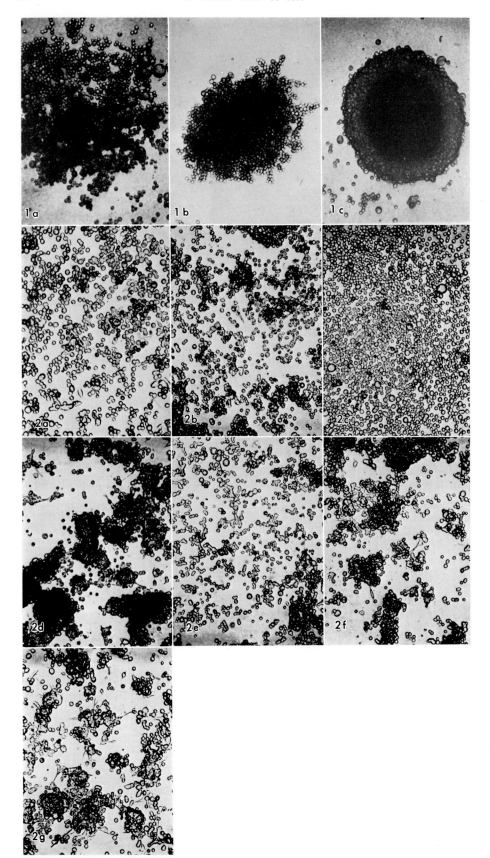

DIAGNOSTIC VALUE OF ALDOLASE AND HEXO-KINASE ISOZYMES FOR HUMAN BRAIN AND UTERINE TUMORS[*1]

Shigeaki SATO,[*2] Yoshihiro KIKUCHI,[*3] Kintomo TAKAKURA,[*4]
Te Chen CHIEN,[*4] and Takashi SUGIMURA[*3]

*Department of Molecular Oncology, Institute of Medical Science, Univeristy of Tokyo,[*2] Biochemistry Division, National Cancer Center Research Institute,[*3] and Neurosurgical Section, National Cancer Center Hospital[*4]*

The aldolase and hexokinase isozyme patterns in human brain and uterine tumors, respectively, were investigated using cellulose acetate membrane electrophoresis. Normal human brain tissue contains aldolase A, aldolase C, and three hybrid molecules, A_3C_1, A_2C_2, and A_1C_3. Gliomas, including glioblastomas, astrocytomas, oligodendrogliomas, and ependymomas, which are derived from glia cells of neuroectodermal origin, also possess both aldolases A and C, and three A-C hybrid sets similar to those in nerve cells. However, in meningiomas, which are of mesenchymal origin, in pituitary adenomas of ectodermal origin, and in metastatic brain tumors, aldolase A and sometimes the A_3C_1 hybrid were detected, but not aldolase C or the other two A-C hybrids.

These observations suggest that brain tumors maintain the same aldolase isozyme patterns as the tissues from which they originated and that study of the aldolase isozyme patterns is helpful in identifying the origin of brain tumors. Hemangioblastomas, which are supposed to be mesenchymal tumors, showed the same isozyme pattern as normal brain and gliomas. The discrepancy between the genetic origin of this tumor deduced from the isozyme pattern and from histological findings is discussed.

In human uterine cervical and corpus carcinomas, a large amount of type II hexokinase is present with some type I enzyme. This isozyme pattern differs from that of control tissues in which type I hexokinase predominates. There was no difference in the hexokinase isozyme patterns of myomas, which are benign tumors of the uterus, and of normal myometrium; in both tissues only type I was present. Accordingly, examination of hexokinase isozymes seems helpful in detecting carcinomas in the uterus.

[*1] This work was supported by grants from the Ministry of Education and the Ministry of Health and Welfare of Japan, and from the Seminar on Metabolic Regulation (Amino Acid and Protein).
[*2] Shirokanedai 4-6-1, Minato-ku, Tokyo 108, Japan (佐藤茂秋).
[*3] Tsukiji 5-1-1, Chuo-ku, Tokyo 104, Japan (菊池義公, 杉村　隆).
[*4] Tsukiji 5-1-1, Chuo-ku, Tokyo 104, Japan (高倉公明, 簡徳珍).

The possible clinical application of these isozyme studies in the diagnosis of human cancers is discussed.

Isozyme studies have contributed to our knowledge of the nature of cancer cells, as described in other parts of this monograph. Phenotypic abnormalities of the isozymes of many kinds of enzymes, the reappearance of some kinds of proteins specific to the cancer cells (*1*), and evidence that the isozyme patterns of some cancer cells can change to other fixed patterns (*23, 31*) suggest that even carcinogenesis may be an epigenetic change of the cells (*31*). It also seems important to consider the clinical application of findings in basic studies on isozymes to human cancers. Fortunately, many methods have been developed in basic studies for separation, purification, and characterization of various isozymes. However, although it is possible to study isozymes in human cancers, it is difficult to obtain results which actually contribute to the conquest of human cancers. Many enzymes and isozymes have been studied in cancer patients (*3–5, 18, 30, 35, 38*), but very few of the data on isozymes obtained so far are conclusive for diagnosis of human cancers or for deciding their genetic origins.

This paper reports the isozyme patterns of aldolase (fructose 1, 6-diphosphate D-glyceraldehyde-3-phosphate-lyase, EC 4.1.2.13) and hexokinase (ATP: D-hexose-6-phosphotransferase, EC 2.7.1.1) in human brain and uterine tumors, respectively. The clinical diagnostic value of these isozyme studies is also discussed.

Aldolase Isozymes in Experimental Brain Tumors

Thorough studies on aldolase isozymes have been made in many laboratories (*7, 9, 14–16, 19–21, 26, 27, 32, 33*) and results are summarized in another part of this monograph. Aldolase has been divided into three types; muscle (aldolase A), liver (aldolase B), and brain (aldolase C).

Type C aldolase has been studied the least, but its crystallization and characterization were reported recently (*19, 20*). There have also been few studies on aldolase C in tumors. The existence of this isozyme which is specific to brain prompted us to investigate the aldolase isozymes in brain tumors.

Photo 1 shows the histology of a mouse tumor transplanted subcutaneously. The original tumor was induced by methylcholanthrene in the brain of a C57BL strain mouse (*11*). It was diagnosed histologically as a glioblastoma, the genetic origin of which is neuroectodermal and is similar to that of nerve cells. It was very difficult to identify the transplanted tumor as being of brain origin from its histological appearance alone because during successive transplantations its morphology changed considerably. Accordingly, we investigated the aldolase isozymes of this transplanted tumor and of normal tissues of the same mouse strain (*25*). Photo 2 shows the aldolase isozyme patterns of the tumor and of brain tissue. Normal brain tissue contained aldolases A and C and the three hybrids. The hybrids were defined as A_3C_1, A_2C_2 and A_1C_3 in order of their electrophoretic mobilities from aldolase A to C (*19*). The transplanted tumor also had aldolase A, three A-C hybrid sets, and weak but definite aldolase C, the pattern being very similar to that of the normal brain tissue. This isozyme pattern was not found in other normal tissues of the same mouse strain or in other types of mouse tumors. From these results this transplanted tumor was thought to be actually derived from brain tissue. Brain tumors generally grow slowly, metastases are rare, and these tumors are more benign than other kinds of tumors. Slowly growing tumors, such as brain tumors, may maintain the same aldolase isozyme pattern as that of the tissue of origin and so studies on their aldolase isozymes should be helpful in identifying their genetic origin.

Aldolase Isozymes in Human Brain Tumors

Brain tumors can be divided into two main groups (2). Gliomas is one group derived from glia cells such as oligodendroglia, spongioblast, astrocyte, or ependyma. These glia cells, like nerve cells, are all differentiated from medullary epithelium of neuroectodermal origin, that is, glioma cells and nerve cells have the same genetic origin. The second group of brain tumors is the meningiomas. These are thought to originate from the arachnoid membrane, arachnoid villus, and pia mater which are mesenchymal tissues.

The aldolase isozyme patterns of these two groups and of some other kinds of human brain tumors were investigated using cellulose acetate membrane electrophoresis (24, 34). Representative isozyme patterns of normal brain, and glioma, meningioma, hemagioblastoma, pituitary adenoma, and metastatic brain tumor tissues are shown in Photo 3. Normal human brain tissue, obtained from an operation, contained aldolases A and C and three A-C hybrid sets, as does brain tissue of various other animals. Glioblastoma, a type of glioma, also had aldolases A and C and three hybrid molecules, the same as normal brain. Other gliomas, such as oligodendrogliomas, astrocytomas, and ependymomas, also had the same pattern as that of normal brain. However, in meningiomas, aldolase A predominated. In some meningiomas a faint band of the A_3C_1 hybrid was detected, but aldolase C and the two other hybrids were always absent. In pituitary adenoma, which is of ectodermal origin, and in metastatic brain tumors of the lung, stomach, and breast cancers, aldolase A and A_3C_1 hybrid were observed but other A-C hybrids and aldolase C were not detected.

These observations, together with the finding that a neurinoma, which is also a brain tumor of neuroectodermal origin, had the same aldolase isozyme pattern as normal brain tissue (13), suggest that brain tumors with the same genetic origin as nerve cells have aldolases A and C. However, in brain tumors of different origins, such as meningiomas, pituitary adenomas, and metastatic brain tumors, aldolase C is absent. Accordingly, aldolase C is a good marker for identifying the genetic origins of brain tumors, especially when a histological diagnosis is inconclusive.

Hemangioblastomas are interesting from this point of view. Aldolases A and C and the three A-C hybrids were found in this type of tumor as shown in Photo 2. The LDH isozyme pattern of this tumor was also the same as that of normal brain (22). These findings suggest that this tumor is of neuroectodermal origin. However, hemangioblastomas are composed of cells similar to hemangioblasts, which form a structure resembling the hemangioendothelium (Photo 4). On the basis of the morphological appearance of this tumor it is called a hemangioblastoma and thought to be of mesenchymal origin. The discrepancy between the genetic origin of this tumor deduced from the isozyme pattern and from histological findings requires further investigation.

Hexokinase Isozyme Patterns of Human Uterine Tumors

Hexokinase in animal tissues is separated into four isozymes by DEAE-cellulose column chromatography (6) or electrophoresis using starch gel (10) or a cellulose acetate membrane (23). Normal liver contains all four isozymes, types I, II, III, and IV, in order of increasing electrophoretic mobility from the origin to the anode (10). Type IV hexokinase has a high K_m value for glucose and is called " glucokinase " (36). The K_m values of types I, II, and III for glucose are lower than that of type IV, and these hexokinases are called " low-K_m hexokinases " (36).

In experimental hepatomas, a decrease or absence of glucokinase and an increase of low-K_m hexokinases were reported (8, 12, 28, 37). These findings were confirmed by electrophoresis (23, 29). In other types of rapidly growing tumors high activity of low-K_m hexokinases was also detected (8, 12).

These findings suggest that abnormal gene expressions of hexokinase isozymes may occur in rapidly growing and highly deviated tumors.

Human uterine tumors are very common so that tumor tissues can readily be obtained during operations. Cervical carcinomas originating in the cervical epithelium are the most common malignant tumors of the uterus. Corpus carcinomas, originating in the endometrium of the body part of the uterus, are rather rate. Myomas, which are benign tumors originating in the myometrium, are also common.

The hexokinase isozyme patterns of these three types of uterine tumors were investigated using cellulose acetate membrane electrophoresis and compared with those of control tissues.

The total hexokinase activity of each type of tissue was also assayed. The methods used for electrophoresis and assay of hexokinase activity were reported previously (23). Photo 5 shows typical hexokinase isozyme patterns of tissues of normal uterus and uterine tumors. In the cervical epithelium, only type I hexokinase was detected. In cervical carcinoma tissue, type II was definitely present in addition to type I. The endometrium showed strong type I and weak type II activity. Corpus carcinoma tissue contained very strong type II and strong type I activity. Thus, unlike control tissues, both carcinoma tissues contained strong type II hexokinase.

However, there was no difference in the hexokinase isozyme patterns in myoma and its control tissue, myometrium. Only type I hexokinase was present in both tissues.

Table I shows the mean values and standard deviations of specific hexokinase activities of tumor and control tissues with the number of specimens examined. The specific hexokinase activities in carcinomas were 3- to 4-fold higher than

TABLE I. Specific Activities of Hexokinase in Human Uterine Tissues and Tumors

Tissue	Activity (units/g protein)	No. of specimens
Cervical epithelium	6.3±2.3	10
Cervical carcinoma	25.1±8.6	10
Endometrium	7.4±3.3	14
Corpus carcinoma	21.2±9.8	5
Myometrium	8.4±5.2	10
Myoma	7.8±4.8	10

in control tissues, but no difference was observed between the activities in myomas and in the myometrium.

The high activity of low-K_m hexokinases, especially type II, may be a characteristic of rapidly growing tumors, as observed also in experimental tumors (*8, 12, 23*). From the clinical standpoint, at least with regard to uterine tumors, strong type II hexokinase activity is a good indicator of the presence of carcinoma tissue. A very little tissue (less than 100 mg wet weight), such as that obtained by biopsy, is enough for separation of hexokinase isozymes by electrophoresis. Study of the hexokinase isozyme pattern will be helpful in detecting uterine carcinomas when histological findings are inconclusive.

DISCUSSION

Human brain tumors, which have the same neuroectodermal genetic origin as nerve cells, contain aldolase C, aldolase A, and three A-C hybrid molecules, as does normal brain tissue. Meningiomas and other brain tumors of non-neuroectodermal origin do not contain aldolase C. Identification of the genetic origins of tumors is important clinically in choosing suitable therapeutic treatments and in determining the prognosis, but identification is sometimes difficult from histological findings alone. In such cases examination of the aldolase isozymes of the tissues seems valuable in identifying the genetic origin of the tumors, especially slowly growing and less deviated tumors.

Hexokinase isozyme patterns of human uterine carcinomas and control tissues were found to be very different. Carcinomas contained strong type II hexokinase as well as type I activity. However, normal tissues of the cervix, endometrium, myometrium, and myomas (benign tumors) contained only type I. Thus the study of hexokinase isozymes is helpful in detecting carcinomas in uterine tissue. These studies can be done using small amounts of tissues and might even be possible with cells collected by vaginal scrapings if the method for detecting hexokinase isozyme activity were improved.

Human primary hepatomas and gastric cancers have also been reported to show different aldolase isozyme patterns from control tissues (*17, 35*). The isozyme patterns of serum and other body fluids may reflect those of tumor tissues as we observed that patterns of the serum of tumor-bearing rats became kinetically similar to those of tumor tissues (*16*). Aldolase and hexokinase isozymes can also be separated in serum and cerebrospinal, ascitic, and pleural fluids. Thus, in the future it may be possible to diagnose human cancers by examination of the isozyme patterns not only of tissues but also of body fluids.

REFERENCES

1. Abelev, G. I. Production of embryonal serum α-globulin by hepatomas: review of experimental and clinical data. *Cancer Res.*, **28**, 1344–1350 (1968).
2. Bailey, P. Cellular types in primary tumors of the brain. *In* " Cytology and Cellular Pathology of the Nervous System," Paul B. Hoeber Inc., New York, pp. 1131–1144 (1932).
3. Bodansky, M. and Bodansky, O. "Biochemistry of Disease," Macmillian Co., New York (1952).
4. Comfort, M. W., Butt, H. R., Baggenstoss, A. H., Osterberg, A. E., and Priestley, J. T. Acinar cell carcinoma of pancreas: report of case in which function of carcinomatous cells was suspected. *Ann. Int. Med.*, **19**, 808–816 (1943).
5. Fishman, W. H., Smith, M., Thompson, D. B., Bonner, C. D., Kasdon, S. C., and Homburger, F. Investigation of glucuronic acid metabolism in human subjects. *J. Clin. Invest.*, **30**, 685–696 (1951).
6. González, C., Ureta, T., Sanchez, R., and Niemyer, H. Multiple molecular forms of ATP: hexose 6-phosphotransferase from rat liver. *Biochem. Biophys. Res. Commun.*, **16**, 347–352 (1964).
7. Gracy, R. W., Lacko, A. G., Brox, L. W., Adelman, R. C., and Horecker, B. L. Structural relations in aldolases purified from rat liver and muscle and Novikoff hepatoma. *Arch. Biochem. Biophys.*, **136**, 480–490 (1970).
8. Gumaa, K. A. and Greenslade, K. R. Molecular species of hexokinase in hepatomas and ascites-tumor cells. *Biochem. J.*, **107**, 22 p (1968).
9. Ikehara, Y., Endo, H., and Okada, Y. The identity of the aldolases isolated from rat muscle and primary hepatoma. *Arch. Biochem. Biophys.*, **136**, 491–497 (1970).
10. Katzen, H. M. and Schimke, R. T. Multiple forms of hexokinase in the rat: tissue distribution, age dependency, and properties. *Proc. Natl. Acad. Sci. U.S.*, **54**, 1218–1225 (1965).
11. Kawai, S. Experimental study on brain tumors (in Japanese). *Trans. Soc. Pathol. Japon.*, **52**, 59–82 (1963).
12. Knox, W. E., Jamder, S. C., and Davis, P. A. Hexokinase, differentiation and growth rates of transplanted rat tumors. *Cancer Res.*, **30**, 2240–2244 (1970).
13. Kumanishi, T., Ikuta, F., and Yamamoto, T. Aldolase isozyme patterns of representative tumours in the human nervous system. *Acta Neuropathol. (Berlin)*, **16**, 220–225 (1970).
14. Lay, C. Y., Chen, C., and Horecker, B. L. Primary structure of two COOH-terminal hexapeptides from rabbit muscle aldolase: a difference in the structure of the α and β subunits. *Biochem. Biophys. Res. Commun.*, **40**, 461–468 (1970).
15. Matsushima, T., Kawabe, S., Shibuya, M., and Sugimura, T. Aldolase isozymes in rat tumor cells. *Biochem. Biophys. Res. Commun.*, **30**, 565–570 (1968).
16. Matsushima, T., Kawabe, S., and Sugimura, T. Serum aldolases of tumor bearing rats (in Japanese). *Proc. Jap. Cancer Assoc. 25th Ann. Meet.*, 14–15 (1966).
17. Nordmann, Y. and Schapira, F. Muscle type isoenzymes of liver aldolase in hepatomas. *Eur. J. Cancer*, **30**, 247–250 (1967).
18. Okabe, K., Hayakawa, T., Hamada, M., and Koide, M. Purification and com-

parative properties of human lactate dehydrogenase isozymes from uterus, uterine myoma, and cervical cancer. *Biochemistry*, **7**, 79–90 (1968).
19. Penhoet, E. E., Kochman, M., and Rutter, W. J. Isolation of fructose diphosphate aldolases A, B, and C. *Biochemistry*, **8**, 4391–4395 (1969).
20. Penhoet, E., Kochman, M., and Rutter, W. J. Molecular and catalytic properties of aldolase C. *Biochemistry*, **8**, 4396–4402 (1969).
21. Rutter, W. J., Blostein, R. E., Woodfin, B. M., and Weber, C. S. Enzyme variants and metabolic diversification. *Adv. Enzyme Regulation*, **1**, 39–56 (1963).
22. Sano, K., Chigasaki, H., and Takakura, K. Diagnostic value of LDH isozyme studies in intracranial tumor. *In* " The 3rd International Congress of Neurological Surgery," Excerpta Medica Foundation, Amsterdam, pp. 575–579 (1966).
23. Sato, S., Matsushima, T., and Sugimura, T. Hexokinase isozyme patterns of experimental hepatomas of rats. *Cancer Res.*, **29**, 1437–1446 (1969).
24. Sato, S., Sugimura, T., Chien, T. C., and Takakura, K. Aldolase isozyme patterns of human brain tumors. *Cancer*, **27**, 223–227 (1971).
25. Sato, S., Sugimura, T., Kawai, S., and Ishida, Y. Aldolase C in experimental brain tumor. *Cancer Res.*, **30**, 1197–1198 (1970).
26. Schapira, F., Dreyfus, J. C., and Schapira, G. Anomaly of aldolase in primary liver cancer. *Nature*, **200**, 995–997 (1963).
27. Schapira, F., Reuber, M. D., and Hatzfeld, A. Resurgence of two fetal-type of aldolases (A and C) in some fast-growing hepatomas. *Biochem. Biophys. Res. Commun.*, **40**, 321–327 (1970).
28. Sharma, R. M., Sharma, C., Donnelly, A. J., Morris, H. P., and Weinhouse, S. Glucose-ATP phosphotransferases during hepatocarcinogenesis. *Cancer Res.*, **25**, 193–199 (1965).
29. Shatton, J. B., Morris, H. P., and Weinhouse, S. Kinetic, electrophoretic, and chromatographic studies on glucose-ATP phosphotransferases in rat hepatomas. *Cancer Res.*, **29**, 1161–1172 (1969).
30. Starkweather, W. H. and Schoch, H. K. Some observations on the lactate dehydrogenase of human neoplastic tissue. *Biochim. Biophys. Acta*, **62**, 440–442 (1962).
31. Sugimura, T. Decarcinogenesis, a newer concept arising from our understanding of the cancer phenotype. *In* " Chemical Tumor Problems," ed. by W. Nakahara, Japan Society for the Promotion of Science, Tokyo, pp. 269–284 (1970).
32. Sugimura, T., Matsushima, T., Kawachi, T., Hirata, Y., and Kawabe, S. Molecular species of aldolases and hexokinases in experimental hepatomas. *GANN Monograph*, **1**, 143–149 (1966).
33. Sugimura, T., Sato, S., and Kawabe, S. The presence of aldolase C in rat hepatoma. *Biochem. Biophys. Res. Commun.*, **39**, 626–630 (1970).
34. Sugimura, T., Sato, S., Kawabe, S., Suzuki, N., Chien, T. C., and Takakura, K. Aldolase C in brain tumor. *Nature*, **222**, 1070 (1969).
35. Tsunematsu, K., Yokota, S., and Shiraishi, T. Changes in aldolase isozyme patterns of human cancerous tissues. *GANN*, **59**, 415–419 (1968).
36. Viñuela, E., Salas, M., and Sols, A. Glucokinase and hexokinase in liver in relation to glycogen synthesis. *J. Biol. Chem.*, **238**, PC 1175–1177 (1963).
37. Weinhouse, S. Glycolysis, respiration, and enzyme deletions in slow-growing hepatic tumors. *GANN Monograph*, **1**, 99–115 (1966).
38. Woodard, H. Q. and Higinbotham, N. L. The correlation between serum phosphatase and roentgenographic type in bone disease. *Am. J. Cancer*, **31**, 221–237 (1937).

EXPLANATION OF PHOTOS

Photo 1. Histological appearance of a subcutaneously transplanted mouse tumor at the 14th generation. ×400
Photo 2. Aldolase isozyme patterns of normal C57BL mouse brain and a transplanted tumor.
Photo 3. Aldolase isozyme patterns of normal human brain and brain tumors.
Photo 4. Microscopic appearance of a human hemangioblastoma. ×400
Photo 5. Hexokinase isozyme patterns of human uterine tissues and tumors.

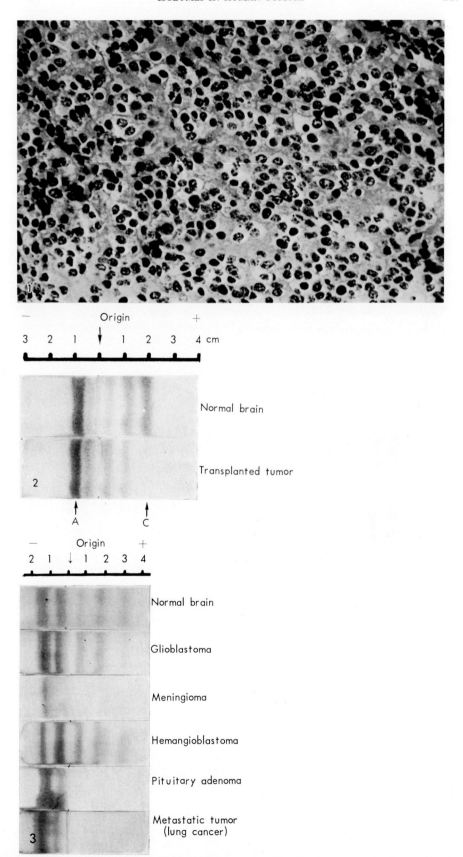

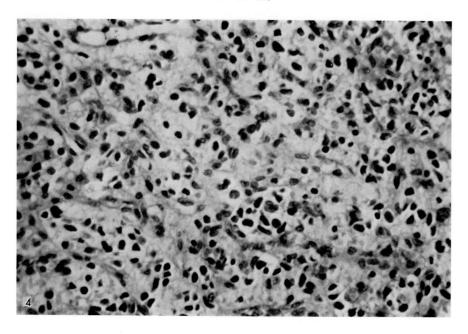

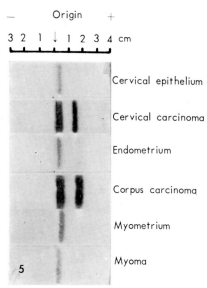

BLOOD OF TUMOR-BEARING ANIMALS AS A CAUSE OF METABOLIC DEVIATIONS

Eiji Ishikawa and Masami Suda

*Division of Protein Metabolism, Institute for Protein Research, Osaka University**

For a long time it has been generally considered that metabolic deviations observed in tumor-bearing animals are brought about by a substance or substances released from the tumor tissue into the circulating blood. Although much effort has been devoted to the isolation of such a substance, little is known as to whether or not it is actually present in tumor-bearer blood except that attempts to demonstrate the presence of causal factors in tumor-bearer blood (using the parabiotic technique and, recently, perfusion of isolated liver) have been successful. The techniques of cross-circulation and blood exchange were used to provide new information on the characteristics of factors possibly causing metabolic deviations in tumor-bearing animals. To follow the effect of blood from rats bearing Walker 256 carcinosarcoma, the activity of histidine decarboxylase in the liver and spleen and the incorporation of L-leucine[U-^{14}C] into tissue proteins were measured. The activity of histidine decarboxylase in normal liver rose to the level in tumor-bearing animals within 5 hr after the cross-circulation began, and the enzyme induction on cross-circulation and partial hepatectomy was apparently inhibited by pretreatment with cycloheximide. It was also shown that about 20 ml of tumor-bearer blood is sufficient to induce histidine decarboxylase in the liver and spleen of normal rats. The incorporation of L-leucine[U-^{14}C] into plasma protein was depressed to one-third of the normal level within 2 hr after infusing about 25 ml of tumor-bearer blood into normal rats. Isotopic incorporation into proteins of the liver, kidney, and spleen was also depressed but to different extents. After depression, the incorporation began to increase gradually returning to the normal level 17 hr after the infusion of tumor-bearer blood. The inhibitory activity of tumor bearer blood appears to reside in the cellular fraction, not in the plasma.

One malignant property inherent in the tumor is manifested and recognized as the cachectic state of the host. This is regarded as a state produced without any metastatic effect of malignant cells and therefore must result from a summation of the metabolic alterations which occur with progressive growth of the tumor. There has been much work done by many investigators on the nature of the metabolic alterations occurring in tumor-bearers, and there is a great deal

---

* 5311 Yamada-kami, Suita, Osaka 565, Japan (石川栄治, 須田正巳).

of information available on these metabolic patterns. However, the mechanism of induction of metabolic alterations in tumor-bearers has not been fully elucidated. The fact that the metabolic alterations take place in tissues remote from the tumor tissue suggests two possible mechanisms by which the tumor could produce its effect on the host; changing the concentrations of materials necessary for continued normal metabolism in the host, and by producing a chemical messenger which alters metabolism.

Nakahara and Fukuoka made the first attempts to extract from tumor tissue an active substance which would cause metabolic alterations comparable to those observed in tumor-bearers (*10*). Subsequently, much effort has been devoted to the purification of this substance. However, little is known as to whether or not such an active substance actually exists in the blood of tumor-bearers, although the presence of a causal factor or factors has been demonstrated in the blood of tumor-bearing rats using the parabiotic technique (*7, 8, 13*) and, recently, perfusion of isolated liver (*14, 16*). An analysis of the causative factors in tumor-bearer blood, together with efforts to purify active substances from tumor tissues, is required for understanding the mechanism causing host metabolic alterations since the active substances extracted from tumor tissues may not all actually or directly act on host tissues. This paper describes approaches that were used to understand the nature of tumor-bearer blood as a cause of metabolic deviations.

Techniques

Parabiosis has been most widely employed in attempts to demonstrate a possible humoral factor. Using this technique, evidence has been obtained that metabolic alterations in tumor-bearers are caused by a factor or factors conveyed through blood circulation (*7, 8, 13*). The technique of parabiosis is simple and easy. However, it can be used only to indicate the existence of humoral factors, since it is not possible with this technique to estimate the amount of humoral factor transferred nor the length of time required for transfer. In order to characterize a humoral factor, it is essential to estimate the volume of blood exchanged between the parabiotic partners and the period of time required for the appearance of an apparent effect and to control the transfer of the factor. In parabiosis, the amount of a humoral factor transferred between parabiotic rats is limited to 0.1 ml/min (*2*). Moreover, using this technique, it may not even be possible to detect the presence of a humoral factor of short life. The technique of perfusion of isolated organs does not have the above limitations since it is readily possible to control and define the experimental conditions; the effects of tumor-bearer blood have actually been observed by perfusing rat liver (*14, 16*). However, the drawbacks to perfusion are that the method is time- and labor-consuming and it cannot be used to detect effects or parameters that appear only after a long period of time, *e.g.*, more than 5 hr. One must also keep in mind the possibility that a humoral agent or agents may be rapidly inactivated in the perfused organs and so be lost from the perfusate.

The third possible method for detecting humoral factors is cross circulation, in which the arterial blood of each partner flows through two cannulas into the vein of the other partner. Using this method it is possible to control the time and duration of blood exchange, and hence the amount of blood exchanged. Moreover, a relatively large volume of blood can be continuously transferred from one partner to the other for as long a period as 2 days (*1*). An inevitable drawback of this method is that it is accompanied and complicated by surgical and immobilizational stresses and probably also by the effects of anesthesia and anticoagulants.

The fourth possible technique is infusion of a certain volume of test blood through a vein while simultaneously withdrawing the same volume of blood from an artery. This enables examination of the effects of various blood fractions by fractionating and reconstituting normal and tumor-bearer blood. However, the method has disadvantages similar to those of the cross-circulation technique.

Effect of Tumor-bearer Blood on Various Parameters

Catalase: Greenstein *et al.* found that the activity of liver catalase in tumor-bearing animals decreases with progressive growth of the tumor (*3*). This has since been confirmed by many investigators. Lucke *et al.* were the first to obtain evidence that a humoral factor may be responsible for metabolic alterations in tumor-bearers (*8*). They demonstrated that catalase activity was lowered in the liver of normal rats connected with tumor-bearers by parabiosis.

Pyruvate kinase: Tanaka *et al.* found two types of pyruvate kinase in rat liver (*15*); type L which changes in activity in response to various physiological conditions and type M which apparently is not under hormonal or dietary control but increases in activity in regenerating liver and in liver of tumor-bearers (*15*). The activity of pyruvate kinase type M also increases in the liver of a normal parabiotic partner connected with a rat bearing a Walker 256 carcinosarcoma (*13*). Suda *et al.* showed that type M activity also increases in normal isolated liver perfused with blood from rats bearing Walker carcinosarcomas (*14*).

Histidine decarboxylase: Johnston reported that histidine decarboxylase activity was abnormally high in tissues of rats bearing Walker carcinosarcoma and in a normal rat connected parabiotically with a rat bearing Walker carcinosarcoma (*7*). As mentioned above, experiments using parabiosis provide little information on the nature of enzyme induction or on the characteristics of a possible humoral factor. Accordingly, in order to study the mechanism responsible for metabolic deviations in the host, we utilized the technique of cross-circulation between normal and tumor-bearing rats. Two cannulas were inserted into the carotid artery and jugular vein of one rat and the jugular vein and carotid artery, respectively, of the partner, so that the arterial blood of each rat flowed into the jugular vein of its partner. Histidine decarboxylase activity was determined by measuring the conversion of L-histidine[U-^{14}C] into ^{14}C-histamine in the livers of normal and Walker carcinosarcoma-bearing rats 2.5 to 5 hr after starting the cross-circulation. The enzyme activity was found to increase remarkably in the liver of normal partners connected with tumor-bearers, while little activity was found when two normal rats were connected with one another

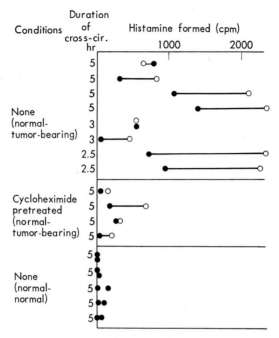

FIG. 1. Induction of histidine decarboxylase in normal rat liver by cross-circulation with Walker carcinosarcoma-bearing rats

After cross-circulation for the indicated periods, liver histidine decarboxylase activity was determined by examining the supernatant obtained by centrifugation of liver homogenates at 80,730 g for 60 min. Activity is expressed as total radioactivity of histamine-^{14}C produced from L-histidine[U-^{14}C]/30 mg protein in 90 min. Cycloheximide was injected i.p. in a dose of 0.1 mg/100 g body weight 30 min before starting cross-circulation (6). ○ tumor-bearing side ; ● normal side.

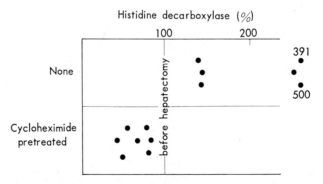

FIG. 2. Effect of cycloheximide on partial hepatectomy-induced increase in histidine decarboxylase activity of Walker carcinosarcoma-bearing rats

Cycloheximide was injected i.p. in a dose of 0.1 mg/100 g body weight 15–30 min before partial hepatectomy. The activity of liver histidine decarboxylase was determined before and 5 hr after removal of two-thirds of the tumor-bearer liver ; the other conditions are the same as for Fig. 1 (6).

(Fig. 1). When using animals that had been treated with cycloheximide, however, much less increase in enzyme activity was observed in normal rats connected with Walker carcinosarcoma-bearing rats. This suggests that enzyme induction is accompanied by *de novo* protein synthesis (Fig. 1). Further evidence of this was obtained in experiments with partially hepatectomized rats. Enzyme activity was measured before and 5 hr after partial hepatectomy of rats with Walker carcinosarcomas. As shown in Fig. 2, the enzyme activity decreased after hepatectomy in all cycloheximide-pretreated rats, whereas it increased after partial hepatectomy of rats that had not had cycloheximide pretreatment.

In order to learn how much blood from tumor-bearers was required for the induction of histidine decarboxylase, normal rats were connected by cross-circulation with rats bearing Walker carcinosarcoma for 10 to 30 min and then allowed to survive for 5 hr. Even 10 min of cross-circulation was found to be sufficient for enzyme induction (Fig. 3). As the volume of blood exchanged

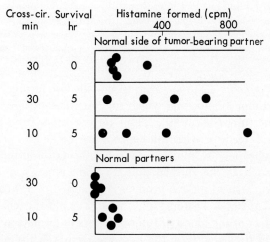

FIG. 3. Short effective duration of cross-circulation for induction of liver histidine decarboxylase

Normal rats were cross-circulated with either normal or Walker carcinosarcoma-bearing rats for 10 or 30 min and then allowed to survive for the indicated time without cross-circulation. Histidine decarboxylase was measured and expressed as described in Fig. 1 (6).

under these conditions was estimated as about 5 ml/min, the volume required for enzyme induction seemed to be less than 50 ml. Tumor-bearer blood was collected from rats with Walker carcinosarcomas; approximately 20 ml each of this blood was infused into normal rats through the jugular vein, while the same volume of blood was removed from the carotid artery. The histidine decarboxylase activities in the liver and spleen were measured 5 hr after blood exchange and the results are shown in Fig. 4. About 20 ml of tumor-bearer blood was apparently sufficient to induce the enzyme. Unfortunately, further analysis has been hampered by the wide variation in the histidine decarboxylase activity of the tissues of rats recently transplanted with Walker carcinosarcoma in this laboratory.

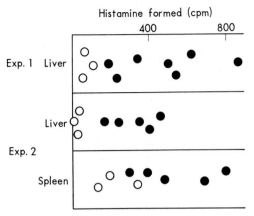

Fig. 4. Induction of histidine decarboxylase by infusion of tumor-bearer blood

About 20 ml of blood was collected from either normal or Walker carcinosarcoma-bearing rats and infused into the jugular vein of a normal rat, while the same volume of blood was being removed from the carotid artery. Five hours after blood exchange, the histidine decarboxylase activity was determined and expressed as described in Fig. 1 (6). ○ normal blood; ● tumor-bearer blood.

Tissue protein synthesis: Tumor-bearing animals lose nitrogen with progressive growth of the tumor in a manner which suggests that nitrogen is translocated from the host to the tumor tissue (9). This nitrogen loss from the host is not prevented by forced feeding with an appropriate diet (12). Parallel to this is the depressed utilization by the host of amino acids for protein synthesis in peripheral tissues such as the skeletal muscles and kidneys (5, 11). Toporek reported that the synthesis of plasma proteins is inhibited in normal isolated liver perfused with blood of tumor-bearers (16). Using the blood exchange technique, we examined the effect of the blood of tumor-bearers on protein synthesis in normal tissues and tried to characterize the humoral factors in the blood of tumor-bearers.

Blood was collected from normal rats and from rats 13 to 18 days after the transplantation of Walker carcinosarcomas. Blood volumes of 20 to 29 ml were infused into normal rats through the jugular vein, and the same volume of blood was removed from the carotid artery. L-Leucine[U-^{14}C] was injected i.p. 1 hr after this blood exchange. The percentage inhibitions of isotope incorporation into tissue proteins were as follows: Total plasma protein, 73%; plasma albumin, 71%; plasma globulin, 74%; liver protein, 41%; kidney protein, 37%; spleen protein, 41%; bone marrow protein, 33%; proteins of pancreas, small intestine, and skeletal muscle showed no significant inhibition (Fig. 5). Isotope incorporation was inhibited to different extents in different tissues, the difference between the inhibitions of incorporation into plasma albumin and into total liver protein being especially significant. Thus the inhibitory effects of the blood of tumor-bearers on different kinds of protein varied even in a single tissue. Therefore, the reduced incorporation was not due to a change in the pool size of leucine in the blood or liver. More trichloroacetic acid-soluble radioactivity was found in

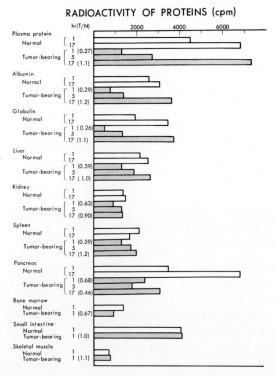

FIG. 5. Inhibitory effect of the blood from Walker carcinosarcoma-bearing rats on the incorporation of L-leucine[U-^{14}C] into tissue proteins

Twenty to 29 ml of blood from either normal or Walker carcinosarcoma-bearing rats was infused into normal rats for blood exchange. After the blood exchange 5 μCi of L-leucine[U-^{14}C] was injected i.p. at the time indicated. One hour after isotope injection the rats were sacrificed to determine the radioactivity of their tissue proteins. Radioactivity is expressed as cpm/0.1 ml of plasma for plasma proteins and as cpm/5 mg of protein for other tissue proteins. The differences between values with blood of normal and tumor-bearing rats on injection of isotope 1 hr after blood exchange were statistically significant for plasma proteins ($P<0.001$) liver, spleen, and bone marrow ($P<0.05$), and kidney ($P<0.01$) (4). T/N: Ratio of radioactivities in rats infused with tumor-bearer blood and normal blood.

the plasma of rats infused with the blood of tumor-bearers than in the plasma of animals infused with normal blood. This is consistent with the above results and suggests that host utilization of amino acids for protein synthesis is inhibited by a factor or factors in the blood of tumor-bearers. Incorporation of L-leucine-[U-^{14}C] into tissue proteins in normal rats tended to increase within 5 hr of its inhibition by the infusion of blood of tumor-bearers and was completely normal again 17 hr after the blood exchange (Fig. 5). This recovery may be explained in two ways; by inactivation of an inhibitory factor in the blood of tumor-bearers or by enhanced protein synthesis in the liver to overcome the action of the blood inhibitory factor, as observed in liver of tumor-bearing animals (5, 11, 16). The results in Fig. 6 show that there was no difference in L-leucine[4, 5-T] incorpora-

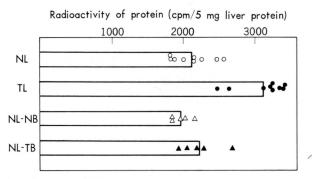

Fig. 6. L-Leucine[4,5-T] incorporation into protein of liver slices from tumor-bearing rats and rats subjected to blood exchange

Twenty-three to 27 ml of blood from either normal or Walker carcinosarcoma-bearing rats was infused into normal rats. The rats were sacrificed 17 to 19 hr after the blood exchange. Liver slices (0.2 g) were incubated with shaking in oxygen and 5% CO_2 in 1.0 ml of normal blood containing 0.2 µCi of L-leucine-[4,5-T] at 37° for 4 hr. NL, normal liver; TL, tumor-bearing liver; NL-NB, liver of normal rats infused with normal blood; NL-TB, liver of normal rats infused with tumor-bearer blood (4).

tion into protein in slices of rat liver which had been infused with normal blood or that from tumor-bearers about 17 hr before sacrifice, while isotope incorporation was stimulated in liver of tumor-bearing rats. These results support the first possibility that an inhibitory factor in the blood of tumor-bearers is inactivated in normal rats within approximately 17 hr.

Blood from tumor-bearing (Walker carcinosarcoma) and normal rats was

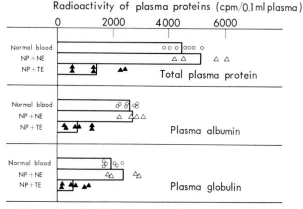

Fig. 7. Inhibitory effect of blood cells from Walker carcinosarcoma-bearing rats on L-leucine[U-^{14}C] incorporation into plasma proteins

Tumor-bearer or normal blood cells were washed 3 times by suspending in saline to the original blood volume and then suspended with normal plasma to give a normal hematocrit. The volume used for blood exchange was 22 to 24 ml. Five µCi of L-leucine[U-^{14}C] was injected i.p. 1 hr after blood exchange. The radioactivity of the plasma proteins was determined 1 hr after the isotope injection. NP+NE, normal plasma+normal erythrocytes; NP+TE, normal plasma+tumor-bearer erythrocytes (4).

separated into plasma and blood cells by brief centrifugation and the following reconstituted bloods were prepared for blood exchange experiments: Plasma of tumor-bearers plus normal blood cells and normal plasma plus blood cells of tumor-bearers. The plasma of tumor-bearers plus normal blood cells did not inhibit L-leucine[U-^{14}C] incorporation into plasma proteins, even when 50 ml of reconstituted blood was exchanged instead of the volume of less than 29 ml, which apparently had been inhibitory when blood from tumor-bearers was used (Fig. 5). However, normal plasma plus blood cells from tumor-bearers was as effective as whole blood from tumor-bearers. Next, blood cells from tumor-bearers were washed three times with saline, suspended in normal plasma, and used for blood exchange. Incorporation of L-leucine[U-^{14}C] into plasma proteins was inhibited by the washed blood cells from tumor-bearing animals (Fig. 7). A preliminary experiment in this laboratory showed that the erythrocyte fraction of blood from tumor-bearers which was essentially free of leucocytes inhibited isotope incorporation into plasma proteins. These results, showing that the cell fraction of blood from tumor-bearers is responsible for the inhibition of isotope incorporation into proteins, differ from those of Toporek (17) who showed that only the albumin fraction of blood from tumor-bearers inhibits ^{14}C-lysine incorporation into plasma proteins by isolated perfused liver. The reason for this discrepancy remains to be investigated.

GENERAL CONCLUSIONS

The period of time and the volume of tumor-bearer blood required for detecting an apparent effect of a humoral factor on histidine decarboxylase and tissue protein synthesis were estimated approximately. From the difference between the immediate effect of blood from tumor-bearers and the state resulting from the continuous action of humoral factors (tumor-bearing state) (Figs. 5 and 6), it seems possible to divide metabolic deviations in tumor-bearers into two categories. The first is the primary effect of a humoral factor or factors and the other is a response of the host to the primary effect of the humoral factor. Thus, the experiments mentioned above on the mechanism of the metabolic deviations in the host provide information on the nature of the humoral factors present in tumor-bearing blood; further information will be obtained when the limitations of the methods have been overcome. It seems important to choose better parameters, although all of the metabolic deviations of the host reported so far could be used. Valuable results should be obtained by work along the lines mentioned above in combination with attempts to obtain active substances from sources other than tumor-bearer blood such as tumor tissue, ascites fluid containing suspended tumor cells, and culture media in which tumor cells have been grown.

Acknowledgments

The present work was supported by grants from the Princess Takamatsu Cancer Research Fund and the Ministries of Education and of Health and Welfare, Japan. We are grateful to Takeda Chemical Industries Co., Ltd., Osaka, for a generous supply of Sprague-Dawley rats.

REFERENCES

1. Alston, W. C. and Thomson, R. Y. Hormonal and local factors in liver regeneration. *Cancer Res.*, **23**, 901–905 (1963).
2. Finerty, J. C. Parabiosis in physiological studies. *Phys. Rev.*, **32**, 277–302 (1952).
3. Greenstein, J. P., Jenrette, W. V., and White, J. The liver catalase activity of tumor-bearing rats and the effect of extirpation of the tumor. *J. Natl. Cancer Inst.*, **2**, 283–291 (1941).
4. Ishikawa, E., Matsuoka, Y., and Suda, M. Inhibitory effect of tumor-bearing blood on the incorporation of labeled leucine into tissue proteins. *GANN*, **62**, 373–380 (1971).
5. Ishikawa, E., Mise, T., and Suda, M. A study on the host-tumor relationship with special reference to protein metabolism. *GANN*, **56**, 543–547 (1965).
6. Ishikawa, E., Toki, A., Moriyama, T., Matsuoka, Y., Aikawa, T., and Suda, M. A study on the induction of histidine decarboxylase in tumor-bearing rat. *J. Biochem. (Tokyo)*, **68**, 347–358 (1970).
7. Johnston, M. Histamine formation in rats bearing the Walker mammary carcinosarcoma. *Experientia*, **23**, 152–154 (1967).
8. Lucké, B., Berwick, M., and Zeckwer, I. Liver catalase activity in parabiotic rats with one partner tumor-bearing. *J. Natl. Cancer Inst.*, **13**, 681–686 (1952).
9. Mider, G. B. Some aspects of nitrogen and energy metabolism in cancerous subjects. *Cancer Res.*, **11**, 821–829 (1951).
10. Nakahara, W. and Fukuoka, F. Toxohormone: a characteristic toxic substance produced by cancer tissue. *GANN*, **40**, 45–71 (1949).
11. Norberg, E. and Greenberg, D. M. Incorporation of labeled glycine in the proteins of tissues of normal and tumor-bearing mice. *Cancer (Philadelphia)*, **4**, 383–386 (1951).
12. Stewart, A. G. and Begg, R. W. Systemic effects of tumors in forcefed rats. II. Effect on the weight of carcass, adrenals, thymus, liver, and spleen. III. Effect on the composition of the carcass and liver and on the plasma lipids. *Cancer Res.*, **13**, 556–559, 560–565 (1953).
13. Suda, M., Tanaka, T., Sue, F., Harano, Y., and Morimura, H. Dedifferentiation of sugar metabolism in the liver of tumor-bearing rat. *GANN Monograph*, **1**, 127–141 (1966).
14. Suda, M., Tanaka, T., Sue, F., Kuroda, K., and Morimura, H. Rapid increase of pyruvate kinase (M type) and hexokinase in normal rat liver by perfusion of the blood of tumor-bearing rat. *GANN Monograph*, **4**, 103–112 (1966).
15. Tanaka, T., Harano, Y., Sue, F., and Morimura, H. Crystallization, characterization and metabolic regulation of two types of pyruvate kinase isolated from rat tissues. *J. Biochem. (Tokyo)*, **62**, 71–91 (1967).
16. Toporek, M. Effect of perfusing blood on serum protein production by isolated perfused livers from normal or tumor-bearing rats. *Cancer Res.*, **29**, 1267–1271 (1969).
17. Toporek, M. Effect of albumin fraction from blood of tumor-bearing rats on serum protein production by isolated perfused normal rat livers. II. *Fed. Proc.*, **29**, 365 (1970).

FERRITIN ISOPROTEINS IN NORMAL AND MALIGNANT RAT TISSUES

M. C. Linder, J. R. Moor, H. N. Munro, and H. P. Morris

Physiological Chemistry Laboratories, Department of Nutrition and Food Science, Massachusetts Institute of Technology,[*1] *and Department of Biochemistry, Howard University School of Medicine*[*2]

The content of ferritin isoproteins in normal and malignant rat tissues was investigated. At least three species of electrophoretically distinct ferritins were found in normal tissues. Adult liver contained one species; kidney, spleen, and newborn liver another; and heart and skeletal muscle a third species as well as the type present in adult liver. Studies of ferritin subunit migration, antibody affinity, and peptide mapping confirmed the differences observed in the electrophoresis of the ferritins of liver, kidney, and heart. Six hepatoma and two kidney tumor lines were also examined. These had growth rates (weight doubling times) spanning at least a four-fold range and varied from highly differentiated to highly anaplastic in their histology. A fairly consistent pattern was found. All but two tumors contained ferritins which were indistinguishable from normal kidney ferritin by electrophoresis or by antibody affinity, and all had ferritins differing from that of normal adult liver. Further examination of the subunits of 3683F hepatoma ferritin, however, suggested the presence of both kidney- and liver-type subunits and suggested that tumor ferritins contain a mixed population of subunits.

All six hepatomas accumulated ferritin protein when iron was admininstered to the rats, just as in several normal tissues. This capacity to accumulate ferritin protein was independent of the growth rate or degree of anaplasia of the tumors. Nevertheless, the iron content of the tumor ferritin was not independent of the degree of malignancy since it showed a clear, inverse correlation with the rate of growth. Together with the observations that regenerating liver ferritin also contains less iron and that regenerating liver has an abnormally high avidity for iron, the results suggest that iron is of some importance in cell proliferation.

The present series of studies are founded on the concept that many of the changes occurring in malignancy are not random and that there is a consistent pattern of biochemical features in all cancer cells. These common biochemical features may be identified by their appearance or persistence after malignant transformation, or by their positive relationship to the degree of malignancy as

[*1] Cambridge, Massachusetts 02139, U.S.A.   [*2] Washington, D.C. 20001, U.S.A.

defined by growth rate and histological differentiation. These ideas are not new and owe much to Greenstein, Weber, Knox, and others for stressing their importance. In earlier work, Knox and Linder (*19, 20, 23*) identified a tumor component which met the criteria mentioned above, namely, the high phosphate-dependent glutaminase of the rat. In transplantable tumors, the concentration of this enzyme exhibited a positive correlation with increasing growth rate and decreasing histological differentiation; this feature was independent of the tissue of origin of the tumors examined. In addition, the study indicated that at least two forms (isozymes) of the glutaminase were present in different normal rat tissues but that the tumors always contained only one of these forms.

In our present work we have examined ferritin, the main site for iron storage in animal tissues, in order to investigate its possible role in the malignant cell. There are several reasons why ferritin is of interest in relation to malignancy. It is widely distributed in animal tissues and in other forms of living matter including fungi and plants, thus suggesting a basic role in cell metabolism. Second, it is an adaptive protein in the sense that it accumulates in various tissues in response to iron. Finally, there was some evidence from the early experiments of Richter (*30*) on human tissues that ferritin might be present in several forms, which we shall call isoproteins by analogy with isoenzymes.

Ferritin Structure

In animals and man, ferritin and hemosiderin comprise about 15% of the total body iron; in rat liver ferritin represents about 75% of the total non-heme iron (*36*). This is also true after iron-loading of the animal, in which case most of the excess injected iron is deposited in ferritin, a large portion being in ferritin of the liver (*11, 13*).

Ferritin is a large molecule (mol. wt., about 440,000 for apoferritin) consisting of a shell of protein subunits within which iron is deposited in varying amounts as a ferric oxyhydroxide salt. When saturated, a ferritin molecule may contain up to 5,000 atoms of iron (*7*). From X-ray diffraction studies and amino acid analyses Harrison and Hofmann (*14, 16, 17*) have concluded that horse spleen ferritin probably consists of 20 identical subunits arranged in the form of a pentagonal dodecahedron. More recently, Crichton and co-workers (*4, 5*) has proposed that there are 24, not 20, subunits, a conclusion which is also compatible with the X-ray data. The proposed structure is based on the results of sodium dodecyl (lauryl) sulfate (SDS) electrophoresis and cyanogen bromide cleavage which suggest a molecular weight for the subunit nearer to 19,000 than to 22,000.

The shell of ferritin protein (apoferritin) has been dissociated with SDS (*5, 34*) and 67% acetic acid (*15*) and analyzed by electrophoresis and ultracentrifugation. No evidence of subunit heterogeneity in horse spleen ferritin has thereby been obtained. On the other hand, the presence of stable oligomers of ferritin in tissue extracts and purified preparations has become generally accepted. These oligomers (dimers, trimers, *etc.*) are separable by electrophoresis (*32, 37*) and repeated Sephadex filtration (*37*); their nature has been confirmed by electron microscopy (*32, 38*).

Ferritin Isoproteins in Normal Tissues

In the last few years, it has become apparent that, in addition to oligomers, normal rat and human tissues contain a variety of different ferritins as evidenced by differences in the electrophoretic migration of the monomer. Since we know that the electrophoretic behavior of the ferritin monomer is not influenced by its iron content (12, 29) these observations suggested that there might be fundamental differences in the composition of the protein shell. Present knowledge of the distribution of these isoferritins with different rates of mobility in man, rat, and rabbit, is displayed diagrammatically in Fig. 1. It is clear that both man and

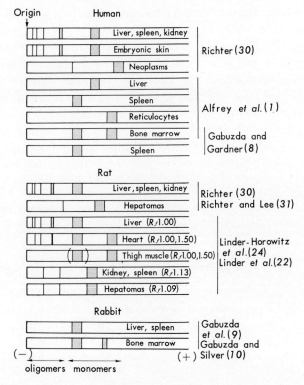

FIG. 1. Summary of electrophoretic studies reported in the literature on ferritins

The distances traveled by the different ferritins towards the anode are presented in relative terms for the individual studies indicated by parentheses and references: Richter (30), Alfrey (1), Gabuzda (8), and their co-workers did the studies on human tissues, Richter (30, 31), Linder (22, 24) and co-workers the studies on rat tissues, and Gabuzda and co-workers (9, 10), those on the rabbit. Ferritin monomers are shown as broad bands and the oligomers as narrow bands to reflect their relative abundance. Except in our studies on the rat (as indicated), no migration rates (Rf's) were actually determined. The Rf's values shown for rat ferritins are those measured relative to control, adult rat liver ferritin.

rat each have at least three electrophoretically distinct species of ferritin monomer in different normal tissues and that the rabbit has at least two. In the rat, the liver and kidney contain different ferritin species and a third type is present in heart and thigh muscle which also contain a type of ferritin indistinguishable electrophoretically from that of liver. The degree of difference in electrophoretic behavior is illustrated by the data in Table I on migration (Rf's) of ferritin monomers by disc electrophoresis using a 7% acrylamide separating gel. Kidney, spleen, newborn liver, and "fast" heart ferritins all migrated significantly faster than ferritin from adult rat liver. The ferritins from kidney, spleen, and newborn liver could not be distinguished from each other, while "fast" heart ferritin was obviously different. The validity of these differences was confirmed when ferritins were partly dissociated into subunits by preincubation with 0.25% sodium lauryl sulfate at 70° for 30 min (34) prior to electrophoresis (Table I, second column). Subunits derived from the ferritins of rat liver and kidney migrated at quite different rates. Although intact liver ferritin migrated more slowly than kidney ferritin, subunits obtained from liver ferritin traveled more rapidly than did the subunits of the kidney ferritin. Spleen ferritin and its subunits both behaved electrophoretically in the same way as kidney ferritin. Heart ferritin preparations yielded only one kind of subunit corresponding to rat liver ferritin, suggesting that the "faster" ferritin found in heart is not so readily dissociated.

The differences indicated by electrophoresis were further substantiated by differences in the antibody affinities of the different ferritins. In earlier experiments we were able to remove preferentially the ferritin corresponding to rat liver

TABLE I. Electrophoretic Migration (Rf) of Ferritin Monomers and Their Subunits Obtained from Holo- and Apoferritins of Different Tissues

Purified ferritins	Migration (Rf) (mean±SD)			
	Monomer		Subunit	
	Holoferritin	Apoferritin	Holoferritin	Apoferritin
Adult liver ferritin	0.085±0.009 (15)	0.084±0.005 (8)	0.88±0.04 (16)	0.82±0.07 (8)
Spleen ferritin	0.106±0.016[b] (4)	0.121±0.006[a] (4)	0.71±0.08[a] (4)	0.69±0.02 (4)
Kidney ferritin	0.114±0.012[a] (7)	0.113±0.008[a] (12)	0.73, 0.76[d] (2)	0.67±0.07 (8)
Newborn liver ferritin	0.113±0.012[a] (4)	0.089±0.012 (5)	0.72±0.02[a] (3)	
Heart				
ferritin I	0.088±0.005[c] (4)			
ferritin II	0.119±0.005[a,c] (4)			
ferritin I		0.084, 0.087 (2)		0.80, 0.80 (2)
ferritin II		0.116, 0.120 (2)		None visible

Purified ferritins were partially dissociated in 0.25% sodium lauryl sulfate at pH 8.3 for 30–60 min at 70° prior to disc electrophoresis in gel tubes containing 7% acrylamide separating gels. Rf's were read relative to a tracking dye (Bromophenol Blue) after staining with Amido Black, with the number of determinations in parentheses. The data are taken from Linder et al. (21).

[a] $P < 0.01$; [b] $P < 0.02$ relative to control rat liver ferritin
[c] Made without SLS preincubation   [d] Subunits seen in only 2 of 7 gels.

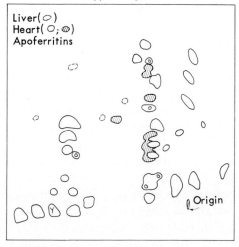
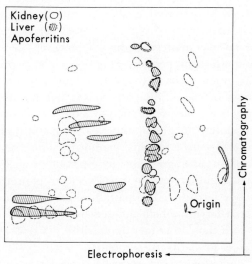

FIG. 2. Peptide maps of tryptic digests of rat liver, heart, and kidney apoferritins

Apoferritins prepared from pure ferritins with 1% thioglycollate (pH 5.4, 6 hr at 25°) were digested for 18 hr with 1/25 weight of trypsin [L-(1-tosylamido-2-phenyl)ethyl chloromethyl ketone] in 0.10 M ammonium carbonate, brought to pH 6.4, lyophilized twice, and mapped on cellulose thin-layer plates (Brinkmann, Polygram Gel 300) using high voltage electrophoresis (1,000 V, 45 min) in one direction and ascending chromatography (pyridine : isopropanol : water, 87.5 : 87.5 : 70) in the other. Maps were stained with cadmium-ninhydrin reagent. The top map shows peptides from rat liver apoferritin (○) and from heart apoferritin (○, ●) which is a mixture of two types (Fig. 1). The "O" and "Y" markings within spots indicate orange and yellow staining, respectively, as compared with the other red stained spots. In the lower map, peptides from kidney apoferritin (○) are superimposed on those from liver apoferritin (◌). The data are taken from Linder et al. (21).

ferritin from heart preparations by precipitation with a particular concentration of specific antiserum (24). The "fast" heart ferritin remained in the supernatant fraction, as shown by electrophoresis after treatment with low amounts of antiserum, and was removed only with much higher amounts of antibody. In the case of rat kidney and liver ferritins, we measured the antibody: antigen ratio of pure ferritin precipitates at the point of maximum protein precipitation. By this method we determined that, for the batch of antiserum used, the maximum ratio of antibody protein: antigen protein in the precipitates was different for the two ferritins, being 3.7 for kidney ferritin and 4.4 for liver ferritin (22).

More recently we have had further confirmation of differences in the protein structure of kidney and liver ferritins from peptide maps made by tryptic digestion of ferritin. As shown in Fig. 2, rat liver apoferritin gave a characteristic peptide map on cellulose thin-layer plates. With high concentrations of peptide, up to 35 different spots spread fairly evenly across the plate could be detected. The peptide spots were characteristic not only in their relative positions, but also in their colors on staining with cadmium-ninhydrin reagent (38). No mapping differences were observed over a wide range of peptide loading (50–1,000 μg) with four independent preparations of highly purified rat liver ferritin. The conditions devised for rat liver apoferritin digests, however, gave a much poorer separation of kidney apoferritin peptides (Fig. 2). The peptides overlapped and were limited mainly to two areas of the plate; some may have migrated off of the plate during electrophoresis. This pattern was not due to overloading of the plate, nor did it vary from one preparation of kidney apoferritin to another.

Purified heart apoferritin preparations gave a pattern similar to that of rat liver apoferritin but had some additional peptides. This was to be expected since heart contained ferritin indistinguishable by electrophoresis from that of rat liver. Heart also contained another ferritin species electrophoretically different from that of liver or kidney ferritin. The pattern obtained is shown in Fig. 2 (top). The peptides of heart ferritin which differed from those of liver ferritin did not appear to coincide with those of kidney apoferritin. A separation of the two heart ferritins is in progress to clarify these points.

In summary, our studies have revealed three kinds of differences in three ferritin species extracted from normal rat tissues. The evidence suggests that these represent differences in protein structure, and therefore amino acid sequence. If this is so, the genetic apparatus of the rat must contain at least three different cistron codings for ferritin peptide subunits. These cistrons must be expressed differently in different tissues, presumably in accord with the functional requirements of each tissue.

Ferritin Isoproteins in Malignancy

In general, cancerous tissues possess certain common biochemical features based on their common malignant characteristics of uncontrolled growth, loss of differentiating elements, etc. Indeed, evidence continues to accumulate that cancers have in common a particular group of enzyme (protein) components and a characteristic pattern of enzyme pathways, whether or not they also contain other enzymes (proteins) which are not characteristic of the malignant process but rather are residual features of the tissue of origin. Components characteristic of cancer cells are recognizable by their appearance or persistence when cancers derived from the same cell of origin are arranged in order of degree of malignancy (20, 23) using rate of proliferation and histological differentiation as criteria. This may also be evident when cancers arising from different tissues are similarly compared (19). The more undifferentiated the malignant cell, the more the enzyme pattern common to other malignant cells will predominate. Isoenzyme (isoprotein) studies present an even more striking picture of the character of the malignant cell, as illustrated by evidence presented in this symposium. What emerges is that certain forms of an enzyme (or nonenzymic protein) prevail when a normal cell is transformed into a malignant one. As the degree of malignancy of the transformed cell increases, those forms of an enzyme or other protein superfluous to the functioning of the cell merely as a malignant cell increasingly tend to be lost. It would seem, therefore, that malignant transformation imposes a malignant program of gene expression on the cell in addition to programs for more specialized cell functions that may have been retained despite transformation. While in no case has it yet been completely proven that the changes in isoproteins seen in malignancy are due to differences in gene expression, the evidence which we are gathering for isoferritins strongly suggests that this is so and adds strength to the theory.

As shown in Table II, we have thus far examined 6 hepatic and 2 kidney tumors of the rat for their content of ferritin isoproteins. These are all transplantable tumors of the Morris series and represent a wide range of histological

TABLE II. Histological and Physical Characteristics of Morris Tumors Studied

Morris tumor	Histology: Differentiation and character of carcinoma	No. of chromosomes (per cell)	Growth rate: Weight doubling time (days)	No. of nuclei (10^8/g)
Hepatomas				
7800	Highly or well-differentiated, trabecular	42	12.3	2.3±0.2 (8)
7793	Well-differentiated, trabecular	45	9.8	3.0±0.5 (8)
5123tc	Intermediate between poorly and well-differentiated, trabecular	49	7.0	5.3±0.6 (4)
7777	Poorly differentiated, trabecular	43	6.6	4.9±0.6 (12)
3924A	Poorly or undifferentiated	73	4.4	3.2±0.5 (8)
3683F	Undifferentiated	39-40	1.5-2.9	3.7±0.6 (7)
Kidney tumors				
MK-1	Highly or well-differentiated tubular cells		62[a]	3.0±0.3 (3)
MK-3	Highly or well-differentiated tubular		12.5	4.4±0.4 (6)

The data for the hepatomas are taken from Linder et al. (22); histological description of the kidney tumors is taken from Morris et al. (25). Weight doubling times of the tumors were determined by regression from the weights of 20 tumors at two different times after implantation when the tumors were between 2 and 8 g in weight. The nuclei were counted in homogenates made with nonnecrotic, nonhemorrhagic portions of tissue.

[a] Small number of tumors sampled; growth rate in doubt.

differentiation, from very highly differentiated (MK-1, MK-3, and 7800) to highly undifferentiated (3683F). The rates at which the tumors doubled their weight (doubling times) correspond to their histological character; as shown, the highly differentiated tumors grow much slower (longer doubling times) than the more anaplastic tumors. The range of growth rates covered was quite large, at least sixfold when comparing the 3683F with 7800 and MK-3. The size of the cells in the different tumors, as reflected in the number of cell nuclei per g, was smaller than for normal rat liver ($2.4 \pm 0.4 \times 10^8$) but showed no consistent variation in relation to growth rate or histological differentiation. All the tumors except 3924A had approximately a diploid complement of chromosomes as compared with the generally-tetraploid complement of normal liver cells (3).

Whether of kidney (MK-1 and MK-3) or liver origin, and whether of differentiated or undifferentiated character, all except one of the tumors examined (3924A) contained ferritins which migrated significantly faster than rat liver ferritin. With the exception of MK-3 ferritin, these more rapidly migrating ferritins could not be differentiated from one another by disc electrophoresis (Table III shows data for gels with 5% acrylamide for separation as compared with the 7% in Table II). Hepatoma 3924A contained a ferritin migrating significantly more slowly than rat liver ferritin and also had twice the number of chromosomes of the other hepatomas and had other unusual traits (8). The ferritin of kidney tumor MK-3 migrated significantly faster than ferritins of the other tumors and is being further investigated. Since, in at least 6 out of the 8 tumors examined so far, the same change in isoprotein was apparent, it may be proposed as a general feature

TABLE III. Electrophoretic Migrations (Rf) of Ferritins Obtained from Hepatomas and Kidney Tumors Relative to Normal Liver and Kidney Ferritins

Tissue		No. of determinations	Migration (Rf) (mean±SD)	
			Absolute	Relative to liver controls
Normal liver	Buffalo, Fisher, Wistar, ACI	41	0.179±0.010	1.00±0.05
Hepatoma	7800	13	0.198±0.008	1.10±0.07
	7793	14	0.194±0.010	1.07±0.09
	5123tc	6	0.200±0.008	1.13±0.04
	7777	6	0.197±0.015	1.06±0.07
	3924A	5	0.170±0.007	0.93±0.04
	3683F	5	0.196±0.012	1.08±0.03
Kidney carcinoma	MK-1	13	0.190±0.009	1.09±0.07
	MK-3	12	0.211±0.013	1.23±0.07
Normal kidney	Fisher, Buffalo, ACl	7	0.196±0.007	1.06±0.04

The data for hepatomas are taken from Linder et al. (22). The ferritins were subjected to disc electrophoresis on 5% acrylamide separating gels. Migrations (Rf) (mean±SD) are given relative to control rat liver ferritin which was run simultaneously, or as absolute migrations relative to tracking dye. All Rf values, absolute or relative, for tumor and kidney ferritins were significantly different from those of control liver ferritin with P values of 0.01 or less. The exception was the absolute Rf for 7777 with a P value of 0.05.

TABLE IV. Affinities of Normal and Malignant Tissue Ferritins for Ferritin Antibody

Tissue	Protein ratio: (antibody+antigen) / Antigen	Protein ratio: (antibody) / Antigen	Fe: Protein ratio of ferritin tested
Liver	5.4	4.4	0.31
7800	4.5	3.5	0.24
7793	4.6	3.6	0.26
5123tc	4.7	3.7	0.21
7777	4.7	3.7	0.10
Kidney	4.7	3.7	0.18

Purified ferritins extracted from normal rat liver and kidney, and 4 hepatomas were titrated with the same specific antiserum to the point of maximum protein precipitation. The ratio of total protein (antibody + antigen) in the precipitate to the amount of antigen protein (ferritin protein) titrated is shown along with the calculated antibody: antigen ratio. The data are taken from Linder et al. (22).

that in malignancy of rat tissues a fairly consistent alteration occurs in the type of ferritin protein made. Antibody-affinity studies for ferritins from four of the liver tumors corroborated the generality of the difference observed by electrophoresis, as shown in Table IV. All four had antibody-protein: antigen-protein ratios of 3.5–3.7 as compared with 4.4 for normal liver. By both of these criteria, electrophoresis and antibody affinity, the ferritins of the liver tumors were indistinguishable from that of normal kidney (Table IV).

Ferritin from one of the hepatomas (3683F) and from the kidney tumors is now being investigated more thoroughly. Just preceding this conference we were able to purify sufficient ferritin to perform preliminary peptide mapping and analysis of subunits. The subunits of the hepatoma apoferritin did not give quite as clear-cut a picture as kidney or liver ferritin. While kidney-type subunits seemed to be present, some protein was also smeared beyond, towards the anode. This smearing had been observed previously when mixed preparations of kidney and liver apoferritin subunits were applied to the same gel (unpublished observations) and suggested that liver-type subunits might also be present in the hepatoma ferritin. In line with this, the undissociated tumor ferritin monomer appeared to migrate slightly more slowly than that of rat kidney when using a 7% rather than a 5% separating gel. The main point to be gained from these studies is that the structure of tumor ferritin differs from that of kidney ferritin despite the implications of the earlier electrophoresis and antibody experiments. Our initial work with peptide mapping further supports this idea. The pattern observed for the tumor ferritin resembles that for liver apoferritin peptides with the streaks of kidney apoferritin peptides superimposed. These findings will have to be confirmed and extended; at present, they suggest that the tumor ferritins may contain both liver- and kidney-type subunits. In malignant rat tissues, therefore, two cistrons for ferritin subunits may be turned on and the resulting ferritin may contain a mixed population of subunits. What this would signify in relation to the functional features of tumor ferritin remains uncertain.

Response to Iron Administration of Ferritin Isoproteins in Normal and Malignant Tissues

The accumulation of ferritin protein in normal tissues following administration of iron is indicative of an adaptive increase in the rate of synthesis of ferritin protein. This has been shown by incorporation studies with leucine-^{14}C and guanido-labeled arginine for ferritins of liver (8, 26) and intestinal mucosa (35). Other normal tissues also accumulate ferritin protein when iron is injected into the whole animal. In Table V the effect on rat heart and kidney is shown relative

TABLE V. Accumulation of Ferritin Protein and Ferritin Iron in Various Tissues One Week after Injection of Iron-Dextran

Tissue	Ferritin protein (μg/g)		Ferritin iron (μg/g)	
	Control	Iron injected	Control	Iron injected
Normal liver	599± 98 (5)	3522±767 (9)	173±27	915±10
Normal kidney	118± 43 (5)	430± 40 (9)	30± 8	66±10
Normal heart	32, 35 (2)	219, 239 (2)	10, 12	55, 56
Hepatoma 3683 F	239±132 (4)	1064±610 (4)	19±15	48±28
3924 A	27± 6 (3)	370±174 (3)	2± 1	35±23
7777	88± 15 (5)	407±199 (5)	8± 1	42±18
5123tc	70± 25 (5)	534±300 (6)	9± 3	73±42
7793	101± 18 (5)	378±176 (4)	18± 5	97±27
7800	255± 64 (4)	492± 74 (4)	38± 9	127±24

Iron (25 mg) as iron-dextran was injected i.p. into normal and tumor-bearing rats. One week later normal or tumor tissues were excised and assayed for ferritin iron and ferritin protein content as described by Linder-Horowitz et al. (24). The tumors were Morris hepatomas of various kinds, as indicated, and the data were taken from Linder et al. (22). Hepatomas are arranged in order of decreasing growth rate and increasing histological differentiation. The results for the other tissues were taken from Linder-Horowitz et al. (24) and represent the combined results for tissues of rats treated with iron 1 and 6 weeks before killing. There were no significant differences for these different periods. Values are given as mean ± SD with the number of determinations in parentheses.

to the effect on liver. While basal levels of ferritin protein were considerably lower in these tissues than in liver, the degree of increase was similar. Iron administration did not change the type of apoferritin made by the tissue. Heart still contained the same two ferritin species in apparently the same proportions, and kidney and liver contained ferritins with the same electrophoretic behavior as before iron administration. Since the ferritins of these three tissues all differ in structure, it seems that the mechanism by which ferritin protein accumulates in response to iron is independent of the species of ferritin protein.

Similar findings were obtained for the six hepatomas tested, as shown on the bottom of Table V. All contained significantly higher concentrations of ferritin after iron treatment than before, indicating no loss of the adaptive mechanism in the malignant tissues. The tumors tested are listed in the order of increasing tumor growth rate and decreasing histological differentiation. The range of

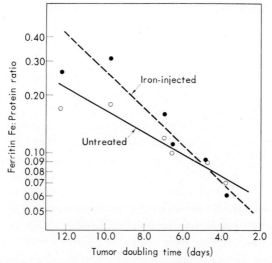

FIG. 3. Relationship between the iron content of ferritin and tumor growth rate for a series of transplantable Morris hepatomas

One-half of the rats carrying tumors were injected with 25 mg of iron as iron-dextran one week before killing (●); the others were untreated (○). The iron: protein ratio for each ferritin was plotted on a logarithmic scale. Tumor growth rate was expressed as weight doubling time, in days, determined by regression. The correlation coefficient between log iron: protein and growth rate was -0.96 (4 d.f., $P<0.01$) for the iron-treated series and -0.94 (4 d.f., $P<0.01$) for the untreated series. Beginning with the most slowly growing tumors (on the left), the hepatomas represented are Morris 7800, 7793, 5123tc, 7777, 3924A, and 3683F. The figure is taken from Linder et al. (22).

growth rates in weight or volume covered by this series was about four fold; histologically the tumors ranged from highly differentiated to undifferentiated (Table II). It is noteworthy that there were no obvious differences in the capacities to accumulate ferritin protein which could be related to growth rate or differentiation. The most rapidly growing, most highly malignant, showed a response similar to that of the most slowly growing and most differentiated.

In the course of these experiments, however, an interesting phenomenon was observed which did reveal a relationship of ferritin to the degree of malignancy of the tumor. As shown in Fig. 3 and by the data for iron and protein in Table V, there was a significant inverse correlation between the growth rate and the iron content of the ferritin present; those tumors with fast rates of growth contained ferritin with less iron than those growing at slower rates. The correlation was statistically highly significant for tumors from both the untreated and iron-treated rats. Since the ferritin isoproteins in these tumors, with one exception, seemed to be identical by several criteria, it seems unlikely that the low iron content of the ferritin in fast-growing tumors is due to a defect in the iron-holding capacity of the protein shell. A more appealing explanation is that the iron is being utilized more rapidly by the more rapidly proliferating tumors than by the slow growing ones. All of the tumors were exposed to the same large iron dose one week before they were excised from the animals. The dose was administered in one injection and was consequently removed from the circulation within a few hours (27), and deposited mostly as ferritin (36) in various tissues. During the week that followed, the tumors grew at different rates. In those tumors growing faster, the iron stored in ferritin may have been utilized more rapidly. The e is other suggestive evidence to support this idea. We have found that regenerating liver exhibits a more avid iron uptake than normal liver. Even 70 hr after removal of 70% of the liver, the uptake of i.v. administered $^{59}FeCl_3$ was 50% higher than normal (22). In addition, as in the proliferating hepatic tumors, the iron content of the ferritin present was reduced from 0.26 (normal liver) to 0.13. More recently, Robbins and Pederson (33) have reported that during proliferation HeLa cells incorporate iron into a substance, possibly a polysaccharide, which can be labeled with glucose-^{14}C. They suggest that this substance may be essential for mitosis since removal of iron from the medium or its chelation results in cessation of DNA (but not RNA) synthesis. Similar observations on the dependence of mitosis on iron have been made for plants (2, 28). Thus, iron may play an important role in cell proliferation.

GENERAL CONCLUSIONS

Our investigations of ferritin in relation to malignancy have borne out several of our premises. Although the number and variety of the tumors studied were limited, the consistency of the findings is already apparent. A particular form of ferritin protein is usually present in the tumors; this is irrespective of the differentiation of the tumor and appears to be similar for liver tumors and some kidney tumors. The ranking of the tumors in the order of degree of malignancy, as quantitatively defined by growth rates for weight and volume, has demonstrated the persistence in malignancy of the capacity to make ferritin protein in response to iron loading. Furthermore, this ranking has indicated that there is a highly significant relationship between tumor growth rate and ferritin iron content. Since ferritin is the site for iron storage rather than iron function, it is logical that the correlation with growth rate should be negative, as has been found. This relationship points to the intriguing possibility that iron utilization is more intimately related to cell proliferation than was previously suspected.

With regard to the isoprotein (isozyme) changes observed by our group and by others and reported in this symposium, most studies support the idea that a particular isomer predominates in malignant cells despite their various origins. Our present understanding of the nature of such isoproteins is that they arise from different genetic cistrons which code for proteins with similar functions. The functional protein is sometimes made up of different proportions of subunits in various tissues (lactic dehydrogenase); sometimes all subunits differ in different tissues (ferritin in normal tissues). In both cases, this implies genetic control of messenger RNA for subunit synthesis. Consequently, the finding of an isoprotein or isozyme predominating in malignant cells points in general to a particular pattern of control of gene expression in these cells which may be related to their malignant nature. It would seem that this pattern of gene expression is of some advantage to the malignant cell in supporting its uncontrolled growth and non-cooperation with the rest of the organism. Thus the preference of the tumor for a particular isomer could be explained on the basis that the tumor selects the one which is under less physiological control. Weinhouse at this symposium pointed out that the adaptive glucokinase form of glucose-ATP-phosphoryl transferases disappears in the more anaplastic tumors and hexokinase II predominates. Glucokinase concentrations vary with diet and insulin levels while hexokinase II concentrations do not.

There may also be another way of understanding the functional significance of the protein changes observed. If we find an isoprotein which is characteristically present in tumors and is present even in the most anaplastic tumors, then, if we know the distribution of this isoprotein relative to its other isomers in a wide range of normal, adult, and embryonic tissues, we may begin to make a judgment about the significance of its presence in the tumors. The isoprotein present in the tumors may well be the one most common in normal tissues (except for liver which is usually different from other tissues). This would suggest that it has a general function in the maintenance of cell viability rather than a specific function in the malignant cell. Aldolase A (Sugimura et al., this symposium), pyruvate kinase M_2 (Suda et al., and Tanaka et al., this symposium), and glutaminase K (*18–20, 23*) would fall into this category. The alternative would be that the uncommon form of the protein is present in the malignant tissue. Here only phosphofructokinase IV (Tanaka et al., this symposium) is available as an example. A study of the characteristics of these uncommon isozymes may be particularly useful in understanding the peculiarities of malignant cell metabolism and function.

Acknowledgments

This work was supported by the American Cancer Society fellowship to Maria Linder, by NIH Grant CA-08893-05 to Hamish N. Munro, and in part by Grant No. 1044 from the Massachusetts Heart Association.

REFERENCES

1. Alfrey, C. P., Jr., Lynch, E. C., and Whitley, C. E. Characteristics of ferritin isolated from human marrow, spleen, liver and reticulocytes. *J. Lab. Clin. Med.*, **70**, 419–428 (1967).
2. Brown, R. and Possingham, J. V. Iron deficiency and the growth of pea roots. *Proc. Roy. Soc.*, **B147**, 145–166 (1957).
3. Bucher, N. Regeneration of mammalian liver. *Int. Rev. Cytol.*, **15**, 245–300 (1963).
4. Crichton, R. R. and Barbiroli, V. Studies on the structure of horse spleen apoferritin. *FEBS Letters*, **6**, 134–136 (1970).
5. Crichton, R. R. and Bryce, C. F. A. Molecular weight estimation of apoferritin subunits. *FEBS Letters*, **6**, 121–124 (1970).
6. Drysdale, J. W. and Munro, H. N. Regulation of synthesis and turnover of ferritin in rat liver. *J. Biol. Chem.*, **241**, 3630–3637 (1966).
7. Fischbach, F. A. and Anderegg, J. W. An X-ray scattering study of ferritin and apoferritin. *J. Mol. Biol.*, **14**, 458–473 (1965).

8. Gabuzda, T. G. and Gardner, F. H. Observations on ^{59}Fe-labeled bone marrow ferritin. *Blood*, **29**, 770–779 (1967).
9. Gabuzda, T. G., Pearson, J., and Melvin, M. Metabolic and molecular heterogeneity of marrow ferritin. *Biochim. Biophys. Acta*, **194**, 50–54 (1969).
10. Gabuzda, T. G. and Silver, R. K. Hemoglobin and ferritin synthesis in erythroid cells in prolonged marrow cell cultures. *J. Cell Physiol.*, **74**, 273–282 (1969).
11. Golberg, L., Smith, J. P., and Martin, L. E. The effects of intensive and prolonged administration of iron parenterally in animals. *Brit. J. Exp. Pathol.*, **38**, 297–311 (1957).
12. Granick, S. Ferritin: its properties and significance for iron metabolism. *Chem. Rev.*, **38**, 379–403 (1946).
13. Hahn, P. F., Granick, S., Bale, W. F., and Michaelis, L. Ferritin. VI. Conversion of inorganic and hemoglobin iron into ferritin iron in the animal body: storage function of ferritin iron as shown by radioactive and magnetic measurements. *J. Biol. Chem.*, **150**, 407–426 (1943).
14. Harrison, P. M. The structure of apoferritin: molecular size, shape and symmetry from X-ray data. *J. Mol. Biol.*, **6**, 404–422 (1963).
15. Harrison, P. M. and Gregory, D. W. Reassembly of apoferritin molecules from subunits. *Nature*, **220**, 578–580 (1968).
16. Harrison, P. M. and Hofmann, T. The structure of apoferritin: evidence for chemical subunits from "finger-prints" of tryptic digests. *J. Mol. Biol.*, **4**, 239–250 (1962).
17. Hofmann, T. and Harrison, P. M. The structure of apoferritin: degradation and molecular weight of subunits. *J. Mol. Biol.*, **6**, 256–267 (1963).
18. Horowitz, M. L. and Knox, W. E. A phosphate-activated glutaminase in rat liver different from that in kidney and other tissues. *Enzymol. Biol. Clin.*, **9**, 241–255 (1968).
19. Knox, W. E., Horowitz, M. L., and Friedell, G. H. The proportionality of glutaminase content to growth rate and morphology of rat neoplasms. *Cancer Res.*, **29**, 669–680 (1969).
20. Knox, W. E., Linder, M., and Friedell, G. H. A series of transplantable rat mammary tumors with graded differentiation, growth rate, and glutaminase content. *Cancer Res.*, **30**, 283–287 (1970).
21. Linder, M. C., Moor, J. R., and Munro, H. N. Unpublished.
22. Linder, M. C., Munro, H. N., and Morris, H. P. Rat ferritin isoproteins and their response to iron administration in a series of hepatic tumors and in normal and regenerating liver. *Cancer Res.*, **30**, 2231–2239 (1970).
23. Linder-Horowitz, M., Knox, W. E., and Morris, H. P. Glutaminase activities and growth rates of rat hepatomas. *Cancer Res.*, **29**, 1195–1199 (1969).
24. Linder-Horowitz, M., Ruettinger, R. T., and Munro, H. N. Iron induction of electrophoretically different ferritins in rat liver, heart, and kidney. *Biochim. Biophys. Acta*, **200**, 442–448 (1969).
25. Morris, H. P., Wagner, B. P., and Meranze, D. R. Transplantable adenocarcinomas of rat kidney possessing different growth rates. *Cancer Res.*, **30**, 1362–1369 (1970).
26. Munro, H. N. and Drysdale, J. W. Role of iron in the regulation of ferritin metabolism. *Fed. Proc.*, **29**, 1469–1473 (1970).
27. Nagarajam, B., Sivaramakrishnan, V. M., and Brahmanandam, S. The distribution of ^{59}Fe in albino rats after intravenous administration of the ionic or chelated form. *Biochem. J.*, **92**, 531–537 (1964).

28. Possingham, J. V. and Brown, R. Nuclear incorporation of iron and its significance in growth. *J. Exp. Bot.*, **9**, 277–284 (1958).
29. Richter, G. W. Electrophoretic and serological properties of the ferritins produced by HeLa and KB cells in culture. *Brit. J. Exp. Pathol.*, **45**, 88–94 (1964).
30. Richter, G. W. Comparison of ferritin from neoplastic and non-neoplastic human cells. *Nature*, **207**, 616–618 (1965).
31. Richter, G. W. and Lee, J. C. K. A study of two types of ferritin from rat hepatomas. *Cancer Res.*, **30**, 880–888 (1970).
32. Richter, G. W. and Walker, G. F. Reversible association of apoferritin molecules. Comparison of light-scattering and other data. *Biochemistry*, **6**, 2871–2881 (1967).
33. Robbins, E. and Pederson, T. Iron: its intracellular localization and possible role in cell division. *Proc. Natl. Acad. Sci. U.S.*, **66**, 1244–1251 (1970).
34. Smith-Johannsen, H. and Drysdale, J. W. Reversible dissociation of ferritin and its subunits. *Biochim. Biophys. Acta*, **194**, 43–49 (1969).
35. Smith, J. A., Drysdale, J. W., Goldberg, A., and Munro, H. N. The effect of enteral and parenteral iron on ferritin synthesis in the intestinal mucosa of the rat. *Brit. J. Haematol.*, **14**, 79–86 (1968).
36. Van Wyk, C. P., Linder-Horowitz, M., and Munro, H. N. Effect of iron-loading on non-heme iron compounds in different liver cell populations. *J. Biol. Chem.*, **246**, 1025–1031 (1971).
37. Williams, M. A. and Harrison, P. M. Electron microscopic and chemical studies of oligomers of ferritin. *Biochem. J.*, **110**, 265–280 (1968).
38. Yamada, S. and Itano, H. A. Phenanthrenequinone as an analytical reagent for arginine and other monosubstituted guanidines. *Biochim. Biophys. Acta*, **130**, 538–541 (1966).

SUBJECT INDEX

A

ACTH in bronchogenic carcinoma	14
Adaptive protein	300
Adenylate kinase	
—— I, II, III, IV	11
—— diabetes	10
—— fasting rat liver	10
—— in hepatoma (9618A, 9633, Novikoff)	11
—— in mitochondria	254
—— insulin	10
—— isoelectric focusing	11
Alcohol dehydrogenase	
—— starch gel electrophoresis	170
Aldolase	
—— A, B, C, A-C hybrid	7, 33, 280
—— amino acid composition	33
—— DEAE-cellulose column chromatography	33
—— diagnostic value	279
—— human brain tumor	279
—— human uterine tumor	279
—— in experimental brain tumor	280
—— in fetal liver	7
—— in hepatoma	7
—— —— AH-130	32
—— —— 7316A, 7793	33
—— K_m	32
Liver ——	7, 33
—— 2'-Me-DAB, 2, 7-FAA	240
Muscle ——	7, 33
—— rat fed 3'-methyl DAB	124, 238, 242
Ali-esterase	
—— starch gel electrophoresis	171
Alkaline phosphatase	13
—— and intestinal metaplasia	38
—— and Regan isoenzyme	13
—— induction by BUdr	27
Allantoin, incorporation of L-serine-U-^{14}C in liver (chick, rat fetal, regenerating) hepatoma (AH-130, 7974, 5123D)	138
Alloenzyme	202
Allosteric enzyme	3
Apoferritin	300
Aspartate transcarbamylase and hepatoma growth rate	67
Azodye, liver DNA, soluble protein,	237
Azoprotein	
—— and alcohol dehydrogenase	178
—— starch gel electrophoresis	171

B

Bile pigment in hepatoma	3
Branched-chain amino acid transaminase	
—— and hepatoma	181
—— and hormone (cortisol, hypophysectomy adrenalectomy)	185
—— DEAE-cellulose chromatography	182
—— fetal, regenerating liver and hepatoma	185
—— in *E. coli*	182
—— in mitochondria	188
—— isozyme I, II, III	182
—— K_m	183
—— tissue distribution	184
—— type III	33
Bromodeoxyuridine (BUdr)	
effect of —— on differentiation	25
——, effect on myogenesis, chondrogenesis, melanin synthesis, hyaluronic acid synthesis	25

C

Cancer
—, blocked ontogeny 34
—, disdifferentiation 34
—, disease of differentiation 34
Carbamylphosphate synthetase 20
— and N-acetylglutamate 147
Carbohydrate metabolism
— key enzyme 4
Carcinoembryonic antigen (CEA)
— and pregnancy 13
— gastrointestinal tract cancer 13
Carcinogen
— of hepatoma 101
Cohnheim alternative 124
Concanavalin A 13
Cross circulation 290

D

dCMP deaminase
— in differentiation and regeneration 59
Decarcinogenesis 31, 34
Dedifferentiation 19, 32, 163
— of enzymes in tumor-bearing animals 79
Derepression 20
Differentiation 12, 19
abnormal — 32, 38
— and extrachromosomal gene 21
— and glutaminase 143
— and growth rate 143
— and isozyme 1
— in hepatoma 1
— of cultured cells 23
Dihydrouracil dehydrogenase
— and growth rate of hepatoma (3924A, 3683F, 44, 47-C, 7787, 9618A, 9618B, 7777, 9618A2) 62
Disdifferentiation 19, 31, 32
— and cancer 34

DNA metabolism 58
— hepatoma growth rate 66
— in hepatoma 68
— 3'-methyl-DAB 129
DNA polymerase in differentiation and regeneration 59
dTMP kinase in differentiation and regeneration 59
dTMP synthetase in differentiation and regeneration 59
Durante alternative 124

E

Ectopic hormone syndromes 53
Epigenetic change 32

F

Ferritin
— electrophoresis 300
— hepatoma (7800, 7793, 5123tc, 7777, 3924A, 3683F) 306
— in malignancy 304
— iron content 309
— isoprotein 299
— kidney tumor (MK-1,-3) 306
— molecular weight 300
— peptide map 303
— response to iron 308
— structure 300
— subunit 300
— tissue distribution 301
Fetal antigen 124
Fetal liver 12
α-Fetoglobulin
— cholangiocarcinoma 13
— hepatoma 13
— liver regeneration 13
— seminoma, chorion-epithelioma, neuroblastoma, nephroblastoma, Wilm's tumor 13
— teratocarcinoma 13
Fructose-diphosphatase 20

Fructose-diphosphatase AMP
　inhibition 154
　—— and differentiation 55
　—— and hepatoma growth rate 57
　—— in Ehrlich ascites carcinoma 154
　—— in hepatoma (7316A, 5123D, AH-35tc$_2$, 3683, 3924A, 108A, 173, 130, 66F) 155
　—— in muscle, kidney, liver 154, 155
Fructose 6-phosphate amidotransferase 153
　—— DEAE-Sephadex column chromatography 160
　—— glucosamine 6-phosphate protection 161
　—— inhibition 160
　—— in liver, brain, Yoshida sarcoma, AH-130, Ehrlich carcinoma 160
　—— in regenerating liver 145
　—— purification 161

G

Ganglioside 168
　—— GD$_{1a}$, GM$_3$, GM$_1$, GD$_1$, GT$_1$, GT$_{1a}$ 177
　—— of Morris hepatoma 174
Gastric tumor 38
Gene expression
　abnormal —— 135
　—— in adult liver 55
　—— in hepatoma 56
　—— in regeneration 60
　—— modulation 55
Glioma (oligodendroglia, spongioblast, astrocyte, ependyma)
　—— aldolase 281
Glucokinase 5, 35
　—— and differentiation 55
　—— in diabetes 6, 81
　—— in diabetic tumor bearer 81
　—— in regenerating liver 6, 87
　—— liver of tumor bearing rat 87

Glucokinase 3'-methyl-DAB 128
　—— of fetal liver 6
Glucose-ATP phosphotransferase (see also hexokinase, glucokinase)
　—— I, II, III, IV 5, 34
　—— multiple form 5
　—— starch gel electrophoresis 5
Glucose 6-phosphatase 20, 33
　—— and differentiation 55
　—— and hepatoma growth rate 57
　—— and insulin 55
Glucose 6-phosphate dehydrogenase
　—— band (TH 1-4, TG1-5) 275
　—— dependency (NADP, NAD) 270
　—— hepatoma (Morris hepatoma, Yoshida ascites hepatoma) 268
　—— H-Group, G-Group 271
　—— in cultured hepatoma 267
　—— induction 272
　—— polyacrylamide gel electrophoresis 269
　—— substrate specificity 271
　—— subunit (A, B, B$_1$, B$_2$, X) 275
Glutamate dehydrogenase 3, 20
Glutamic-oxalacetic transaminase (GOT)
　—— and adrenalectomy 99
　—— and adrenal hormone 99
　—— isozyme 96
　—— peak A, B 96
Glutamic pyruvic transaminase (GPT)
　—— in hepatoma (3924, 5123TC) 99
Glutaminase 33
　—— and development 145
　—— differentiation 300
　—— growth rate 300
　—— in hepatoma (AH-130, Morris hepatoma, 3924A, 5123C, 7777, 7316A, 7793, 9618A) 146

Glutaminase in kidney tumor
    (9789K, 9786K, 8997K)   146
    —— in regenerating liver   145
    —— isozyme   145
    —— kidney type PI   144
    —— N-acetyl-glutamate   147
    —— phosphate dependent (PD), phosphate independent (PI)   144
    —— product inhibition   148
    —— —— in hepatoma   149
Glutamine synthetase
    —— of E. coli   3
Glutathionase   13
Glycogen
    —— in hepatoma   3
    —— phosphorylase   3
    —— synthetase   3, 153
    —— —— D and I   157
    —— —— in liver, muscle, hepatoma (AH-130, 66F)   156
    —— —— K_m 158
Glycolytic/gluconeogenic enzyme and hepatoma growth rate   57
Glycosyl transferase
    —— of liver and hepatoma   176

H

Hemangioblastoma
    —— aldolase   281
Hemosiderin   300
Hepatoma (see also Morris hepatoma)
    —— and fetal liver   12, 73
    —— and regenerating liver   73
    —— biological characteristics   102
    —— carcinogen   101
    —— chromosomes   4, 104
    —— cultured (HTC, 7288tc)   268
    —— differentiation   4, 50
    —— glycolysis   4
    —— growth rate   50, 103
    —— histology   102
    —— spectrum   100

Heteroglycan of hepatoma   167
Hexokinase (see also glucose-ATP phosphotransferase)
    —— I, II, III, IV   34, 282
    —— DEAE cellulose chromatography   282
    —— diagnostic value   279
    —— electrophoresis   282
    —— hepatoma growth rate   57
    —— human brain tumor   279
    —— —— uterine tissue (cervical epithelium, cervical carcinoma, endometrium, corpus carcinoma, myometrium myoma)   283
    —— —— —— tumor   279, 282
    —— in Yoshida sarcoma   35
    —— K_m 282
    —— 3'-methyl-DAB   128
    —— tumor bearing animals   81
Histidine decarboxylase
    —— induction   294
h protein   169
Humoral factor   88
Hyaluronic acid production   24
Hybridization
    cell ——   23
    RNA-DNA ——   22

I

Infusion   291
Insulin in gastric tumor   14
Isoprotein   299
Isozyme   202
    —— change during hepatocarcinogenesis   121
    fetal ——   124
    identification of ——   2
    —— in selected hepatoma   95

K

Key enzyme   48
    —— of carbohydrate metabolism   4
    —— of gluconeogenesis   55

Key enzyme of glycolysis 55, 79
—— of nucleic acid metabolism 59
—— of sugar metabolism 82

L

Lactase 39
Lactate dehydrogenase (LDH)
—— in Yoshida sarcoma 35
—— isozyme in hepatoma 98
—— starch gel electrophoresis 170
Leucine-aminopeptidase 38
Liver marker enzyme 4
—— and growth rate 20

M

Malic dehydrogenase (MDH)
——isozyme A, B 99
Malignoenzyme 205, 215
Meningioma
—— aldolase 281
Messenger DNA 25
Metaplasia
intestinal —— 38
Minimal deviation hepatoma
—— FDP aldolase 125
—— glucokinase 125
—— hexokinase 125
—— pyruvate kinase 125
Mitochondria
—— ATPase 257
—— —— gel electrophoresis 258
—— branched amino acid transaminase 188
——, catalase genome 22
—— compartmentation of enzyme 253
—— component (outer membrane, inner membrane, intermembrane space, matrix) 252
—— cytochrome oxidase, malate dehydrogenase, adenylate kinase, monoamine oxidase 254
—— extrachromosomal gene 21

Mitochondria from Morris hepatoma (9618A, 7800, 3924A) 254
—— phosphorylation 256
—— respiration 255
Molecular correlation concept 47, 49, 214
—— kidney tumor 74
—— mammary cancer 74
Morris hepatoma (see also hepatoma)
—— chromosome karyotype 50
—— —— number 50
—— differentiation 4, 50, 305
—— ganglioside 174
—— glycolysis 50
—— growth rate 4, 50
—— histology 305
—— mitochondria 251
—— respiration 50
Mouse mammary carcinoma (FM3A) 23, 36
Multiple forms of enzyme
——, a new classification 202
—— of hexokinase 79, 83
—— of pyruvate kinase 83
——, molecular basis 3

N

N-Acetyl glucosaminyltransferase in liver and hepatoma (7777, 7800, 5123D) 173
Neurinoma
——aldolase 281
Nucleic acid metabolism 58
—— in hepatoma 70
Nucleotide diphosphokinase in liver, hepatoma 69

O

Oncogeny
—— and ontogeny 123
—— as blocked ontogeny 121
Ornithine carbamoyltransferase 72
—— in hepatoma (8999, 7787,

7800, 7777, 3683F, 3924A,
9618A2)   71
Ornithine decarboxylase
—— in hepatoma   72
Ornithine metabolism
—— in liver, hepatoma   69
Ornithine transcarbamylase   20
—— chick liver   140

P

Parabiosis   290
Perfusion of liver   290
Phosphoenolpyruvate carboxykinase   55
—— and hepatoma growth rate   57
Phosphofructokinase   33
—— against ATP   230
—— and hepatoma growth rate   57
——, cyclic AMP, AMP, ADP   231
—— immunochemistry   227
—— in tumor (LY, AH-66F, -130, Walker carcinosarcoma, gastric cancer)   229
—— kinetic property   229
—— multiple form   219, 226
—— TEAE-cellulose chromatography   226
—— tissue distribution   228
—— tumor bearing animals   81
—— type I, II, III, IV   226
Phytohemaggultinin (PHA)   37
Pituitary adenoma
—— aldolase   281
Precancerous liver   125
Preneoplastic state   235
Pyrimidine metabolism   58
—— and growth rate   61
—— in hepatoma   68, 70
—— in liver   68
Pyruvate carboxylase
—— and differentiation   55
—— and hepatoma growth rate   57
Pyruvate kinase   219
—— adrenalectomy   82
—— and differentiation   55

Pyruvate kinase, cellular location   92
—— hepatoma   83
—— —— (Novikoff, 3924A)   10
—— —— (9618A, 7800, 5123C)   126
—— —— growth rate   57
—— in diabetes   55, 81
—— in liver, muscle   8, 83
—— in liver of tumor bearing rat   83
—— in regenerating liver   86
—— isoelectric focusing   10
—— kinetic property (K_m, FDP activation, ATP inhibition, PCMB inhibition, alanine inhibition)   223
—— M_1, M_2, L   85, 222
—— 2'-Me-DAB, 2, 7-FAA   240
—— 3'-methyl-DAB   127, 238, 242
—— multimolecular form   221
—— newborn rat muscle, brain, and liver   85
—— polyacrylamide gel electrophoresis   81, 84
prototype of ——   86
—— starch block electrophoresis   81, 83, 126, 221
—— synthesis   90
—— tissue distribution   223
—— type I, II, III   8, 126

R

Regan isoenzyme   13
Regulation of enzyme synthesis   19
Ribonucleotide reductase
—— and cell proliferation   205
—— and development in liver   208
—— and growth rate   206
—— enzyme synthesis   210
—— hepatoma growth rate   67
—— in differentiation and regeneration   59, 210
—— in *E. coli*, *Lactobacillus*   206
—— in hepatoma (3683F, 3924A, 7777, 7228C, 5123C,

7800, 7793, 7794B, 7787, 38A, Novikoff) 207
Ribonucleotide reductase in spleen 209
—— natural inhibition 213
—— regulation by ATP and deoxyribonucleotide 206
—— RNA messenger 215
—— subcellular distribution 213
—— thioredoxin (reductase) 206

S

Serine dehydratase 191
—— I, II 193
—— amino acid composition 193
—— DEAE cellulose chromatography 195
—— in diabetes 81
—— in diabetic tumor bearer 81
—— in liver and hepatoma 195
—— leucine-^{3}H incorporation 196
—— peptide map 194
—— polyacrylamide gel electrophoresis 193
Serological (S) antigen 13
Sialyltransferase
—— in liver and hepatoma (7777, 7800, 5123D) 173
Sorbitol dehydrogenase
—— starch gel electrophoresis 170
Sucrase 38
Switched on and off 14, 20
Synzyme 202

T

Tdr kinase
—— aggregation and disaggregation 113
—— bone marrow cell 107
—— DEAE-cellulose column 109
—— embryonic liver 107
—— growth rate 63
—— in differentiation and regeneration 59, 107
—— molecular weight 112

Tdr kinase, a neoplastic tissue 107
—— peak I, II 111
—— potato 107, 115
—— purification 109
—— tetrahymena 107
Tdr to DNA
—— in differentiation and regeneration 59
Thioredoxin-thioredoxin reductase 208
Threonine dehydrase 20
Transplantability
loss of —— 38
Tumor (T) antigen 13
Tumor-bearing animal
—— catalase 291
—— histidine decarboxylase 291
—— metabolic deviation 289
—— pyruvate kinase 291
—— tissue protein synthesis 294
Tumor reversal
—— of neuroblastoma, myeloid leukemia erythroblastic leukemia 35
—— of plant tumor 36
Tumor-specific transplantation antigen (TSTA) 13
Tyrosine-aminotransferase 191
—— adrenalectomy 198
—— development 201
—— distribution 198
—— hormone 200
—— hydroxylapatite chromatography 197
—— in Morris hepatoma (9618A, 9121, 7800, 5123) 201
—— multiple form 197

U

UDP-N-acetylglucosamine-2'-epimerase 163
Uric acid
—— L-serine [U-^{14}C]incorporation in hepatoma (AH-130, 7974), in liver (chick, rat, fetal

Uricase (regenerating) 137
—— in hepatoma (AH-130, 7974, 5123D) 139

X

Xenozyme 202